Techniques in Ecology and Conservation Series

Series Editor: William J. Sutherland

Bird Ecology and Conservation

A Handbook of Techniques

William J. Sutherland,
Ian Newton,
and Rhys E. Green

OXFORD
UNIVERSITY PRESS

OXFORD

UNIVERSITY PRESS

Great Clarendon Street, Oxford OX2 6DP

Oxford University Press is a department of the University of Oxford.
It furthers the University's objective of excellence in research, scholarship,
and education by publishing worldwide in

Oxford New York

Auckland Bangkok Buenos Aires Cape Town Chennai
Dar es Salaam Delhi Hong Kong Istanbul Karachi Kolkata
Kuala Lumpur Madrid Melbourne Mexico City Mumbai Nairobi
São Paulo Shanghai Taipei Tokyo Toronto

Oxford is a registered trade mark of Oxford University Press
in the UK and in certain other countries

Published in the United States
by Oxford University Press Inc., New York

© Oxford University Press, 2004

The moral rights of the author have been asserted
Database right Oxford University Press (maker)

First published 2004

A catalogue record for this title is available from the British Library

Library of Congress Cataloging in Publication Data

(Data available)

ISBN 0 19 852085 9 (Hbk)
ISBN 0 19 852086 7 (Pbk)

10 9 8 7 6 5 4 3 2 1

Typeset by Newgen Imaging Systems (P) Ltd., Chennai, India
Printed in Great Britain on acid-free paper by
Biddles Ltd.www.biddles.co.uk

Preface

We decided to produce this book, as we were aware of the difficulties of providing information on the methods available to field ornithologists or conservationists. We also believed that the difficulties of accessing methodology were hindering the development of the science and the enactment of effective conservation. Thus when we were asked how to conduct a project involving say foraging behavior or breeding biology, we could point out a few papers and describe some methods, but had no source that would outline the major techniques in a comprehensive manner. Our target audiences are young biologists starting a research or conservation project involving birds, or more established researchers who may be familiar with some but not all of the more useful methods available.

This is the first in an intended series of books devoted to methods in ecology and conservation. Each book will either treat a taxonomic group, as this book does, or a broad subject area. We also thank the authors for their efforts and, not least, for putting up with our various idiosyncrasies.

We are donating two hundred copies of this book to ornithologists and libraries outside Western Europe, North America, Australia, New Zealand and Japan who would otherwise be unable to obtain a copy. We will donate another hundred with each reprinting. We thank Ian Sherman at OUP for organizing this, the British Ecological Society for funding the postage and the nhbs.com bookstore for coordinating the distribution. Suggestions of recipients for copies can be made at the Gratis books website http://www.nhbs.com/gratis-books.

We thank Ian Sherman of Oxford University Press for his enthusiasm and efficiency. Working with him has been a pleasure.

William J. Sutherland, Ian Newton and Rhys E. Green

Contents

3. Breeding biology   57

Rhys E. Green

10. Diet and foraging behavior   233

William J. Sutherland

13. Exploitation   303

Michael C. Runge, William L. Kendall, and James D. Nichols

List of Contributors

Susanne Åkesson Department of Animal Ecology, Lund University, Ecology Building, SE-223 62 Lund, Sweden

Malcolm Ausden RSPB, The Lodge, Sandy, Bedfordshire, SG19 2DL

Colin J. Bibby Birdlife International, Wellbrook Court, Girton Road, Cambridge CB3 ONA, UK

John E. Cooper School of Veterinary Medicine (Mount Hope), University of the West Indies (UWI), St Augustine, Trinidad, West Indies

Alistair Dawson Centre for Ecology & Hydrology, CEH Monks Wood, Abbots Ripton, Cambridgeshire, PE28 2LS

Paul F. Donald RSPB, The Lodge, Sandy, Bedfordshire, SG19 2DL

David W. Gibbons RSPB, The Lodge, Sandy, Bedfordshire, SG19 2DL

Rhys E. Green RSPB and Conservation Biology Group, Department of Zoology, University of Cambridge, Downing Street, Cambridge CB2 3EJ, UK

Richard D. Gregory RSPB, The Lodge, Sandy, Bedfordshire, SG19 2DL

Andrew Gosler Edward Grey Institute of Field Ornithology, Department of Zoology, South Parks Rd., Oxford OX1 3PS UK

Anders Hedenström Department of Animal Ecology, Lund University, Ecology Building, SE-223 62 Lund, Sweden

Carl G. Jones Durrell Wildlife Conservation Trust, Les Augres Manor, Trinity, Jersey JE3 5BP, Channel Islands and Maurition Wildlife Foundation, Grannum Road, Vacoos, Republic of Mauritius

William L. Kendall US Geological Survey, Patuxent Wildlife Research Center, 12100 Beech Forest Road, Laurel, MD 20708, USA

Robert Kenward Centre for Ecology & Hydrology, CEH Dorset, Winfrith Technology Centre, Dorset, DT2 8ZD

Ian Newton Centre for Ecology & Hydrology, CEH Monks Wood, Abbots Ripton, Cambridgeshire, PE28 2LS

James D. Nichols US Geological Survey, Patuxent Wildlife Research Center, 12100 Beech Forest Road, Laurel, MD 20708, USA

Michael C. Runge US Geological Survey, Patuxent Wildlife Research Center, 12100 Beech Forest Road, Laurel, MD 20708, USA

William J. Sutherland Centre for Ecology, Evolution and Conservation, School of Biological Sciences, University of East Anglia, Norwich NR4 7TJ, UK

1

Bird diversity survey methods

Colin J. Bibby

1.1 Introduction

The perfect bird census would locate and identify every individual bird at an instant in time. This is the approach used every 10 years to count humans in the United Kingdom. But individuals in many bird species are dispersed and hard to find. Rigorous counting might be achieved in a small area and is easier for some species than others. Bibby *et al.* (2000) give the whole subject of counting birds a more detailed treatment and many more references than is possible in two chapters in this book. A shorter practical guide book (Bibby, Jones, and Marsden1998) is freely available at www.conservation.bp.com/advice/field.asp

There is no rigid distinction between survey (this chapter) and census (Chapter 2), but some general gradients can be perceived singly or in combination (See Table 1.1).

Multispecies surveys may be appropriate where the aim of study is to describe the birds of a relatively poorly known area. In this case, an assessment of the species complement of the avifauna may be sufficient, with abundance estimates being at best qualitative or relative. Questions might be biogeographic, asking about the factors affecting the distribution and diversity of birds. Atlas studies are a powerful way of describing the distribution of birds over large areas. Here the primary aim is to produce distribution maps species by species and quantification, though possible, is relatively less important. A study aiming to compare the conservation importance of known sites might similarly be satisfied with a survey approach. Questions may concern the impact of habitat variation on the fauna where the habitat variation is either natural or human-induced. Studies of the impact of logging in forests would be a typical example. Assessment of the number of species in an area is often important in studies looking at habitat fragmentation, using an island biogeographic model. Formal monitoring programs have to date tended to use more sophisticated census approaches. It is possible

Table 1.1 *Comparisons of survey and census*

	Survey	Census
Prior knowledge	Limited	Moderate to good
Area covered	Large	Restricted
Species covered	All or many	Selected few or one
Methods	Simple	Some technicality
Effort	Extensive	Intensive
Quantification desired	Qualitative/relative	Quantitative/absolute

that simple survey approaches would be sufficiently good in some cases for monitoring. They would certainly be an improvement on having no monitoring information at all, as is currently the case for the majority of the world.

The common feature of these classes of question is that the effects being investigated are bigger than simply changing the abundance of a few species. Important differences between habitats or sites are generally determined by the presence or absence of quite a few species. The same could be said of long-term monitoring as well. Sophisticated schemes might measure population changes to within a few percent year on year, and this detail is valuable for analytic approaches to diagnosis of causes of change. On the other hand, the kinds of changes that trigger any practical response will rarely be less than a 50% decline. This was the case for all the most impacted common farmland birds in the United Kingdom, so that they were listed as national conservation priorities (Gregory *et al.* 2002). Changes of this magnitude lead to widespread gaps in the occurrence of species on scales of a few kilometers or more, and could be detected by simple relative methods.

Effectively, you get the results (absolute or relative abundance) that you pay for, in terms of intensity and technical sophistication of effort. Absolute counts, returning an unbiased estimate of the real number of birds in a specified time and place, are an ideal. They can be obtained with high effort and technically appropriate methods. As a result, they are usually reported for a selected range of species or sometimes only one. The intensity of effort required also means that absolute results are normally described for relatively restricted areas. The main cost of technical census approaches comes from the difficulty of removing bias from the results, for instance by measuring distance to registrations to allow for differences of detectability across habitats.

Relative abundance data tend to return larger numbers for more abundant species if, but only if, other things such as habitats or the behavior of species are similar. They differ from absolute counts in the following important respects.

The relationship between relative and absolute counts may not be linear, so the fact that one number is twice the size of another does not necessarily mean a twofold difference in absolute abundance. Such counts cannot be relied on for comparative purposes across species. Noisy or conspicuously perching or flying species will appear to outnumber those that are quiet, and otherwise cryptic. Nor can relative counts be fully relied on within one species across habitats. Birds are easier to find in open than closed habitats because they are easier to see at greater ranges and can be detected even if silent. Most detections in closed forests have to be made by ear. Thus even if absolute numbers are the same; an open country survey is likely to return higher relative counts.

In many circumstances, relative results are good enough for the purpose of a study. They are an inevitable consequence of seeking results applicable over many species and large areas at low cost. They might be especially appropriate in many tropical studies because of generally poorer knowledge combined with larger differences between habitats in species complement than are found elsewhere.

1.2 Designing the fieldwork

Survey fieldwork consists of going to selected places and following a recording protocol at each. Choosing where to go is a critical element of survey design. Such sampling is not always properly acknowledged, but it can influence results as much as the choice of survey or census method. Many studies do not even mention how the field locations were chosen.

Any well-designed survey has a boundary; the area to which the conclusions will apply. This should be explicitly identified. In some cases, the whole area within the boundary will need to be sampled. In other cases, the target might be just one habitat such as forests. In this case, the location of forests will first need to be identified from maps or satellite images.

One way to locate study plots is to select them by the generation of randomly distributed coordinates within the study boundary. This will give the best unbiased estimates for the whole study area. It will sample individual features of habitat in proportion to their overall abundance. For many purposes, this might not be appropriate. If the intention is to study the variation of bird communities with a feature such as habitat type or human impact, then you will want a sample that well covers the range of variation. This will mean more intensive sampling of the rare features and less of the common ones than a random sample would deliver. The way to do this is by stratification. All possible sample locations (grid cells or habitat blocks) are allocated to a particular stratum. Strata might be one of several predefined habitat types, patch sizes, management histories, or whatever is the target of study. Sample locations are then selected at random within each stratum.

One problem with the random selection of sample locations is that the overall distribution of effort may well look geographically patchy. While random points are the best way to get a good overall result, they may not give the best chance of picking up any geographic patterns of potential interest. A common way round this is to use a systematic selection, visiting plots uniformly distributed on a grid every kilometer or hundred kilometers or whatever scale is appropriate. Hybrids between systematic and random are possible too. It might be sensible to visit every 100-km square to get a good spatial spread but to pick actual sample points within each at random.

Another difficulty with random selections can be cost and access. In large and remote areas the time and cost in getting even a short distance away from a road could be considerable. What is good from a statistical viewpoint can look ridiculously impractical. For this reason, expedience often plays a large part in selecting where to go in the field. The best advice that can be given is to acknowledge and understand what is going on. What makes for easy access for an ornithologist does the same for other people too. Species that are hunted or trapped may be scarcer or absent in more accessible areas, which will cause those species to be underestimated if only the accessible areas are sampled. Many other edge effects are possible. If sampling does not penetrate very far from edges, then forest interior species will be underestimated.

1.3 Finding the birds

Finding all the species in an area depends primarily on good bird-watching skills. The observer needs to get to the right place at the right time and to identify every bird species there. Fast and accurate identification is essential. Few people can identify absolutely everything from a brief sight or sound but if more than about 10% of contacts are unidentified, you need to improve your identification skills to collect worthwhile data. In an unfamiliar area, especially in the tropics, this might take several days, but they are days well spent. Most detections will be by ear, but to begin with you may need to see the bird to identify it. Help from a local expert can be invaluable and increasingly it is possible to get sound recordings to practice with. Observers can work in pairs and compare notes as to what they are recording to see that identification is consistent and everyone is up to standard before serious data collection begins. Sounds can be recorded and brought back for subsequent expert identification.

A target list of likely species can help to remind you of what might be found but is still missing, and this can double as a standardized recording form. Lists can be derived from field guides or local faunal studies. Endemics of restricted range (Stattersfield *et al.* 1998) or threatened species (BirdLife International 2000) are likely to be of particular interest. Local people may often know something of their birds and can be prompted with pictures from field guides but beware of the common tendency to want to please, which can result in optimistic or mistaken evidence of occurrence.

Working at the right time is fairly obvious as far as time of day is concerned. Some species are only active and vocal early in the morning. Especially in the tropics, this period of activity may be extremely short for some species; perhaps just one call per day before first light. Other species require evening visits. At temperate latitudes with a distinct breeding season, visits can be made early enough in the year to catch the resident species when they are singing, and continue through the period of peak activity for the migrants. In the tropics, breeding may follow the seasonality of rainfall but different species may have different seasons. Experience will be needed to understand the effects of surveying at different seasons and for some purposes, year round surveys may be required if residents and non-breeding migrants are involved. In many places in the tropics, the patterns of non-breeding movements and habitat requirements are not well known. So year round surveys are much needed.

The third key to finding everything in a sample area is looking in all the right places. Some species will only be found in particular habitats such as wet areas, streamsides, or bamboo thickets. Many species will be attracted to particular fruiting or flowering trees. Bird flocks are worth watching carefully to make sure that all member species have been identified. Some species are most easily detected in flight over a forest canopy and can best be found by watching from a good vantage-point. Especially in the tropics, altitude is a major determinant of distribution and all altitudes in a study area need to be checked. Competitive birdwatchers will be well familiar with these techniques, which are those used to collect the longest list as quickly as possible. Playback of pre-recorded calls can be used to check for the presence of particular likely species, but this procedure is unlikely to produce different results from surveys not using this technique. Playback of field recorded sounds can also be used to help identify unidentified calls although there should not be too many of these.

Mist nets are sometimes used as a bird survey tool, but there is not much to recommend them. While they may catch some skulking and hard-to-see species, they will fail to catch a large part of the avifauna especially larger species, agile aerial foragers and anything that lives in the higher canopy. Safe use of netting requires a high level of experience and care and effort to visit nets frequently. In pure bird survey terms, the return is poor in relation to the effort required, but there might, of course, be other reasons for catching birds and the bird in the hand may be needed for reliable identification.

1.4 Standardizing the effort by time and space

The most obvious source of bias in multispecies surveys is the variation of effort put into different sampling units. This can be standardized in several ways. The simplest is time. Species lists or counts are accumulated for a fixed period and when this has elapsed a new one is started. The hour is a commonly used and practical unit. It is a simple quantum of effort and many can be completed in any single field session. It is common to collect data from dawn to the point in the morning when bird activity subsides, and there may be another good period in the late afternoon. The day is another commonly used unit. If several observers are involved and perhaps other work is being done, it is convenient to be able to tally a day's records every evening. The length of a day is often less explicit. Daylength varies, as does the duration and intensity of the quiet patch in the middle. Also, the longer the recording period, the less likely it is that observers are working at full effort to cover more ground and find more species.

Smaller time units can be used. This approach makes particular sense if the search area is also constrained. The second Australian atlas used 20 min within 2 ha, which is equivalent to a circle of radius about 80 m. This makes each list rather like a point count of all the birds within detection range from a fixed point. The critical difference is that you walk around to try to flush or otherwise locate birds that you would not necessarily find by standing still for 20 min. Hewish and Loyn (1989) found that this method particularly appealed to observers. It overcomes a frustration of point counts of knowing, from what you see while walking, that there are birds out there that you are missing while keeping still for the actual count. Two hectares is a practical minimum area to use and is roughly equivalent to a point with a search of the area around it. Twenty minutes is a plausible time to search this much ground in an average habitat. At a slow walking pace you might cover some 500 m in this time and thus get to within 20 or 30 m of everywhere in the imagined plot.

If area is to be constrained, then shorter time periods go with smaller areas. An advantage of this approach is to generate results with finer spatial resolution, if that is important for the study aim. Also, more lists can be collected in a given time, enabling frequencies to be measured with more precision. The disadvantage is that more time is needed for subsequent data handling and writing.

Pomeroy (1992) developed a rather more elaborate 1-h procedure in which birds are listed according to which of the six 10-min divisions they are first recorded in. He then weighted them (6,5,4,3,2,1) on the grounds that those detected in the first 10 min will on average be commoner and called the resulting statistic a Timed Species Count Score. In a later development (Freeman *et al.* 2003), the distribution of timing of first detections was used to estimate encounter rates. Encounter rates in this case mean number of detections per unit time rather than number of individuals. Encounter rates can of course be measured directly, but this involves more writing time in the field than the timed species count approach because every detection has to be logged rather than just the first detection of a species. There is also the difficulty in recording encounter rates of having to work out which detections are of a bird already recorded and exclude them. The estimated encounter rate is probably superior to the timed species count in lacking the arbitrary weighting coefficients and coming closer to being proportional to relative abundance.

With listing methods, the end result for each species is expressed as the proportion of lists in which it occurs. Commoner species will clearly be recorded more frequently than rare ones but there is no reason why the relationship between frequency and absolute (unknown) abundance should be linear. Indeed it almost certainly would not be. No effort is made to deal with the fact that species vary in how detectable they are, even when present. Thus the paucity of owls might not be due to the fact that they are rare so much as the fact that they are hard to detect, especially during daylight. In addition, very common species are likely to occur on most or all lists (the more so the longer the recording time). As a result, the method is not so good at separating the relative abundances of the most common species. An obvious way round this is to shorten the recording time and collect more lists.

Encounter rates (total number of individuals divided by time) are rather better than frequencies for separating the relative abundances of the more common species which will tend to having frequencies close to one (they occur on most lists). The disadvantage of encounter rates is that the field recording is greater. There is a natural tendency to prefer to give time to finding new birds than spending a lot of time writing in the field.

1.5 Standardizing the effort by McKinnon's list method

McKinnon has proposed an alternative method of standardizing effort by repeated accumulation of fixed length species lists (McKinnon and Phillips 1993). The observer writes down each new species occurrence until a target number of species has been recorded. At that point, a new list starts with all species again being available to count as new. This is replicated several times. Target list lengths might be as small as 10 species in poor habitats or perhaps 20 in richer ones. The advantage of this method is that it allows to an extent for the fact that some people will accumulate lists faster than others. If hourly lists are being recorded, any time given to trying to catch sight of a strange call to identify it will detract from finding other species in the time. With McKinnon lists, you can take as long as you need to identify individual birds. A more skilled observer will simply collect more lists in a given time period.

Not many people have actually taken up this attractive method. Perhaps the idea of frequency on lists from a fixed time period has the inherent appeal of greater simplicity. Additionally, it could be argued that the benefit of helping less skilled observers is unnecessary or inappropriate. It is possible to stop the clock and to identify a mystery bird during a fixed time sample (provided the clock really is stopped and you do not record anything else). If your skills in identifying the birds of a study area leave you taking a lot of time on identification then perhaps they need to be improved with more practice.

1.6 Atlas studies

Atlas studies have proved to be a very effective way of documenting the avifauna of a region. The aim is simply to map the distribution of occurrence of species. Target areas are divided into grid cells. Squares of side 2, 10, or 50 km on the UTM grid have been used in Europe, with the smaller being used to cover smaller areas. Elsewhere, latitude and longitude cells have been used with sides of one-eighth, quarter, or half a degree. Half a degree at the equator is approximately 55 km but cells become narrower toward the poles. An important practical consideration is that the grid needs to be marked on commonly available maps, although with cheap access to Global Positioning Systems (GPS) this constraint may not last much longer. So far, it seems that a few thousand grid cells is plenty of work to administer which is probably why bigger regions have used a less fine resolution.

At its simplest, fieldwork consists of accumulating lists of species in grid cells. Most atlases have classified some information on the breeding status of the species. Broadly similar criteria for doing this have been standardized in Europe and North America (summarized in Table 1.2). Some of the details look subtle but there are significantly different interpretations between different atlases. The second atlas in the United Kingdom, for instance, did not distinguish records of summering according to whether birds were in suitable breeding habitat or not on the grounds that this is an arbitrary decision especially if habitat boundaries are poorly known or a new habitat use is emerging.

Table 1.2 *Categories of evidence of breeding in European and American bird atlases*

Observed	Species observed in a block during its breeding season, but no evidence of breeding. Not in suitable nesting habitat May include birds such as raptors, waders or gulls far from any breeding area. Used in America but not Europe
Possible breeding	Species observed in breeding season in possible nesting habitat including single records of song
Probable breeding	Pair observed or permanent territory presumed from records at least a week apart in suitable nesting habitat in breeding season. Courtship and display or agitated behaviorseen or heard. Nest building or brood patch observed
Confirmed breeding	Nest with eggs or young seen or heard including those not accessible. Recently fledged young or downy young or distraction display. Adult carrying food for young or fecal sacs. Recently used nest or eggshells

See sources for fuller treatment (Hagemeijer and Blair 1997; Smith 1990).

The simplest atlases have two weaknesses. First, they provide no measure of abundance. A rare species that is found to breed once in the study period, which is usually several years, is mapped just the same as one with thousands of pairs in the grid cell. Second and more problematic is any bias due to difference of intensity of coverage across grid cells. Part of the study area inevitably will have more birdwatchers or easier access, so a higher proportion of the breeding birds will be found and mapped. This means that you cannot be certain of the degree to which a mapped range is biased by indicating apparently thinner occurrence in a region that is actually just less well surveyed. This is a particular weakness when it comes to comparing maps made at different periods of time. People will surely want to use any particular atlas for this purpose at some time in the future and it cannot legitimately be done if there is bias in coverage.

Both these weaknesses can be overcome at the same time by using effort standardization approaches as described above while remaining within the target grid cells. Different atlases have used different units (see Table 1.3).

Most atlas studies have chosen a study region, divided it into a suitable grid and then attempted to cover every grid cell to an adequate level. The outputs are generally presented as maps with one of several different kinds of dot either reflecting relative abundance or level of proof of breeding. There is no inherent reason why atlases should not be constructed from a sampling approach aiming only to visit a selection of grid squares. Sampling could be random, stratified or systematic, or an appropriate blend. This approach would have the advantage of considerable saving of effort compared with the traditional approach, which

Table 1.3 *Examples of the range of units used to standardize effort in different atlas projects*

Southern Africa (Harrison *et al.* 1997)	Checklists for any time period between 1 day and 1 month (but mainly 1 day) within a quarter degree cell
UK (Gibbons, Reid, and Chapman 1993)	Species list from 2 h in a 2 km². At least 8 of the 25 such squares were targeted within a 10 km²
Australia www.birdsaustralia. com.au/atlas/index.html	20 min counts within 2 ha (preferred method) or: up to 5 km from a central point in a time over 20 min but no more than 1 week

is labor intensive. The largest atlases so far attempted have covered Australia, 6 countries of Southern Africa and Western Europe. A sampling approach would make possible the mapping of large areas of the species-rich tropics, even where the density of potential recorders is not high.

An example of an atlas based on sampling has been produced for the United States (Price, Droege, and Price 1995). In this case, the fieldwork was not designed to produce an atlas but was the Breeding Bird Survey, which is designed for population monitoring. Sample units consist of routes of 20 roadside point counts. The distribution of survey routes is designed within a bioregional framework but also allows for differing numbers of contributors in different parts of the country. Distribution maps showing relative abundance were generated by interpolation. There are several ways of inferring distributions from individual locality records or from the rough maps in field guides and knowledge of habitats. At atlas of the distribution of larger African mammals is an excellent example (Boitani *et al.* 1999).

1.7 Estimating species richness

Species diversity is a common focus of study. Notwithstanding the ranges of indices available, the primary dimension of species diversity is the single figure of the number of species occurring in an area—species richness. The critical question is how to tell when to stop looking for more. At a first visit, all the species listed will be new. Over time, the number of new ones gradually diminishes. A plot of total species seen against accumulated effort will rise at an ever-decreasing rate until reaches an asymptote. The full list is practically impossible to achieve. Even after a 100 years of very intensive searching, the list of birds that occur in Britain is not complete. One or two new ones are still added in most years though of course these are vagrants of highly irregular occurrence and marginal to the regular avifauna. The same thing happens in a smaller study area where there is a risk that later additions to a list might be irregular transients.

To demonstrate that a list is virtually complete, it is necessary to show that the rate of accumulation of new species with further effort has reduced to an acceptably small number. By this point, the number of species recorded only once will have fallen to a very low level. In the field, it is helpful to have a list of the likely or possible species so as to focus attention on checking for the missing ones that might occur. Do you know these species well enough to have a good chance of picking them up? Have you looked in the best habitats at the right time of day and season to find them if they were there?

Ideally, species number would be compared across plots that have either received equal effort or sufficient to obtain a near full list for each. It is possible to make a retrospective estimate of the number of species for a fixed effort that is less than the total put in by resampling the data. In this way, all plots can be compared at a standardized effort level of that which received the least but this is rather wasteful of data. Alternatively, it is possible to estimate the number of missing species from the total numbers observed and the numbers found only once or twice (see Colwell and Coddington 1994 and Boulinier *et al.* 1998). There are programs at www.mbr-pwrc.usgs.gov/software/comdyn.html

A much used study design looks for the effects of fragmentation and habitat modification by comparing species richness across a range of sites. Species lists are accumulated, usually by timed visits to study plots, which might vary in size by several orders of magnitude from a few hectares up to tens or hundreds of square kilometers. It might take only a few hours to estimate the species richness in a small plot but considerably longer to complete the list in a large plot. It is tempting to under sample the smaller plots when the species list is slow to grow and put more effort into the larger ones where new finds continue to look likely. A sound quantification requires the demonstration that the species lists for all plots are comparably complete. Ideally, each would have received several visits beyond the point where the list ceases to grow. This can be done with a stopping rule such as stop when the number of species seen only once is less than or equal to the number recorded twice.

1.8 Conclusion

The development of bird survey and census methods really got going about 40 years ago. There was a major conference on the subject 20 years ago (Ralph and Scott 1981). Reading the proceedings, one could be forgiven for concluding that no method produces consistent and reliable results. The list of potential biases and problems is considerable. On the other hand, the literature on bird numbers, habitats, distribution, and trends is increasingly full of interesting and practically important findings. Many have been produced by some of the simpler methods described in this chapter. Superficially this is paradoxical. I believe the explanation is that many of the things we need to know about for practical conservation purposes are sufficiently plain as to be revealed by studies with quite simple methodology.

It is clear to me, on a global scale, we have so far done insufficient to document the basic parameters of distribution and relative abundance of birds. This is especially true in the species-rich tropics where there is a great need for better data to understand and respond to change. It is also clear that simple methods can be very powerful. I suspect that some of the more sophisticated methods, especially those used for monitoring in wealthy northern countries, have deterred people from seeing the potential of simpler approaches for primary exploration and documentation. The earlier we lay down such baseline information, the quicker future generations will be able to use it to promote conservation.

It would be nice if there were a standard approach, because this would make data more comparable across studies and capable of being amalgamated for different purposes. While individual studies will continue to need specific design elements, some general standards can be recommended (Table 1.4).

Table 1.4 *Key points for designing bird surveys*

1	The selection of fieldwork locations should be designed (systematic, random, or stratified) and the design should be a documented part of any publication or database
2	The recommended basic field method is the collection of multiple complete species lists each within a defined period of time and area of ground
3	The time period should be 20 min, 1 h, or 1 day with the smallest possible chosen to be consistent with aims, observer acceptability and data handling consequences
4	Individual timed lists should be collected from the smallest reasonable area (2 ha, 1 km^2, 10 km^2, or quarter degree cell), which should be defined and located by map reference, or GPS coordinate. Separate lists are preferable to amalgamation across major changes of habitat or altitude
5	If evidence on breeding status is collected it should follow EOAC standards (above)
6	The completeness of lists should be assured by use of a stopping rule whereby single occurrences are as frequent or less than doubles. Completeness should be explicitly considered and declared in published analyses
7	For dispersed species, the frequency of occurrence on lists is a measure of relative abundance. This does not make sense for congregatory species where numbers should be estimated
8	Survey data should be deposited in appropriate secure and persistent electronic archives to maximize their potential value

Methods need to suit both purpose and available resources (money, manpower, and skill levels). To improve the rate of data generation, methods need to be acceptable to observers. In countries well endowed with amateur and professional ornithologists, it may not matter that most do not contribute to surveys because there are plenty who will. In places with far fewer ornithologists, it becomes more important that methods are simple enough to be acceptable to them all.

Overall, there is a convergence on the value of species tick lists that are really the simplest kind of recording that could be imagined. They are also the most similar to the kinds of notes that almost all birdwatchers routinely collect. Such lists improve in value with narrower time boundaries. One month, 1 day, 1 h, and 20 min have all been used to good effect. They also improve in value with tighter area constraint. Areas of several hundred square kilometres (quarter degree grids), 10 km^2, 2 km^2 down to 2 ha have again all been used to good effect. Smaller time boundaries are appropriate for smaller area boundaries. If data are collected at a finer resolution of time and space, there are more analytical possibilities. Such records can always be aggregated to a coarser resolution if needed. The converse cannot be done if the original recording was at a coarse scale. An example might be the fact that a record of presence in a quarter degree cell cannot necessarily be attributed to a major habitat type. Nor can it even be relocated although this might be required if it is an uncommon species.

Tick lists are not an efficient recording method for congregatory species such as breeding seabirds or non-breeding waders or wildfowl. Confronted with a sight of tens or even thousands of individuals, it is wasteful not to record a number even if it is only an order of magnitude.

To date, there has been something of a separation of design of studies looking at spatial or temporal patterns. Atlas studies have attempted to achieve complete coverage of grid cells. They tend to last for several years—five is typical—and to be repeated after tens of years. Atlases generally use simple methods. Temporal or monitoring studies have tended to use more sophisticated methods, such as mapping, transects, or point counts. Plots are distributed as observers choose although increasingly with an element of sampling design. Plots are visited annually for as long a run as the individual observers can sustain. It is clear from those atlases that have been repeated that quite large changes in range (and presumably numbers) can be found over periods as short as 20 years. It is also clear, from the American bird atlas or the African mammal atlas, that valuable distribution maps can be drawn from studies not based on uniform coverage of a grid. There seems to be an as yet under-exploited potential to design studies that measure spatial and temporal patterns within the same design.

Data handling is a considerable cost in large bird surveys, such as atlases. This applies both to individual recorders and to the process of centralizing, checking, and computerizing the complete data set. There is clearly a potential for many of these processes to be automated. Field recorders could log data electronically or computerize their own records before submitting them. Locations of plots could be logged, with great precision and less chance of map reading and transcription errors, from a GPS. Centralized archives could harvest data over the Internet and respond to enquiries with real time analyses and presentations of results. The Australian atlas maps can be viewed at www2.abc.net.au/birds/mapviewer.html. This site will also return lists for any one degree square. Observers can add their new data electronically. Because the process is efficient (once set up), Birds Australia intends to collect records beyond the formal period of atlas study as the start to a monitoring program. Birdsource at www.birdsource.org is another excellent example of a site which collects and displays bird survey data. A well designed system could be much more efficient than anything seen to date because individual records could be used for a variety of analyses both spatial and temporal. This would contrast markedly with current approaches where most records are used only within the framework of the study design in which they were collected, or they reside in birdwatchers' notebooks and are not used at all. Such a future will be aided by the further development of electronic archives where basic data can be deposited independently or as annexes to (electronic) journal publication.

Acknowledgments

My thanks to Rhys Green, Ian Newton and Bill Sutherland for critical comments on an earlier draft.

References

Bibby, C., Jones, M., and Marsden, S. (1998). *Expedition Field Techniques: Bird Surveys*. Royal Geographical Society, London.

Bibby, C.J., Burgess, N.D., Hill, D.A., and Mustoe, S.H. (2000). *Bird Census Techniques*, 2nd ed. Academic Press, London.

BirdLife International (2000). *Threatened Birds of the World*. Lynx Edicions and BirdLife International, Barcelona and Cambridge.

Boitani, L., Corsi, F., Reggiani, G., Sinibaldi, J., and Trapanese, P. (1999). *A Databank for the Conservation and Management of the African Mammals*. Istituto di Ecologia Applicata, Rome.

Boulinier, Nichols, Sauer, Hines, and Pollock. (1998). Estimating species richness: the importance of heterogeneity in species delectability. *Ecology*, 79, 1018–1028.

Colwell, R.K. and Coddington, J.A. (1994). Estimating terrestrial biodiversity through extrapolation. *Phil. Trans. R. Soc. Lond. B.*, 345, 101–118.

Freeman, S.N., Pomeroy D.E., and Tushabe, H. (2003). On the use of Timed Species Counts to estimate avian abundance indices in species-rich communities. *African J. Ecol.*, 41, 337–348.

Gibbons, D.W., Reid, J.B., and Chapman, R.A. (1993). *The New Atlas of Breeding Birds in Britain and Ireland: 1988–1991.* T. and A.D. Poyser, London.

Gregory, R.D., Wilkinson, N.I., Noble, D.G., Robinson, J.A., Brown, A.F., Hughes, J., Proctor, D.A., Gibbons, D.W., and Galbraith, C.A. (2002). The population status of birds in the United Kingdom, Channel Islands and the Isle of Man: an analysis of conversation concern 2002–2007. *Br. Birds*, 95, 410–450.

Hagemeijer, E.J.M. and Blair, M.J. (eds.) (1997). *The EBCC Atlas of European Breeding Birds: Their Distribution and Abundance.* T. and A.D. Poyser, London.

Harrison, J.A., Allan, D.G., Underhill, L.G., Tree, A.J., and Brown, C.J., (eds.) (1997). *The Atlas of Southern African Birds.* SABAP, University of Cape Town, Cape Town.

Hewish, M.J. and Loyn, R.H. (1989). *Popularity and Effectiveness of Four Survey Methods for Monitoring Populations of Australian Land Birds.* Royal Australasian Ornithologists' Union, Moon Ponds, Victoria.

McKinnon, J. and Phillips, K. (1993). *A Field Guide to the Birds of Borneo, Sumatra, Java and Bali.* Oxford University Press, Oxford.

Pomeroy, D.E. (1992). *Counting Birds.* African Wildlife Foundation, Nairobi.

Price, J., Droege, S., and Price, A. (1995). *The Summer Atlas of North American Birds.* Academic Press, San Diego, San Fransisco, New York, Boston, London, Sydney and Tokyo.

Ralph, J.C. and Scott, J.M. (eds.) (1981). *Estimating numbers of Terrestrial Birds. Studies in Avian Biology No. 6.* Cooper Ornithological Society, Kansas.

Smith, C.R. (ed.) (1990). *Handbook for Atlasing North American Breeding Birds.* North American Ornithological Atlas Committee. Available at www.americanbirding.org/norac/atlascont.htm.

Stattersfield, A.J., Crosby, M.C., Long, A.J., and Wege, D.C. (1998). *Endemic Bird Areas of the World: Priorities for Biodiversity Conservation.* BirdLife International, Cambridge.

2

Bird census and survey techniques

Richard D. Gregory, David W. Gibbons, and Paul F. Donald

2.1 Introduction

In Chapter 1, we saw how it was possible to use simple methods to assess the species composition in an area and to give an idea of their relative abundances. Here, we consider methods that will allow us to derive estimates of population size or density or, where this is unnecessary or impossible, population indices. Armed with such information over a number of years, we can then track changes in population levels and, where appropriate, compare population levels between different sites. As described in Chapter 1, the distinction between a *census* and a *survey* is somewhat artificial, but here we use *census* to describe a particular type of *survey* that counts the total numbers in an area (Figure 2.1).

2.1.1 What are bird surveys and why do we need them?

If we need a reliable estimate or index of the population size of a particular species in a given area, then we must undertake a survey. There may be a number of reasons for wishing to do this. It may simply be that, as the owner of a nature reserve, we wish to know how many individuals of a particular species of bird are present, or we may need baseline information for an area, or a species, that is poorly known. If repeated at regular intervals, the counts allow us to track changes in bird populations. Alternatively, it may be because a piece of land is being developed (e.g. turned into an industrial area) and we need to undertake an assessment of the likely impact of the development on the nature conservation value of the land. Frequently, bird survey data are used to assess whether a piece of land should receive legal protection from governments and their agencies; such designations are important to conservation because they are intended to constrain potentially damaging activities. Information on population sizes of individual species can also be used to set priorities, allowing conservation effort to be focused on those species most in need of attention. In general, smaller population size is

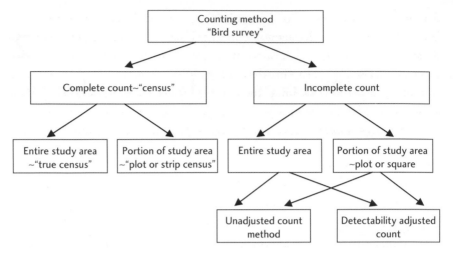

Fig. 2.1 Distinctions between surveys and censuses. Census counts, by their nature, require no correction for detectability. All other counts, here termed "incomplete counts," can be used in their unadjusted, raw form, or preferably with adjustment for detectability (adapted from Thompson 2002).

associated with greater risk of extinction locally, regionally, or globally. Such information is collected by undertaking surveys over varying geographical areas. The lists of globally threatened bird species (BirdLife International 2000) or of species of conservation concern in individual continents, countries or regions (e.g. Carter *et al.* 2000; Gregory *et al.* 2002; www.partnersinflight.org), are based largely on information on population size. In addition, surveys can be used to collect information on where birds are in relation to different habitats, and so assess habitat associations.

2.1.2 What is monitoring and why do we need it?

Monitoring is a simple step on from a survey, in that by undertaking repeat surveys we can estimate the population trend of a particular species over time. Here consistency of method is crucial to measuring genuine population fluctuations. Trend data are central to setting species conservation priorities. All other things being equal (e.g. population, range size and productivity), a species whose population is declining will be of higher conservation priority than one that is not. Monitoring has more uses than this, however. If a monitoring program is well designed, it can be a research tool in its own right providing that suitable environmental data (e.g. habitats, predators, food supplies, weather) are collected, or are available elsewhere. Frequently, such analyses provide early pointers towards the underlying causes of trends in species numbers. The monitoring of demographic parameters, considered in Chapters 3 and 5, can also yield clues about the underlying demographic mechanisms, for example, declining productivity or declining adult survival that may drive a decline in numbers. Monitoring also plays a role in ascertaining the success or failure of conservation actions by faithfully recording their outcomes—these actions might be the acquisition of land to protect particular species, the adoption of new management practices, species recovery programs, or the success of government environmental policies. Sadly, such monitoring is often neglected and the true efficacy of conservation actions is then hard to evaluate.

In some circumstances, birds can be excellent barometers of wider environmental health, particularly when such assessments use summarized data from a wide range of species (Bibby 1999, see also Niemi *et al.* 1997). Two of the best examples of such indicators are WWF's Living Planet Index (Loh 2002, www.panda.org/news_facts/publications/general/livingplanet/index.cfm), and the UK Government's headline indicator of wild bird populations (Figure 2.2; Gregory *et al.* 2003, www.sustainable-development.gov.uk/indicators/headline/h13.htm).

2.1.3 Useful sources of information

This chapter is an introduction to survey design. The following publications give more detail: Ralph and Scott (1981), Ralph *et al.* (1995), Bibby *et al.*

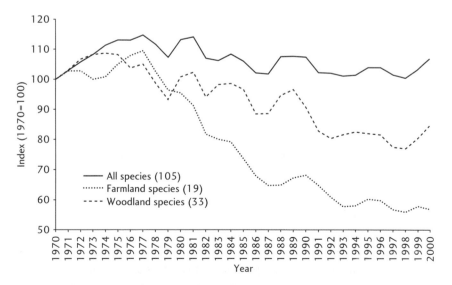

Fig. 2.2 The UK Government's *Quality of Life* indicator showing population trends among common native breeding birds.

(1998, 2000), and Bennun and Howell (2002). In addition, Gilbert *et al.* (1998) and Steinkamp *et al.* (2003) outline species-specific methods for many types of birds, while Greenwood (1996) introduces the underlying theory. Finally, Buckland *et al.* (2001) describe special methods for density estimation, known as *distance sampling* (see below), which use data from line or point transects.

2.1.4 Begin at the beginning

Before rushing into undertake a survey or set up a monitoring program, we first need to clarify our objectives and review our resources. This is a key stage in planning, and any ambiguity or uncertainty at this point could be fatal—wasting time and money, and limiting the usefulness of the results. A common mistake is to be overambitious and try to collect much more information than is strictly required to the point where this compromises quality and other activities. A useful technique here is to list your goals, the data required to fulfill them, the time required to collect these data, and then revisit and prioritize your aims. It is always tempting to ask a whole range of interesting questions, but in attempting to do so, you may fail to answer the key ones. This section outlines how to go about planning a survey; information on sampling strategies and field methods are developed in later sections.

The key decisions to take are:

- Do we want to estimate population size accurately or will an index meet our needs? In other words, are we interested in absolute or relative abundance?
- Where will we undertake the survey?
- Should we cover the whole area of interest, or only sample part of it?
- If we plan to sample, how should we select the study sites?
- What geographical sampling units will we use? Mapped grid squares, forest blocks, or other parcels of land?
- What field method will we use?
- What are the recording units for the birds: individuals, singing males, breeding pairs, nests or territories?
- How will the subsequent data analysis be carried out?
- How will the results be reported and used?

A useful way of planning a survey is to try to envisage clearly the finished product, even down to the details of what tables of data you wish to include in your report. This will clarify the various stages that you need to go through to collect these data.

2.1.5 Population size or index?

If the aim of our survey is to determine accurately the *population size* (=total numbers) of a species in a particular area, then a population index is insufficient for our needs. If, for example, we want to estimate the global population of the Raso Lark *Alauda razae* on its tiny island home, or the numbers of Sharpe's Longclaw *Macronyx sharpei*, on a particular grassland, then we must choose a method that yields an absolute measure of population size and where error can be estimated. If, however, we are not interested in having population size *per se*, only whether a population is increasing, decreasing or stable, then a *population index* would meet our objectives. The implicit assumption here is that there is a direct correlation between the population index and the true, but unknown, population size. A population index is a measure of population size in which the precise relationship between the index and population size is often not known. The index, however, should ideally be directly proportional to changes in population size, such that if the population doubles then so does the index. Population monitoring can be achieved by obtaining, over a period of years, repeated measures of population size or index; frequently the latter is much less resource-intensive than the former and a reliable index is preferable to a poor count. As we saw in the previous chapter, because we are often interested in quite large changes in populations to trigger conservation action (such as 25–50% declines: Gregory *et al.* 2002), then simple methods are often more efficient.

In truth, the distinction between an estimate of population size and an index may be less we think, because in neither case do we actually know the real population size.

2.1.6 Survey boundaries

The decision on where to undertake the survey again depends on its objectives, which should guide the setting of *survey boundaries*. These boundaries are largely self-evident if we want to obtain an estimate of the numbers of one or more species in a discrete habitat area, such as a forest or marsh, or in a particular geopolitical (e.g. country) or geographical (e.g. island) area.

Survey efficiency, however, can be greatly improved if we further refine the boundaries within the area of interest, as it is likely that the species will not be present everywhere. It would be inefficient to cover large areas of clearly unsuitable habitat, but conversely little confidence could be placed on a study that excluded areas or habitats in which the species might be present. Boundary setting should be based on existing information, ideally previously available distributional data. If the general distribution of the species has been mapped by an atlas project (see Chapter 1), then set the boundaries of the survey to those shown by the atlas—but be aware of any limitations to the original data collection. If such information is not available—and for most parts of the world it will not be, or it is of uncertain provenance—then set your boundaries based on factors that you think might affect the species distribution, for example, altitudinal or habitat preferences. For example, Arendt *et al.* (1999), set the boundaries for their survey of the critically endangered Montserrat Oriole *Icterus oberi* on the known distribution of its favored habitat, humid and wet tropical forest, supplemented by knowledge of the bird's distribution from local foresters. Some areas outside this boundary were also checked, but no orioles were found.

Frequently, decisions on where to set survey boundaries, and on how to design the survey within those boundaries are closely linked. In many situations, our knowledge of a species' distribution and ecology is based on relatively scant and sometimes uncertain information. In this instance, we need to be more careful in defining our survey boundaries and be cautious of the received wisdom. The areas or habitats with uncertain information become particularly important when they are large in extent. The practical implication is that we will often need to collect data over a wider area than is apparent at first sight, although it is sensible to sample at a much lower intensity in peripheral areas. This is the basis of *stratification*, which will be discussed in more detail later. It is also sensible to count over a larger area when a bird is known, or suspected, to be expanding its range. Paradoxically, it can be as important to confirm that a bird does not occur in an area (and record a nil count), as it is to count it where it does occur.

2.1.7 Census or sample?

The next decision is whether to undertake a *true census* by attempting to count all birds, pairs or nests within the survey boundary, or to count in only a *sample* of areas within the survey boundary. While it might be tempting to census the whole area for the sake of completeness, it is often considerably more effective to census or survey representative sample areas and to extrapolate the results to obtain a figure for the total population with estimates of the likely error. Highly clumped and conspicuous species, such as breeding seabirds or non-breeding waterbirds, may be more amenable to counting most of the population at a limited number of sites. Where numbers are extremely large, however, within-site sampling may also be advisable. Rare birds with restricted ranges are often easier to count using a true census, because sampling might record too few birds to produce a reliable estimate. For more common and widespread species, it may be expensive and unnecessary to count the whole area, and it might be more cost-effective to census or survey a representative selection of areas.

It is possible to mix sample and census approaches within the same survey. Thus, in some areas or habitats a census of all birds is used, for example, where densities are high in limited geographical areas, yet in others only a sample of areas or habitats is counted, for example, where densities are low over wide areas.

2.1.8 Sampling strategy

If we decide to undertake a sample survey, we need to be very clear about the *sampling strategy*. We need to ensure that the areas in which we count are truly representative of the area within the survey boundaries. If they are not, our final estimate or index may be biased in an unknown manner. Strategies based on *random, random stratified* or *regular sampling* (also known as systematic sampling) are likely to be most robust. As this is such an important topic, it is outlined later.

2.1.9 Sampling unit

In tandem with our sampling strategy, we need to decide upon our *sampling unit*, the bits of the whole survey area we actually count birds in. This might be a grid square, the precise location and boundaries of which are available from maps. The area encompassed within the survey boundary can be subdivided into a large number of grid squares on a map, and a sample of these squares chosen at random for survey. While this approach is simple and statistically sound, it may not always be practical. It might be difficult to use, for example, when surveying birds living in fragmented forest plots of variable size surrounded by farmed land. In such circumstances, individual plots can become the sampling unit. In this case, unlike the grid squares, the individual sampling units are likely to vary in size.

2.1.10 Field methods

We now need to consider what *field method* we will use to count the birds. There are a variety of options and the one we choose will depend upon the species or group of species being counted, the habitats involved and the level of detail required. For some species, it is necessary to develop specially tailored methods (see Gilbert *et al.* 1998; Steinkamp *et al.* 2003). If we are trying to survey a number of species together, however, then we need a generic method that will encompass most species well. There are two principal methods for generic or single species surveys; *mapping* and *transects*. These methods, plus others with specific uses, are outlined below.

2.1.11 Accuracy, precision, and bias

The terms *accuracy, precision, and bias* have specific meanings when applied to scientific data, such as bird surveys, though *accuracy* and *precision* are generally interchangeable in common use. It is extremely important to understand these terms at the outset and to use them appropriately when we report survey results. As we will see, survey design essentially revolves around the twin aims of increasing accuracy and precision and reducing bias, but this is easier said than done.

Accuracy is a measure of how close our estimate is to the true population. For example, if our estimate is 510 parrots and the true population is 500, most people would accept that our estimate was quite accurate. If our estimate is 510 but the true population is 2000 parrots, then our estimate is patently inaccurate. Of course, the problem is that we usually do not know the actual numbers and so it is extremely difficult to measure accuracy. In most circumstances, it is practically impossible to count every last individual in a population, and even if it were technically possible, it would be prohibitively expensive. The only practical way to measure accuracy would be to carry out very intensive work in small areas and to calibrate the findings with a wider survey—but such studies are very time-consuming (e.g. DeSante 1981).

Precision is a measure of how close replicated estimates are from each other (and so it is unrelated to the true population size). This is the same as asking how much error is there around a mean estimate. Take the parrot example above; suppose that we have five counts during a period when the true population stayed the same, and we get estimates of 490, 495, 500, 505, 510. Because these estimates are close together, the difference between the extreme counts being just 4% of the mean, most people would accept that the estimates were relatively precise. Five counts of 300, 400, 500, 600, 700, with a difference between extreme counts of 80% of the mean, are imprecise. Coincidentally, the average of both sets of counts is accurate because it is close to the actual number of parrots, though of course this would not be known. A final set of counts of 990, 995, 1000, 1005, 1010, is exactly as precise as the first set, with again a difference of 4% of the mean between extreme counts, but hopelessly inaccurate. Hence, precision is independent of the true population size.

Unlike accuracy, precision can be measured in statistical terms (e.g. as a range, variance, standard error, 95% confidence limits, percentage error etc.) by looking at the differences in counts between the different sampling units. Be aware, however, that standard methods of calculating confidence limits assume that the counts follow a normal (or Gaussian) distribution, which is unlikely to be the case for bird counts. The way around this is to use distribution-free methods, such as bootstrapping, to derive confidence limits (see later). Precision is determined by two factors: the number of sample units visited (=numbers of sites and hence birds counted) and the degree of variation in the counts made in those sampling units.

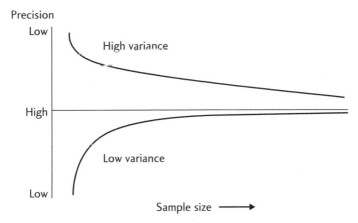

Fig. 2.3 The relationship between precision and sample size with high and low variance. Larger samples sizes deliver greater precision, but as sample sizes increase the benefit declines.

Multiple counts can be obtained by counting the same study site repeatedly in the same season, or by counting multiple study sites once. The first option tells us about temporal variation at sites within a season, the second about spatial variation across the sites—both may be important depending on the study aims.

The relationship between precision, sample size and variance is shown in Figure 2.3. This shows that precision rapidly increases with increasing sample size and that it does so more rapidly where there is little variance in counts between sampling units. As a good rule of thumb, the width of the confidence intervals is related to the number of sampling units N, as $N^{-0.5}$. With smaller sample sizes, great increases in precision can be achieved for relatively small increases in sample size. However, as sample sizes increase, so the additional precision gained declines, and when sample sizes become very large, we gain little in precision, even for very large increases in sample size. From a practical perspective, this tells us that if we wish to increase precision we need to take a larger sample of sites, but beyond a certain point, which we could think of as the optimum sample size, this produces diminishing returns. We can use pilot data to make an informed decision about the optimum sample size but, of course, there are often other more practical considerations (e.g. individuals and time available for fieldwork, survey time required within each plot, or the terrain), and the ultimate decision about sample size will be based as much upon these as on the theory.

The other element to influence precision is the variation in counts between sampling units. If a bird is widely and evenly distributed, occurring in roughly

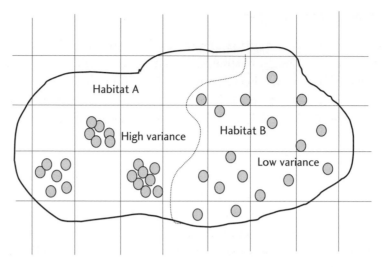

Fig. 2.4 An illustration of uneven variance between two habitats. The filled symbols represent birds.

similar numbers in different sampling units (as in Habitat B in Figure 2.4), then counts from different squares are likely to be similar and the estimates of population relatively precise. If a bird has a more clumped distribution, giving lots of variation between sampling units (as in Habitat A), then counts from different squares are likely to be dissimilar and have lower precision. Note that differences in the ability of the observers to make the counts can also lead to high variance even when the birds are actually evenly distributed.

Bias occurs when our estimates are either systematically larger or smaller than the true value. Put another way, inaccuracy is brought about by bias, which can arise from a poor sampling strategy (e.g. by only surveying the best areas) or an inappropriate field method (e.g. by counting around midday when a species is most active in the morning), or a combination of factors. A whole range of factors could lead to bias, for example, the field method, effort and speed of surveying, the habitat, the bird species and their density, the time of day, the season of the year, the weather conditions, double counting, the observer's skills, etc. The challenge is, first, to recognize all the potential sources of bias and, second, to standardize survey methods and improve standards where appropriate, to reduce bias as much as possible. That said, bias is an unpleasant and often unavoidable fact—and surveys should always consider the likely sources of bias and how they might influence the findings. We should never assume that our survey is free of bias.

2.2 Sampling strategies

We saw in the previous section that, if we are to obtain an unbiased measure of bird abundance (e.g. an estimate of absolute or relative population size), we will often need to count birds in a number of sampling units that are representative of the area within the survey boundaries. This raises two important questions; how many sampling units should we visit to count birds? And, crucially, which ones?

2.2.1 How many sampling units?

As we have seen, the larger the sample size (=number of areas and hence birds counted) the more precise our estimate. Sample size will therefore depend largely on the reliability we want to place in our estimate. If we want a very precise estimate, we need to have a larger sample of sites than if we just want a good approximation. Statistical methods, requiring the collection of some pilot data, are available for calculating sample sizes necessary to achieve predetermined levels of precision (Snedecor and Cochran 1980). In the real world, however, our sample sizes are generally influenced by financial and human resources, and, as these are generally low, we will rarely be at risk of having sample sizes that are much higher then we actually need. Instead, we need to ask ourselves whether our sample size will be sufficient to meet the objectives that we set ourselves at the outset.

2.2.2 Which sampling units to count?

Next, we need to determine which sampling units, out of all those available, should be visited. In other words, what is our *sampling strategy*? This is probably the most critical decision in a sample survey, as failure to use an appropriate sampling strategy could invalidate the results. Only when we are certain that our sampling strategy is appropriate should we start to think about how we will actually count the birds when we get into the field.

There is a tendency for fieldworkers to visit areas they expect to be good for their target species or for their particular study. *Free choice* of this kind can lead to a bias toward higher quality sites, or particular types of site. Remember that our sample must be representative of the whole area of interest if we are to extrapolate the results to areas that are not visited. So how can we select our sample without falling into this trap? The most frequently used methods, and the best, are *random sampling* and *regular sampling*. A definition of truly random sampling is that each sampling unit has an *exactly* equal chance of being selected. Contrast this to free choice, where better areas are far more likely to be selected than less good areas.

Sampling randomly is not as straightforward as it might seem. One might think that closing ones eyes and sticking a pin in a map would be random, but it

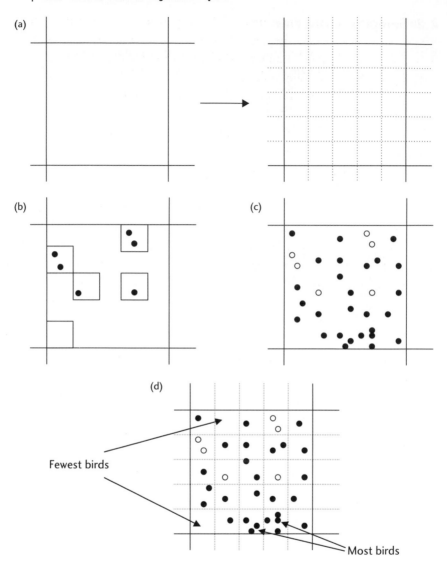

Fig. 2.5 Choosing the right sampling units to count from a grid. (a) First, break the whole area down into bits that can be counted—these are *sampling units*. In this example, we have the resources to count 5 of the 25 sampling units. (b) Next select your squares randomly (see text), count the birds (filled symbols) in these specially selected sampling units (and no others), and estimate the population. The estimate = number of birds counted divided by number of squares counted (= average density of birds per square) multiplied by the total number of squares. Thus, for example, population estimate = 6/5 × 25 = 30. Or, more correctly, add your census count to an estimate of the number of birds in the remaining un-surveyed squares = 6 + (6/5 × 20) = 30. This *extrapolates* data from areas

is not—squares toward the center of the map would be more likely to be selected than those around the edges. Trying to pick "random" squares by eye, or trying to guess "random" numbers, will be similarly biased. If we deliberately select squares we think might hold "average" numbers, this also biases our estimate of precision. There are a number of ways that sampling units can be selected randomly. Assigning each a different number, or using a grid in which each cell has unique coordinates, allows us to select sampling units using random numbers. Random numbers can be selected using random number generators from scientific calculators, from most database packages (such as Excel), or from statistical tables. Alternatively, bits of paper each with the grid coordinates of 1 square can be put into a hat and drawn out blind (this is only random if every square has a corresponding piece of paper). This low technology alternative is perfectly acceptable and scientifically robust. The power of random selection is that it does not matter if we miss the squares with most birds. In the example in Figure 2.5, the two "best" squares were missed, and one of only two squares where the species was absent was selected, but the estimate was still extremely close to the real population size.

The procedure for randomly sampling non-regular units, such as nesting colonies, lakes, forest blocks, etc., is similar. The key is to number or label each of the individual entities and then randomly sample from the whole set (so that each has an exactly equal chance of being picked). Note that for irregularly distributed sampling units, picking a point at random and selecting the nearest sampling unit does not produce a random sample, since sampling units that are more isolated from others are more likely to be selected using this method than sampling units close to others.

2.2.3 Using stratification

We can often use prior knowledge about a species or an area to be surveyed in order to sample more effectively. An important refinement is *stratification*, where

---

where we count (our sample) to those we do not count. (c) Random selection of sampling units almost always provides a good estimate of the true population. In this hypothetical example, our estimate was 30 and the "real" population was 33. Here, open circles represent birds that were counted and filled circles those that were not. (d) It may seem odd that our random sample has missed both the "best" areas for birds, (i.e. with most birds in them), and actually counted one of only two squares with no birds, but this does not matter. As we have seen above, the information we collect from our random sample allows us to estimate the population accurately. Had we based our counts on the best areas, our overall estimate would be a hopeless overestimate.

the area of interest is broken down into different sub-areas, known as *strata* (singular *stratum*). Two simple examples of stratification are shown in Figures 2.6 and 2.7. In the first case, there is prior information from a bird atlas that the species is largely absent, or at least very rare, in the southern part of our region. Randomly sampling across the whole region might, quite by chance, result in us selecting a high proportion of our samples in the area where the species is largely absent (Figure 2.6(a)). This would lead to an imprecise and inaccurate estimate and might lead to other problems, such as reluctance by fieldworkers to visit these areas because they expect to see so little. As an alternative, we could predetermine that, for example, 80% of our samples are drawn at random from the area we think is largely occupied, and only 20% of our samples from that thought to be largely unoccupied. In the second example, our area of interest is known to comprise two distinct habitats, which we expect to hold different densities of the species of interest. Once again, we can get a more precise estimate by using stratification, this time to allocate a predetermined 50% of our samples to each habitat (Figure 2.7(b)). Selection of strata clearly depends upon some knowledge or well-founded assumptions about the distribution of the study species.

We can stratify by habitat, climate, altitude, land use, bird abundance, accessibility of survey sites, administrative or geopolitical boundaries, and so forth. From

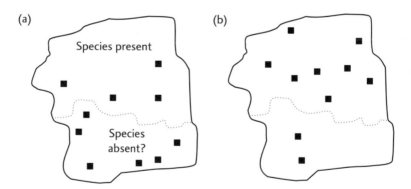

Fig. 2.6 Imagine we are surveying a bird in an area divided into two distinct habitats. (a) A pure random sample of the whole area could, by chance, result in 60% of our samples falling in the southern habitat—which we have reason to believe has very few, if any birds. The filled squares represent survey plots. This would be wasteful of time and resources. (b) Far better would be to use prior knowledge to stratify our sample and, say, take 80% of our random samples from the occupied habitat, and 20% from the habitat that is likely to be unoccupied (see text for further details). Note that, although the sample is smaller in the unoccupied area, it is still vital that it is surveyed.

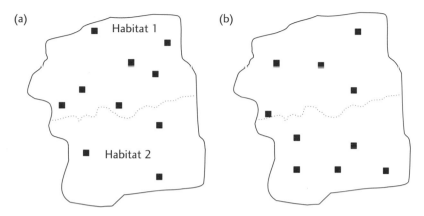

Fig. 2.7 Next, imagine our study area comprises two distinct habitats of roughly equal area, within which our chosen study species lives but at quite different densities. (a) A random sample across the whole area is quite likely to result in an uneven split of survey squares between the two habitats. If 70% of the squares happen to fall in one habitat then the population estimate for the whole area based on the 10 squares would inevitably be dominated, or biased, by that habitat. (b) The solution to this problem is to stratify so, for example, half the samples fall in each habitat—the data are then analyzed by strata and the results combined to give an unbiased estimate of population size (see text for further information on sampling within strata).

what we know about the ecology of birds, it will often make sense to stratify our sample by obvious factors, such as habitat and altitude. Where surveys rely on local observers, it might also make sense to stratify by their availability. Stratification by observer density might seem odd at first sight, but it provides an efficient way of maximizing the use of skilled volunteers when their distribution is uneven, as it often is. Stratification is strongly recommended because it can improve both precision and accuracy, and it ensures proper habitat coverage. Thankfully, there are simple rules that help us choose the most appropriate strata—and it turns out that, even when our prior assumptions about strata prove to be wrong, there is no detrimental effect.

In those situations where we have little information about the habitats used by a species, it makes sense to sample in proportion to the area of the different habitats. For example, if 80% of the area is forest and 20% farmed land, then 80% and 20% of our samples should be in forest and farms, respectively. When we know more about species density in different habitats there are some simple rules designed to improve precision. For example, Sutherland (2000) suggests that sampling should be proportional to the likely proportion of the species occurring in a habitat—so if preliminary information suggests 60% of a population lives in forest, then 60% of our sample should be in that habitat. Of course, there is an element of circularity in this, and it depends on the reliability of the original information. There is the added complication that numbers may be much more variable in one habitat than in another, requiring many more counts there to achieve the same level of precision.

In general, we can improve precision by choosing strata that minimize the variation between sampling units within a stratum while maximizing the variation between strata. This is quite easily achieved because birds generally occur at different densities in different habitats. As we have seen above, the simplest choice is proportional allocation of sampling units within strata, but if the costs of counting sampling units differs across strata, or the counts are more variable in some strata, we can adjust our sampling to optimize allocation (Box 2.1: Snedecor and Cochran 1980). The basic rule is to take smaller samples, compared to proportional allocation, in a stratum where sampling is expensive, and to take bigger samples in a stratum where the counts are more variable. Even rough estimates of variability and cost can help to improve sampling design.

Problems can arise if the number of strata is large relative to the total number of study plots (so that only a few sampling units are selected in each stratum). We recommend using a small number of strata; 2–6 is generally sufficient. One of the reasons for this is that a separate population estimate should be calculated for each stratum and these estimates must be added together to get an overall estimate of the total population. Likewise, confidence limits on these estimates have to be found by combining information from the strata (Box 2.2; see Wilkinson *et al.* 2002, Wotton *et al.* 2002).

In the real world, it may be very difficult to sample totally at random, for example, because you are unable to travel long distances to remote areas to count

---

Box 2.1 Choice of sample sizes within strata

1. Proportional allocation: Take the same fraction of sampling units from each stratum; that is, make n_h/N_h the same for all strata

2. Optimum allocation: Make n_h proportional to $N_h S_h/\sqrt{C_h}$. This delivers the smallest standard error around an estimate for a given cost.

Where: n_h is the sample size chosen in the hth stratum, N_h the total number of sampling units in the hth stratum, S_h the standard deviation of sampling units in the hth stratum, and C_h is the cost of sampling per sampling unit in the hth stratum.

Box 2.2 Analyzing stratified samples

The simple rule in analysing stratified samples is that each step of calculation needs to be carried out at the level of the stratum and the estimate then combined with those from all other strata. If we want to estimate the size of a bird's population and had collected data from three strata (e.g. low, medium, and high abundance, or farmland, scrub, and forest habitats), we would calculate the bird's density in each stratum separately based on our field counts, then multiply up by the area of each stratum, and then add these numbers together to give an overall population estimate. All very simple—and the same approach holds when calculating confidence limits using the bootstrap procedure, but here we add counts from the sampling units we visited to an estimate of the numbers from the remaining area of that stratum that was not visited. Thus, we re-sample at random with replacement from sample sites within strata, calculate an estimate of density and multiply by the area of the habitat that was not surveyed, and add to this the actual number of birds counted. We repeat this process to create 999 unique estimates of the number of birds within each stratum. For each replicate, (1,2,3, ... ,999) the number of birds would then be summed across the strata (strata 1, replicate 1 + strata 2, replicate 1 + strata 3, replicate 1, etc.), to give 999 "bootstrapped" estimates of the overall population size. These totals are then sorted or ranked in size and the 25th and 975th values taken as the 95% confidence intervals.

birds. A more pragmatic approach is *semi-random* sampling, where sampling units are randomly selected within a predefined area. If, for example, you are able to travel a maximum of 50 km from your base to count birds, it is possible to select count sites at random from those available within this radius. An alternative is to define a larger area (which does not need to be contiguous) within which you are able to count, comprising say 5 or 10 km^2, and randomly select smaller sample squares from within this area. This is, however, liable to introduce bias. For example, a semi-random approach is likely to over-sample areas close to human population centers if that is where you live. Nevertheless, semi-random is better than just visiting areas that seem good for birds. By sampling a small number of genuine randomly chosen squares, it is also possible to check on the nature and degree of bias.

A potential problem with random sampling, particularly when sample sizes are low, is that, just by chance, our samples might be concentrated in one part of the survey area that is particularly good for a species, or might miss an area in which we were particularly interested (Figure 2.8(a)). If we are using stratification, this is less of a problem; we can, for example, stipulate that every grid square, or every stratum, contains a fixed number of sampling units (Figure 2.8(b)). An alternative to random sampling that gets around this problem is *regular* or *systematic sampling*. This involves selecting the sampling units by choosing them in a regular pattern (Figures 2.8(c) and 2.9(a)). We can use random numbers to help us do this. If we want a 10% sample from 100 squares, we can select a random number, say 7, then take every 10th square from a list in standard order; 7, 17, 27, 37, . . . , 97. Alternatively, we could simply decide to sample every 1-km square in the north-east corner of every 10-km square and so forth to achieve a predetermined sample size. There are advantages to regular sampling compared to a random design:

- Regular samples are easier to select—a single random number is all that is required.
- It samples evenly over the area of interest; there is 'built-in' stratification that ensures that samples are taken from across the whole area of interest.
- In consequence, it is often more accurate.
- It can be used to create maps and atlases.
- It is easy to understand and explain to others.

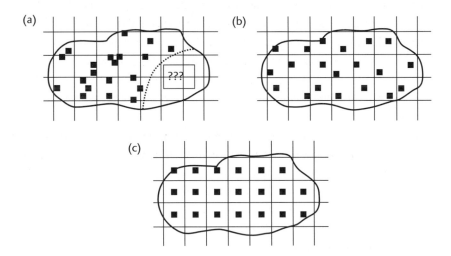

Fig. 2.8 There are certain situations, in which a pure random sample can, by chance, miss an important part of the study area, which could lead to serious under- or over-estimation of a population depending on its distribution. In this example, a random sample (a) under-samples the southeast corner of the study area. A stratified random approach (b) could alleviate this problem by requiring a survey point in every grid square in the study area. Similarly, a regular sample (c) overcomes this problem because survey points are located in the center of every grid square. Here the filled circles represent sampling units ($n = 20$) within a study area defined by the bold border.

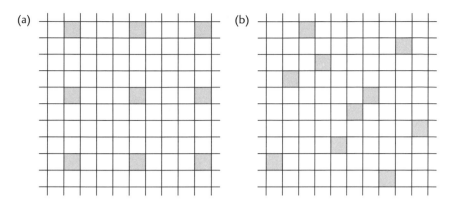

Fig. 2.9 (a) An example of a regular sampling method, and (b) a randomized Latin square design. Survey squares are shaded.

Stratification can be used alongside regular sampling too. For example, we could take every seventh square from a stratum where a bird is thought to be common, but every fourteenth square from a stratum where it is thought to be rare (see Nemeth and Bennun 2000 for a similar approach).

There is, however, a possible bias in systematic sampling, in that this method might over- or under-sample certain features that are regularly distributed in the landscape. For example, it might be that parallel roads are the same distance apart as our lines of samples, leading to over- or under-sampling of areas near roads. In reality, however, such biases are very rare, although we need to be aware of them. In summary, regular sampling has much to recommend it and it has probably been under-used in the past.

An attractive alternative is to integrate the strengths of random and regular sampling by using a *randomized Latin square* design (Figure 2.9(b)), in which each column and each row holds one, and only one, sampling unit. Sampling units are drawn randomly from the rows and columns on the condition that every row and column can only contain a single square, which ensures balanced coverage of the area. This pattern of sampling can be repeated across the study area and within larger sampling units.

2.3 Field methods

In the section above, we considered the key question of how we choose where to make our counts. Now we must consider how to choose between counting methods. Although we have presented survey design as a linear process, in reality, there should be a strong feedback loop in which the sampling strategies and

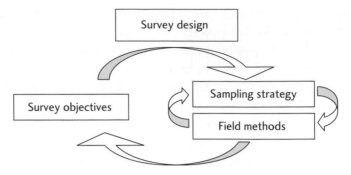

Fig. 2.10 Feedback loops operating in survey design between the survey objectives, sampling strategy, and field methods.

field methods influence and alter each other, and they will in turn influence and potentially alter the survey objectives (Figure 2.10). For example, if the required survey method for a particular species, or habitat, is labor intensive, this might dictate that a smaller number of census plots could be covered. Equally, if the sampling strategy dictated that survey effort needed to be spread across several potential habitats because of uncertainty over the true habitat requirements of a scarce species, this might lead us to re-define and simplify our survey objectives.

There are some general issues to consider in planning fieldwork:

- The season and the time of day the survey is to be carried out.
- The size of the survey plots.
- The number of visits to be made to each sample plot or area (commonly around 10 visits for territory mapping, 2–4 for transects, see below).
- The recommended search effort, for example, walking speed (this is particularly important for line transects) or count duration (for point counts), and general counting protocol for the observers.
- The recording units and behavior of the birds to be noted (ages, sexes, nests, singing, calling males, etc).

The three most common field methods are mapping, and line and point transects; each of these is discussed in turn below.

2.3.1 Mapping

During the temperate zone breeding season, many individual birds are restricted to relatively small areas, actively defending a territory or spending much time around a nest. If a number of visits are made to an area, and the exact location of birds plotted on maps, it becomes possible to identify clusters of sightings and so to estimate directly the total number of pairs or territories of each species present.

An essential component of this method is the use of activity codes to describe bird behavior in the field. These allow observers to record simultaneous observations of territory-holding birds, different forms of territorial behavior and other factors that later allow an analyst to approximate the boundaries between adjacent bird territories. This is the method of *territory* or *spot mapping*. Examples of these codes, and of the way that maps can be analyzed, are given in Marchant *et al.* (1990), Gibbons *et al.* (1996), and Bibby *et al.* (2000). At first sight, this would appear to be an extremely accurate and precise method, but this is not always the case and one needs to be aware of the underlying assumptions about territoriality. An obvious advantage of the method is that it produces a detailed map of the distribution and size of territories, allowing us to link bird distribution with habitats. For certain purposes, for example, habitat management on a nature reserve, such information can be invaluable. The method does, however, have a number of disadvantages:

- It requires high quality maps of the study area.
- It is time consuming, requiring up to 10 visits to each site to be able to identify territories (though fewer visits could be made if only one species is being surveyed—a minimum is around four). The time required for mapping can be up to seven times that of transects.
- Because of the intensity of recording, only small areas can normally be covered, generally 1–4 km^2 (though again this depends on whether a single species is being studied and its ecology, and how much time is available).
- Mapping requires a high level of observer skill in identifying and recording birds.
- Interpretation of the results can be difficult, subjective, and requires the application of consistent rules, particularly when territory densities are high. Territories at the edge of a plot are troublesome and require arbitrary rules.
- It is an inefficient method for recording non-territorial species, semi-colonial species, those that sing for brief periods, or those that are not monogamous.
- It is difficult to use in dense or featureless habitats (e.g. thick forests, flat deserts) or when bird densities are high.
- It is difficult to compare results across studies unless common standards of territory analysis have been applied.

Despite these limitations, territory mapping has proved a useful method of surveying birds in temperate situations and the results have proved a valuable data source for ecological research. In those situations where it is critical to map individual territories, and sufficient resources exist to do this, it is the method of choice. When used appropriately, it allows fine-scale habitat associations to be studied and probably provides relatively accurate estimates of population size (although precision, and especially accuracy, are not easily measured). Mapping methods can also be usefully combined with nest finding, radio telemetry, mist netting etc. in research projects. Mapping has seldom been used in the tropics, largely because breeding is more asynchronous and many species have complex social behaviors.

2.3.2 Transects

There are two types of transect most commonly used in bird surveying, *line transects* and *point transects*. The latter are often termed point counts. Both are based on recording birds along a predefined route within a predefined survey unit. In the case of line transects, bird recording occurs continually, whereas for point transects, it occurs at regular intervals along the route and for a given duration at each point. There are a number of variations on this theme where birds are recorded to an exact distance (variable distance) or within bands (fixed distance) from the transect point or line. The two methods can also be combined within the same survey. While there are important differences between the line and point transects, and choosing between them is an important decision in survey design, there are also many practical and theoretical similarities.

Line and point transects are the preferred survey methods in many situations. They are highly adaptable methods and can be used in terrestrial, freshwater, and marine systems. They can be used to survey individual species, or groups of species. They are efficient in terms of the quantity of data collected per unit of effort expended, and for this reason they are particularly suited to monitoring projects. Both can be used to examine bird–habitat relationships (though generally less well than territory mapping), and both can be used to derive relative and absolute measures of bird abundance. Transects can be usefully supplemented and, to some degree, verified in combination with other count methods such as sound recording, mist netting, and tape playback (e.g. Whitman *et al.* 1997; Haselmayer and Quinn 2000).

There are a series of issues to consider when using transects in the field. The recommended walking speed is particularly important for line transects, as are the counting instructions for the observers. A further important consideration is whether to use full distance estimation, that is, estimating distances from the center of the point count or from the transect line, to all birds heard or seen, or to use estimation within distance bands or belts. In the latter case, one needs to decide on the specific distance bands.

We would always recommend recording some measure of the distance to each bird seen or heard because this provides a useful measure of bird detectability in the habitat concerned and allows species-by-species density estimation (see *Detection probabilities*). It is always preferable to record the exact distance to birds, or failing this, distance within many belts, but in reality, this will often prove to be impractical. As range-finders become increasingly affordable, they open the way for simple and accurate distance estimation, especially for single species surveys.

2.3.3 Line transects

At its simplest, a line transect involves traveling a predetermined route and recording birds on either side of the observer. The distance a bird is seen or heard from the transect line is normally recorded as an absolute measure, or in distance bands. Distances should be estimated perpendicular to the transect line (rather than the distance from the bird to the observer). Distance estimation of this kind is key to the estimation of bird densities. Perpendicular distances can be estimated in a number of ways:

1. Distance is estimated by eye from the line, given practice and periodic checking against known distances; fixed distances can also be marked unobtrusively in the field using marker posts or colored tape to aid recording.
2. Observers may be able to visually mark the position of a bird when detected and then use a tape or range finder to measure the distance when they are perpendicular to where the bird was recorded.
3. Bird observations can be plotted on to high quality maps and the distance measured subsequently. This requires good mapping skills and is helped by having fixed markers in the field.
4. Observers can use a sighting compass to estimate the angle (θ) between the transect line and a line from the observer to the bird, and use a tape or range finder to measure the distance (d) from that point to the bird. The perpendicular distance is then calculated as $d \cos \theta$.

The sampling strategy chosen for a particular survey determines the sample square or unit to be surveyed, but there is still the choice of line transect routes within this area. There are several options, and some flexibility is advisable. For example, a regular or systematic approach could be used with parallel transects orientated north to south, or a series of transects oriented along the long axis of the study area. A random approach, for example, with starting points and directions of transects selected randomly, could be used. One could even use a stratified random approach, for example, with the starting points and direction of transects selected at random, but where each lies within an individual habitat stratum. In reality, topography, watercourses, roads, certain land uses, and access permissions, might all limit access, so that the actual routes counted will differ to some degree from the ideal routes—but such deviation cannot be avoided. In some cases, it might be necessary to substitute a piece of transect for one that cannot be covered, providing it is equivalent in habitat.

The survey design of the Breeding Bird Survey in the United Kingdom, which uses a line transect approach, provides a useful model that can be adopted elsewhere for breeding birds (Gregory 2000; Gregory and Baillie 1998, www.bto.org/survey/bbs.htm). This survey is based on two counting visits to a square each breeding season, with one previous visit to set up a route, and uses three distance bands, 0–25, 25–100, and over 100 m. In general, and for ease of comparison across studies of terrestrial breeding birds, we recommend a minimum of two visits to a plot each season and a maximum four visits. We recommend, as a minimum, 2 distance bands, 0–25 and over 25 m for line transects, and preferably three (as above) or more.

Observers often differ in their ability to record birds and other data. If more than one observer is available, bias can be reduced by matching observers to particular tasks they suit (e.g. one spotting and identifying birds, one estimating distances, one acting as data recorder), and by incorporating training. Inter-observer differences in bird identification can be monitored and compared (e.g. by plotting the decline in the percentage of bird records unidentified through time).

Line transects are highly adaptable; they have been used to survey seabirds from ships, and waterbirds and seabirds from the air, although these are specialized and expensive applications.

2.3.4 Point transects

Point transects differ from line transects in that observers travel along the transect and stop at predefined spots, allow the birds time to settle, and then record all the birds seen or heard for a predetermined time, ranging, at the extremes, from 2 to 20 min. Again, we have three choices in deciding where to site point counts within the study plot. There are, of course, many variations on this theme and the counting stations do not need to follow a set route. One could select individual points at random, or by a stratified random design, and access each of them individually—in fact, this is one of the strengths of point transects because they do not require access across the whole survey area. As with line transects, practical barriers might limit the degree to which the ideal routes can be followed, but equivalent points can be substituted with a little care.

If the point transect is the chosen method for a particular survey, then the same set of considerations outlined above would apply. In addition, for point counts one needs to decide on a settling time once the counting station is reached, and on the duration of the count itself. For ease of comparison across studies of terrestrial breeding birds, we recommend the minimum number of visits to a plot is two and a maximum four. We recommend a 5- or 10-min count period plus an initial settling time of 1 min. For the longer period, we suggest that birds recorded in the first and second 5 min are noted separately (allowing some check on double counting, on whether birds are attracted to the observer, and allowing comparison with 5-min counts). We recommend a minimum of two distance bands, 0–30 m and over 30 m, better still would be 3 bands, 0–30, 30–100 and over 100 m. Lastly, we suggest a minimum of 200 m between counting stations. Ralph *et al.* (1995) review point count methods and provide practical recommendations for their use.

The North American Breeding Bird Survey, which is a continent-wide survey, involves point counts along randomly selected road transects (Sauer *et al.* 2001; www.mbr-pwrc.usgs.gov/bbs/).

2.3.5 Rules for recording birds in the field

The aim is to record all birds identified by sight or sound with an estimate of distance when first detected. It might be helpful to indicate whether a bird is detected by sight or sound on a recording form. Birds that are seen flying over the census area (aerial species) are recorded separately because they cannot be included in standard density estimation. For such mobile species, it is best to make an estimate of their numbers along each section of transect, or at each point. If birds fly away as you are counting, record them from the point you first saw them. We recommend that birds flushed as you approach a point count station should be recorded from that point and included in the point count totals (but you must make this plain in the write-up). Try to avoid double-counting the same individual birds at a point count or within a transect section by using careful observation and common sense. It is, however, correct to record what are likely to be the same individual birds when they are detected from subsequent point counts or transect sections.

2.3.6 Choosing between line and point transects

There is little to choose between line and point transects because they are so adaptable to species and habitats, but each is better suited to particular situations (Table 2.1). The strengths and weaknesses of the methods need to be matched against your survey objectives.

Both methods require a relatively high level of observer skill and experience because a large proportion of contacts and identifications will be by song or call.

Table 2.1 *A comparison of line and point transects*

Line transects	Point transects
Suit extensive, open, and uniform habitats	Suit dense habitats such as forest and scrub
Suit mobile, large or conspicuous species, and those that easily flush	Suit cryptic, shy, and skulking species
Suit populations at lower density and more species poor	Suits populations at higher density and more species rich
Cover the ground quickly and efficiently recording many birds	Time is *lost* moving between points, but counts give time to spot and identify shy birds
Double counting of birds is a minor issue, as the observer is continually on the move	Double counting of birds is a concern within the count period—especially for longer counts
Birds are less likely to be attracted to the observer	Birds may be attracted to the presence of observers at counting stations
Suited to situations where access is good	Suited to situations where access is restricted
Can be used for bird–habitat studies	Better suited to bird–habitat studies
Errors in distance estimation have a smaller influence on density estimates (because the area sampled increases linearly from the transect line)	Errors in distance estimation can have a larger influence on density estimates (because the area sampled increases geometrically from the transect point)

Some thought needs to be given to surveying birds that are non-territorial, semi-colonial species, those that sing for brief periods, and those that have unusual mating systems; but this is less of a concern than in territory mapping. A potential disadvantage of both transect methods for some purposes is that they tend to follow paths, tracks, or roads and so may not be representative of the area as a whole. A practical way around this using point counts is to establish counting stations at right angles to the transect, and say 30 or 50 m into the habitat.

2.3.7 Detection probabilities

Having conducted a survey of a species in a particular habitat, it makes sense to compare the results with those of other similar studies in order to place your findings in context. This is often easier said than done, however, because to do so using the raw, or "unadjusted counts," you must assume that the probability of detecting birds is the same for each data set that is compared. It is an inescapable fact that some birds present in your study area will go undetected regardless of the survey method and how well the survey is carried out. Detectability is a key concept in wildlife surveys and we neglect it at our peril. Thus, comparison of "unadjusted counts" will only be valid if the numbers represent a constant proportion of the actual population present across space and time. This assumption is often questionable and has been a matter of much debate (Buckland *et al.* 2001; Rosenstock *et al.* 2002; Thompson 2002). To be clear, this could affect comparisons between different habitats surveyed at the same time, and between the same or different habitats surveyed at different times.

The solution is to "adjust" counts to take account of detectability, and a number of different methods have been proposed (Thompson 2002). For example, the "double-observer" approach uses counts from primary and secondary observers, who alternate roles, to model detection probabilities and adjust the counts (Nichols *et al.* 2000). The "double-sampling" approach uses the findings from an intensive census at a subsample of sites to correct the unadjusted counts from a larger sample of sites (Bart and Earnst 2002). The "removal model" assesses the detection probabilities of different species during the period of a point count and adjusts the counts accordingly (Farnsworth *et al.* 2002). Finally, "distance sampling" models the decline in the detectability of species with increasing distance from an observer and corrects the counts appropriately.

Distance sampling is a specialized way of estimating bird densities from transect data and of assessing the degree to which our ability to detect birds differs in different habitats and at different times (Buckland *et al.* 2001; Rosenstock *et al.* 2002). The software and further information to undertake these analyses are freely available at: www.ruwpa.st-and.ac.uk/software.html. *Distance sampling* takes account of the fact that the number of birds we see or hear declines with distance from the observer. The shape of this decline, the distance function, differs among species, among observers and, importantly, among habitats—birds within open grassland are detectable over greater distances than those within dense forest—even when they occur at the same densities. *Distance sampling* models the "distance function" and estimates density taking into account both the birds that were observed, plus those that were likely to be present but were not detected. This method is strongly recommended.

Distance sampling provides an efficient and simple way of estimating bird density from field data. It allows for differences in conspicuousness between habitats and species (though not observers), enabling comparisons to be made between and within species, and across different habitats at different times. Density estimates improve with the number of birds recorded—a minimum of about 80 records is recommended. The method relies on a number of assumptions which need to be evaluated carefully in the field and steps taken to lessen their effects (Buckland *et al.* 2001). The key assumptions of distance methods are that all the birds actually on the transect line or at the counting station are recorded (for cryptic and shy species this may not be true), and that birds do not move in response to the observer prior to detection.

2.3.8 Colonial birds

Around 15% of bird species nest in colonies, either on cliffs, in trees, on the ground, in caves or in burrows. In some ways, this makes them easy to count, since birds are concentrated in generally conspicuous aggregations. However, counting birds in colonies also poses problems:

- Numbers may be huge, making counting difficult; it may be necessary to sample parts of the colony (using strategies described above) and extrapolate.
- Breeding may not be synchronous. At any time, part of the population might be elsewhere, and the birds present on the second visit might not necessarily be those present on the first; individual marking of birds may be necessary.
- There may be large numbers of non-breeders or "helpers" present, or birds might be absent from the colony for long periods; it may be better to count nests rather than individuals.
- Old nests might appear to be active; it might be advisable to count apparently active or occupied nests only.
- Colony attendance might vary greatly during the day and over the year; it may be necessary to make a number of counts at different times.

A critical step is to decide what it is that you want to count. Is it the total number of birds present, the number of breeding pairs, the number of apparently active nests, or the number of occupied burrows? This decision will help to determine the count method used.

Counts of large colonies often involve breaking the colony down into smaller units for ease of counting. In the case of cliff colonies, photographs can be used to divide the cliff into counting units, or even to count the birds directly. Cliff colonies should always be counted from opposite the colony rather than from above when nests are more easily missed. Aerial photography has been used to estimate numbers of large colonial birds, such as Gannets *Morus bassanus*. Tree-nesting colonies can be counted in a similar fashion, with nests in either all trees being counted or just a sample of trees. Large colonies of ground-nesting birds can be subdivided into smaller counting units by using a grid system marked out with string. The counters can then visit all, or a random stratified or regular sample of grid squares. Alternatively, densities of nests can be estimated using *distance sampling* (see above) and extrapolated for total colony area. Burrow-nesting seabirds are particularly difficult to count, many of them return to land after dark, and burrows may be occupied by more than one pair, or they may be unoccupied. It is possible to assess whether burrows are occupied using playback methods (although you need to know or measure the response rate), endoscopes, smell, or by planting toothpicks around the entrance to the burrow and seeing whether these get knocked over (but beware pre-breeding birds that are prospecting for nest sites). Multiple occupancy of burrows is difficult to detect and remains a problem. Steinkamp *et al.* (2003) provide a practical and detailed review of survey methods for seabirds and colonial waterbirds.

2.3.9 Counting roosts and flocks

Counting large aggregations of birds away from breeding colonies poses many of the same problems as counting birds in colonies, but with some additional considerations:

- If disturbed by the counter, birds are unlikely to return to the same place; observers need to maintain a distance.
- Birds may be closer together than when they are in nesting colonies where they tend to space themselves out, so great care is needed to count those present.
- Flocks often contain several species; it is necessary to count each separately.
- Some aggregations, such as roosting flocks, form for only short periods, often when light conditions are poor. Counts of nocturnal roosts often require the use of photography or of counts of groups of birds joining the roost.

Stationary flocks of up to 500 birds can be counted directly with relative ease if conditions are good. For larger flocks, and for rapidly moving flocks, photography or estimation methods are needed. A common method when estimating very large flocks is to count, say, 10, 20, 50, 100, or 500 birds and then estimate what proportion of the flock this represents. An important consideration when using this method is that birds in flocks do not tend to be evenly distributed, with higher densities in the center of the flock and lower densities at the periphery. Alternatively, for wading birds feeding on open mudflats and waterbirds on lakes, the flock can be broken down into smaller counting units using natural features of the habitat or distant landmarks. When birds are in dense groups, accurate counts are only possible by counting from above, or by counting them as they enter or leave an area. Care is needed so that counting does not disturb the birds; count from concealed or raised positions. The exception to this rule is the *flush* method in which birds are deliberately flushed into the air in order to get a better count of numbers (see Steinkamp *et al.* 2003). Coastal birds might be more easily counted at particular stages of the tide, for example, at high tide roosts, than when more dispersed over a larger area. Photography is a useful method, but in tightly packed flocks, many birds may be obscured. For larger birds, aerial or even satellite photography gets around this problem, although identification may be difficult. A general consideration when counting flocks is that observers show a natural tendency to overestimate small flocks and underestimate large flocks, although the extent to which different observers do this varies greatly. Furthermore, most observers estimate the size of larger flocks far less accurately than smaller flocks. It is always helpful for individual counters to repeat their own section counts and compare them with those from another observer.

For flocking species that disperse to feed over wide areas, it is often advisable to count the birds as they enter or leave roost sites at dawn or dusk, particularly where the sites are used traditionally and predictably.

2.3.10 Counting leks

In a small proportion of birds (around 150 species), males gather in communal gatherings, known as leks, to display and compete for females during the breeding season. At this time, a high proportion of males can be detected at a relatively small number of often traditionally used sites. One or two counts of the leks may be sufficient to give a reasonable and efficient census of the local population. There are downsides to this method however. For example, you need to be sure that all the leks present in an area have been detected, as birds can move between leks, and the smaller they are, the harder they are to find. Counts restricted to the largest traditional leks may well sample a specific group of birds and we do not know the area from which the birds came. In addition, some males may not choose to visit leks and this is particularly true for younger males. Finally, lek counts provide a poor means of surveying female birds.

2.3.11 Counting migrants

Counting large, diurnal migrants, such as raptors, cranes, storks, and pelicans, where they pass through migration bottlenecks, is often more efficient and easier than trying to count them when dispersed over huge breeding or wintering grounds, although this only samples birds that are low enough to be seen. In Israel, counters are arranged in a line across the front of migration and use radios to ensure that no more than one observer records each large flock of migrating birds. As migration can take place at great heights, observers often count in teams, continually scanning the sky and working together. Similar coordinated raptor counts occur across North America where their potential for population monitoring has been explored (Lewis and Gould 2000).

Estimation of the numbers of smaller nocturnal migrants is particularly difficult, but considerable progress has been made in this field (www.birds. cornell.edu/brp). Many smaller migrants call as they migrate, allowing at least minimum numbers to be assessed and species to be identified. Recently developed methods use microphones and complex computer programs to try to estimate total numbers of calling birds passing, as well as their height and speed (Evans and Rosenberg 2000, www.birds.cornell.edu/brp). Radar has been used to not only detect passing flocks, but also to estimate their numbers, direction of flight, speed, altitude, and even wing beat rate, but not their specific identity. This method requires access to extremely sophisticated, and usually militarily sensitive equipment and is generally beyond the reach of most researchers. Counts of migrants passing in front of the moon, or passing through the beams of bright lights, are of limited use, because only a small proportion of birds can be seen and most cannot be identified. A further indirect method of measuring changes in numbers of migrants, although not the absolute numbers, is ringing (banding), and a high proportion of ringing effort is concentrated at migration stopover points (Dunn et al. 1997). These methods are described in detail in Chapter 7.

2.3.12 Capture techniques

Because most species of bird tend to be visible and vocal, methods to survey them generally rely on observers seeing or hearing them. Occasionally, however, this may not be the case, as in species that live in dense undergrowth, or in the forest canopy, which may be rarely seen or heard. Under such circumstances, one way to census them is to catch them using mist nets. Capture techniques have been widely used in the tropics where they can be usefully combined with other census methods (e.g. Whitman et al. 1997). Broadly, two separate approaches can be used; either *capture-mark-recapture* (also known as mark-release-recapture, MRR) which allows estimations of population size, or *catch per unit effort* which can be used to produce population indices.

Capture methods can be time consuming and require substantial training to develop the skills necessary to catch, handle, and mark birds. The safety and welfare of the birds are always of paramount importance. In many countries, these techniques are licensed, and anyone considering using them should apply to the relevant authority well in advance. As we have seen in the previous chapter, mist netting is a relatively poor method for surveying birds. Further information on methods of capture and marking are given in Chapter 4. Despite these disadvantages, capture techniques yield much information besides population size and trend estimation. In particular, they can provide valuable information on demographic parameters, such as survival and breeding success, in addition to information on bird movements. Chapter 5 covers these issues in detail.

The principle behind standard effort *capture-mark-recapture* is that, if birds are caught and individually marked (e.g. with rings or bands), then from the ratio of marked to unmarked birds subsequently recaptured, population size can be estimated. Imagine that on the first day of capture at a site, 100 birds of a particular species were caught in nets, marked, and released. A week later, the nets were put back up. This time 50 of the same species were caught, 25 of which had been marked on the first day. If we assume that the population is closed and the original 100 birds caught had become fully mixed back in the population over the intervening week, then the total population size of the species on the site is 200. That is, we assume that the proportion of birds caught on the second date that were marked (25/50 or 50%) is the same as that in the total population on the site. Because we know that the number of marked birds is 100, then the total population is twice that, that is, 200. Expressed mathematically: the total population size, $P = n_1 n_2 / m_2$ where n_1 is the number caught, marked, and released on the first date, n_2 the number caught on the second date, and m_2 is the number of those caught on the second date that were marked. In practice, there is no need to actually catch birds on the second date, as they could be recorded by walking around the site trying to see as many birds as possible and recording those that were marked.

While the capture-mark-recapture approach may seem simple, it is in practice fraught with problems because it relies on a suite of assumptions, many of which may be untrue. For example:

- It assumes that birds mix freely within the population and this may rarely be the case.
- It assumes that the population is closed and that no birds enter or leave the population, either through births, deaths, or movements.
- It assumes that marking does not affect the probability that a bird will be recaptured, and that marked birds have the same probability of survival as unmarked birds.
- It assumes that marks do not fall off or become less visible.

While many of these assumptions may be broken, it is possible to plan fieldwork to minimize their influence on the results. For example, if the first and second capture dates are reasonably close together, the study site is well defined, and the study is undertaken outside of the breeding and migration periods, then the population will more approximate a closed one.

An array of mathematical models has been developed to analyze data from capture-mark-recapture studies. While it is not within the scope of this chapter to go into these methods, a range of approaches is available. The simplest of these, which is known as the Lincoln index (or Petersen method) assumes one capture and one recapture (or re-sighting) event only, and that the population is closed. The calculations for this model are essentially those described above. More complex models allow for multiple capture (re-sighting) events, and for open populations. The latter types of model, generally known as Jolly-Seber models, provide information on both population size and survival rates. Further information on these models is given in Chapter 5.

The principle behind *standard effort capture* is that populations of birds can be reliably monitored by capture methods if capture effort is kept constant over time, and done at the same season each year. Several programs for monitoring birds with this method exist, but perhaps the best known is the Constant Effort Sites scheme of the British Trust for Ornithology (Peach *et al.* 1996, www.bto.org/ringing/ringinfo/ces/index.htm), which is being followed by an increasing number of European countries. The Monitoring Avian Productivity and Survival (MAPS: www.birdpop.org/maps.htm) program is a similar initiative in North America.

Catch per unit effort data can be used to:

- Monitor population trends of adult birds, based on the numbers caught.
- Estimate absolute population size using the capture-mark-recapture methods outlined above.
- Monitor changes in productivity using the ratio of juveniles to adults caught late in the season.
- Estimate adult survival rates from between-year re-traps of ringed (banded) birds (see Chapter 5).

For the Constant Effort Sites scheme, the capture method involves placing the same types (e.g. mesh size) and lengths of mist nets (see Chapter 4), in the same positions, for the same length of time (about 6 h per visit) over a series of 12 visits during May to August. These methods are held constant from year to year. All birds caught are identified, aged, and sexed, and all un-ringed birds are ringed. While it might be tempting to vary net lengths from visit to visit, particularly if the number of fieldworkers varies from visit to visit, this could influence the catches. Simply calculating the number of birds per 10 m of net is insufficient, because doubling net lengths does not necessarily double the number of birds caught. Similarly, catching for twice as long with half the length of nets on some visits is not advised as capture success may vary with time of day.

Constant effort ringing is commonly used in dense habitats (scrub, reed beds, undergrowth, etc), but it can also be used in forest canopies, with nets raised high above the ground using pulleys or telescopic poles. Because some dense habitats, such as scrub and reed bed, can be successional, care needs to be taken to ensure that population trends reflect real changes in bird numbers rather than local habitat change around the nets. Although the Constant Effort Sites scheme uses mist nets, any accepted capture technique (Chapter 4) can be used, providing that effort is standardized (same number of traps, places, time periods, etc).

As a general survey or monitoring tool, catch per unit effort has some limitations, such as requiring specialist equipment and training and thus being expensive to maintain.

2.3.13 Tape playback

Some species of bird are particularly difficult to see or hear. Examples of such species are those that have skulking behavior, live in dense habitats, are nocturnal or crepuscular or nest down burrows. The probability of detecting these species can sometimes be increased by the use of tape playback, in which the taped call or song of a bird is played, and a response listened for. Recordings of the calls and songs of many species are now commercially available, and can be copied to tape. Ideally, use a tape loop, so that a short length of call can be repeated continuously for as long as is required. The call can be broadcast from a simple hand-held loudspeaker but care is needed to keep disturbance to a minimum and not to affect the bird's natural behavior.

The results from census work involving tape playback need careful interpretation. If the aim is simply to determine whether a given species is present in an area, then tape playback may simply increase the chance of finding it. If, however, the aim is to estimate population size or to produce a population index, then more care is needed. To generate a reliable population index, the probability of birds responding to the tape needs to be held as constant as possible. This can be helped, for example, by standardizing the manner in which the tape is played (same volume, recording, playback length, time of day, season, etc), and ensuring that the tape is not played to any one individual too frequently, causing it to habituate and respond less frequently. Tape playback has been used widely for monitoring populations of marsh birds, owls and raptors (Gibbons *et al.* 1996; Newton *et al.* 2002; Lor and Malecki 2002).

Estimating absolute population size from tape playback is more complex, as the probability of the average bird in the population responding to playback needs to be known. Frequently, detailed additional work will be required to determine response probabilities. Such work has been undertaken on owls and nocturnal burrow-nesting seabirds. For example, Brooke (1978) has shown that responses to playback of their call were obtained only from half of all occupied Manx

Shearwater *Puffinus puffinus* burrows. Detailed observations on incubating birds showed that this was because males and females shared incubation equally, but that only males responded to playback. Playing the tape into numerous burrows, counting the number of responses, and doubling this number could thus yield an estimate of the overall population. Unfortunately, response probabilities are not always constant. In their studies of Storm Petrels *Hydrobates pelagicus* Ratcliffe *et al.* (1998) have shown that response probabilities vary among years and colonies, and the cause of this variation is unknown. To estimate population size, it is thus necessary to determine year-specific and colony-specific response probabilities.

2.3.14 Vocal individuality

The songs and calls of many bird species are unique and often identifiable at the level of an individual, if not by ear, then from a sonogram. Acoustically distinct calls of this kind have considerable potential in monitoring and conservation, particularly for birds that occur in dense vegetation or are otherwise difficult to observe, but this potential has not always been realized (McGregor *et al.* 2000). The method involves recording songs or calls with a directional microphone and examining sound spectrograms using freely available software. The spectrograms from an individual bird are often recognizable by eye and discrimination can be formalized using statistical techniques.

Work on Bitterns *Botaurus stellaris*, in Britain has shown that their booming calls are individually quite distinct. This has allowed their numbers to be monitored more accurately and their year-to-year survival to be estimated (Gilbert *et al.* 2002). In a study of the Corncrake *Crex crex* information gained from vocalizations increased census estimates by some 20–30% (Peake and McGregor 2001), and showed that males called less frequently than was previously thought. The churring call of male European Nightjar *Caprimulgus europaeus*, a mainly nocturnal and mobile species, has been shown to differ between individuals (Rebbeck *et al.* 2001). The pulse rate of calls and the phase lengths together allow identification of nearly 99% of males. Interestingly, males were shown to move some distance within a breeding season, but return to the same territory year after year. It is hard to see how these insights could have been gained by other methods. One can also apply *capture-mark-recapture* methods to re-sightings based on vocalizations to estimate population size. In contrast, although the calls of Black-throated Diver *Gavia arctica* are distinct, the method proved impractical as a monitoring tool because calls are infrequent and difficult to record (McGregor *et al.* 2000). In each case, quantitative rules were developed to help discriminate one bird from another, but this is not always straightforward and, in some cases, ambiguity remains.

An advantage of this method is that it is non-intrusive, which might be particularly useful in studying rare and endangered species. The disadvantages are: that it requires high quality recording of birds that often live at low densities across scattered sites; ideally, one needs an independent means of identification, such as marking or radio tracking, to corroborate the findings; it requires specialist and quite expensive equipment; it often tells us only about breeding males; and it can be time-consuming, unless the analysis is automated (see Rebbeck *et al.* 2001).

2.4 Conclusions

A whole variety of different approaches can be used in surveying birds, but a series of questions need to be asked before work can begin. For example, are we interested in relative or absolute abundance, or a population index instead of a population estimate? As we have seen, it is vital to establish the objectives of the survey at the outset and consider their practicality and relative priority. The survey objectives will interact with, and be influenced by, the sampling strategy (choosing where to count) and the field method (how to count); these taken together define our survey design. A number of generic rules help us decide how to select our survey plots; random stratified and regular sample designs are best. Stratification should always be considered. Furthermore, a number of rules allow us to choose between survey methods and apply them in an appropriate fashion. We recommend line and point transects as the two most adaptable and efficient methods for most surveys. While each survey must be tailored to a particular situation, the common application of field methods will greatly enhance our ability to compare across studies; and we make some practical suggestions. A number of specialized and often more intensive techniques are available for survey and research purposes.

References

Arendt, W.J., Gibbons, D.W., and Gray, G. (1999). Status of the volcanically threatened Montserrat Oriole *Icterus oberi* and other forest birds in Montserrat, West Indies. *Bird Conserv. Int.*, 9, 351–372.

Bart. J. and S. Earnst. (2002). Double sampling to estimate density and population trends in birds. *Auk*, 119, 36–45.

Bennun, L. and Howell, K (2002). African Forest biodiversity: a field survey manual for vertebrates, eds. G. Davies, and M. Hoffmann, Earthwatch Europe, Earthwatch, Oxford.

Bibby, C.J. (1999). Making the most of birds as environmental indicators. *Ostrich*, 70, 81–88.

Bibby, C., Jones, M., and Marsden, S. (1998). *Expedition Field Techniques: Bird Surveys*. Royal Geographical Society, London.

Bibby, C.J., Burgess, N.D., Hill, D.A., and Mustoe, S.H. (2000). *Bird Census Techniques*, 2nd ed. Academic Press, London.

BirdLife International (2000). *Threatened birds of the world*. Barcelona and Cambridge, Lynx Edicions and BirdLife International, UK.

Brooke, M. de L. (1978). Sexual differences in voice and individual recognition in the Manx shearwater (*Puffinus puffinus*). *Anim. Behav.*, 26, 622–629.

Buckland, S.T., Anderson, D.R., Burnham, K.P., Laake, J.L., and Borchers, D.L. (2001). *Introduction to Distance Sampling: Estimating Abundance of Biological Populations*. Oxford University Press, New York.

Carter, M.F, Hunter, W.C., Pashley, D.N., and Rosenberg, K.V. (2000). Setting conservation priorities for land birds in the United States: the partners in flight approach. *Auk*, 117, 541–548.

DeSante, D.F. (1981). A field test of the variable circular plot censusing technique in a California coastal scrub breeding bird community. In *Estimating Numbers of Terrestrial Birds*, eds. C.J. Ralph, and J.M. Scott. Studies in Avian Biology, 6, 177–185.

Dunn, E.H., Hussell, D.J.T., and Adams, R.J. (1997). Monitoring songbird population change with autumn mist netting. *J. Wildlife Manage.*, 61, 389–396.

Evans, W.R. and K.V. Rosenberg. (2000). Acoustic monitoring of night-migrating birds: a progress report. In *Strategies for Bird Conservation: The Partners in Flight Planning Process*. Proceedings of the 3rd Partners in Flight Workshop; Cape May, NJ. October 1–5, 1995.

Farnsworth, G.L., Pollock, K.H., Nichols, J.D., Simons, T.R., Hines, J.E., and Sauer, J.R. (2002). A removal model for estimating detection probabilities from point count surveys. *Auk*, 119, 414–425.

Gibbons, D.W., Hill, D., and Sutherland, W.J. (1996). Birds. In *Ecological Census Techniques: A handbook*, ed. W.J. Sutherland, pp. 227–259.

Gilbert, G., Gibbons, D.W., and Evans, J. (1998). Bird Monitoring Methods—a manual of techniques for key UK species. RSPB, Sandy.

Gilbert, G., Tyler, G.A., and Smith, K.S. (2002). Local annual survival of booming male Great Bittern *Botaurus stellaris* in Britain, in the period 1990–1999. *Ibis*, 144, 51–61.

Greenwood, J.J.D. (1996). Basic techniques. In *Ecological Census Techniques: A handbook*, ed. W.J. Sutherland, pp. 11–110.

Gregory, R.D. (2000). Development of breeding bird monitoring in the United Kingdom and adopting its principles elsewhere. *The Ring*, 22, 35–44.

Gregory, R.D. and Baillie, S.R. (1998). Large-scale habitat use of some declining British birds. *J. Appl. Ecol.* 35, 785–799.

Gregory, R.D., Wilkinson, N.I., Noble, D.G., Brown, A.F., Robinson, J.A., Hughes, J. Procter, D.A., Gibbons D.W., and Galbraith, C.A. (2002). The population status of birds in the United Kingdom, Channel Islands and Isle of Man: an analysis of conservation concern 2002–2007. *Br. Birds*, 95, 410–448.

Gregory, R.D., Noble, D., Field, R., Marchant, J.H, Raven, M., and Gibbons D.W. (2003). Using birds as indicators of biodiversity. Ornis Hungarica (in press).

Haselmayer, J. and Quinn, J.S. (2000). A comparison of point counts and sound recording as bird survey methods in Amazonian southeast Peru. *Condor*, 102, 887–893.

Lewis, S.A. and Gould, W.R., (2000). Survey effort effects on power to detect trends in raptor migration counts. *Wildlife Soc. Bull.*, 28, 317–329.

Lor S., Malecki, R.A. (2002). Call-response surveys to monitor marsh bird population trends. *Wildlife Soc. Bull.*, 30, 1195–1201.

Loh, J. (ed.) (2002). Living Planet Report 2002. WWF Gland Switzerland.

Marchant, J.H, Hudson, R., Carter, S.P., and Whittington, P.A. (1990). Population trends in British breeding birds. British Turst for Ornithology, Thetford.

McGregor, P.K., Peake, T.M., and Gilbert, G. (2000). Communication behaviour and conservation. In Behaviour and Conservation, eds. L. Morris Gosling and W.J. Sutherland, pp. 261–280.

Nichols, J.D., Hines, J.E., Sauer, J.R., Fallon, F., Fallon, J., and Heglund, P.J. (2000). A double-observer approach for estimating detection probability and abundance from avian point counts. *The Auk*, 117, 393–408.

Niemi, G.J., Hanowski, J.M., Lima, A.R., Nicholls, T., Weiland, N. (1997). A critical analysis on the use of indicator species in management. *J. Wildlife Manage.*, 61: 1240–1252.

Nemeth, E. and Bennun, L. (2000). Distribution, habitat selection and behaviour of the East Coast Akalat *Sheppardia gunningi sokokensis* in Kenya and Tanzania. *Bird Conser. Int.*, 10, 115–130.

Newton, I., Kavanagh, R., Olsen, J., and Tyalor, I. (2002). *Ecology and Conservation of owls*. CSIRI Publishing, Collingwood, Victoria, Australia.

Peach, W.J., Buckland, S.T., and Baillie, S.R. (1996). The use of constant effort mist-netting to measure between-year changes in the abundance and productivity of common passerines. *Bird Stud.*, 43, 142–156.

Peake, T.M. and McGregor, P.K. (2001). Corncrake *Crex crex* census estimates: a conservation application of vocal individuality. Anim. Behav. Conserv. 24, 81–90.

Ralph, C.J. and Scott, M. (1981) *Estimating numbers of terrestrial birds*. Studies in Avian biology No. 6. Cooper Ornithological Society, USA.

Ralph, C.J. Sauer, J.R., and Droege, S. (1995). Monitoring Bird Populations by Point Counts. www.rsl.psw.fs.fed.us/projects/wild/gtr149/gtr_149.html.

Ratcliffe, N., Vaughan, D, Whyte, C., and Shepherd, M. (1998). The status of storm petrels on Mousa, Shetland. *Scott. Birds*, 19, 15–163.

Rebbeck, M., Corrick, R., Eaglestone, B., and Stainton, C. (2001). Recognition of individual European Nightjars *Caprimulgus europaeus* from their song. *Ibis*, 143, 468–475.

Rosenstock, S.S., Anderson, D.R., Giesen, K.M., Leukering, T., and Carter, M.F (2002). Landbird counting techniques: Current practices and an alternative. *Auk*, 119, 46–53.

Sauer, J.R., Hines, J.E., and Fallon, J. (2001). *The North American Breeding Bird Survey, Results and Analysis 1966–2000*. Version 2001.2, USGS Patuxent Wildlife Research Center, *Laurel, MD*.

Snedecor, G.W. and Cochran, W.G. (1980). *Statistical Methods*. The Iowa State University press, Iowa.

Steinkamp, M., Peterjohn, H., Bryd, V., Carter, H., and Lowe, R. (2003). Breeding season survey techniques for seabirds and colonial waterbirds throughout North America. www.im.nbs.gov/cwb/manual/.

Sutherland, W.J. (2000). *The Conservation Handbook: Research, management and Policy*. Blackwell Scientific, Oxford.

Thompson, W. L. (2002). Towards reliable bird surveys: accounting for individuals present but not detected. The Auk 119:18–25.

Whitman, A.A., Hagan, J.M., and Brokaw, N.V.L. (1997). A comparison of two bird survey techniques used in a subtropical forest. *Condor*, 99, 955–965.

Wilkinson, N.I., Langston, R.H., Gregory, R.D., Gibbons, D.W., and Marquiss, M. (2002). Capercaillie *Tetrrao urogallus*, abundance and habitat use in Scotland, in winter 1998–99. *Bird Stud.*, 49, 177–185.

Wotton, S.R., Carter, I., Cross, A.V., Etheridge, B., Snell, N., Duffy, K., Thorpe, R., and Gregory, R.D. (2002). Breeding status of the Red kite *Milvus milvus* in Britain in 2000. *Bird Stud.*, 49, 278–286.

References

3

Breeding biology

Rhys E. Green

3.1 Introduction

Collecting information on breeding biology and performance is an important part of many studies of the population ecology of birds and is often essential in identifying effective conservation measures for threatened and declining species. As with all scientific research, care is needed in designing an appropriate program of fieldwork on breeding birds. It is important to think through in detail the questions you wish to answer. For example, do you want to estimate the mean number of young reared per female or pair in the breeding season or is it also important to know about the success of individual breeding attempts and the causes of failure? Make a preliminary model of the population you are studying or the sequences of events in the breeding cycle of the individuals within it. This might be a simple flowchart describing how you think the events in the breeding season of an individual female relate to one another and to her environment. What determines whether or not she attempts to breed, when she begins her first breeding attempt, and how many eggs she lays? What influences the chance that her nest fails? Does she renest after failure? How long does that take? What influences whether she re-nests or not? It may seem strange to do this before you begin your study. Many people think you should collect a lot of information first and only then use it to construct a model. However, building a preliminary model based upon whatever is already known about the birds you are studying or, failing that, what is known about other species in the scientific literature, will help you to organize existing knowledge for many bird species. There is much useful information on basic aspects of breeding biology, especially clutch size, duration of incubation, and nestling period, and time of breeding, and this is well summarized in ornithological handbooks. Most importantly, constructing a preliminary model will force you to make explicit the assumptions you are making about how the things you can observe and measure relate to the reality you want to know about.

3.2 Choosing study areas

Where will you carry out your study? For a small population of a rare, threatened species there may be no choice, but most studies of breeding birds are conducted in particular areas that cover only a small part of the species' total or regional range. Often the study area is chosen by the researcher because it is easy to get to, because the owner of the land is friendly, because it has many of the birds that you wish to study or because they are easy to observe there. For many applications, this is not a good way to choose a study area. If you want to make estimates of breeding parameters that are typical of the population as a whole, then it would be better to have several study areas chosen in such a way that they are representative. These might be a sample of sites chosen at random or at least from representative habitats or regions. If your study concerns the causes and correlates of rarity or population decline, then it may be appropriate to select study areas with a range of population densities or recent population trends. Comparisons among these areas may then help to identify factors that influence population status as well as breeding performance.

3.3 Measuring the success of individual breeding attempts

3.3.1 Finding and selecting nests

Studies of the success of individual breeding attempts usually involve, as the first step, locating nests that are being used by birds. In some species with durable nests it may be possible to find them at the end of the breeding season and assess whether they have been used and whether they have fledged young from signs. However, pilot studies would be required to confirm that the probability of finding a nest was not related to its outcome and that the signs gave a reliable indication of use and outcome.

The ideal stage at which to find nests is before egg-laying has occurred. If some nests fail, and if the probability of failure is correlated with attributes of the environment or parents, then nests found at later stages will be a nonrandom sample because those that fail early will have been selected out. There may even be some pairs or females that make a single breeding attempt and then leave the study area if it fails. To overcome this problem it is not necessary for all nests to be found pre-laying, but it is desirable to have as high a proportion in the sample as possible.

Some species have nests that are easy to find and access, but if finding or accessing nests is difficult, then there is a risk that you will be studying an unrepresentative sample which may have a different success rate and differences in causes of failure. Cold searching, that is searching visually for nests in all potential nesting habitat in the study area, is a frequently used method for finding nests. For some species of woodland raptors that re-use nests for several years or build new nests near to old ones, it can be productive to search for old nests in winter when deciduous trees have shed their leaves and nests are easier to see. This gives a good indication of where there are likely to be active nests in the spring. There is a danger that cold searching can give a biased sample of nests for study because some habitats are easier to search than others. For example, several bird species that nest on farmland in Britain build hidden nests in hedgerows, the lines of shrubby vegetation on uncultivated field margins, but also nest in field crops. Cold searching for nests by carefully looking in trees, bushes, and ground vegetation is a practical, though time-consuming, method for finding nests in hedgerows; but it is unproductive in field crops because, even though there may be as many nests in fields as in hedgerows, the total area of the fields is much larger and farmers dislike researchers walking through their crops. Nests in hedgerows are probably at greater risk of being located by predators, but nests in crops are at risk from farming operations.

For species that can easily be watched, potential bias of this kind can be overcome by first finding a bird and then watching it back to its nest during visits for nest building, incubation changeovers, or feeding nestlings. The use of a hide, or using a car as a hide, makes this practical for many species. Some ground nesting birds can be watched back to the nest by an observer who has climbed a nearby tree and are especially unwary if the observer has a companion who leaves the area. Providing that birds can be located and watched with similar ease in different habitats, then watching back may well yield a less biased sample of nests than cold searching. The efficiency with which nests can be found in this way can be increased if the observer is aware of the significance of special clues provided by behavior or signs. In many species of galliform birds, females that are foraging during a break from incubation peck at food much more rapidly than normal and they also produce unusually large droppings because they accumulate fecal material during long incubation stints. Other signs that can draw the observer's attention to a nest include anxiety calls, carrying of nestling droppings away from the nest by passerines, carrying food or nest material and displays used by male birds when leading their mate to a nest site. Changes in the height or density of foliage during the breeding season can also affect the ease with which nests can be found and may lead to undersampling of late nests. This can lead to serious bias in estimates of nest success if success varies markedly with time of year. As with habitat differences, it may be that finding nests by watching birds back to them is less susceptible to the effects of vegetation changes than is cold searching.

For species that are secretive and conceal themselves in vegetation, another possible solution to finding an unbiased sample of nests is to radio-tag adult birds and later track them to their nests. Radio-tagging is expensive and time-consuming, so choosing this method is likely to put severe limits on the number of nests you can study. However, this approach can work well for difficult, secretive species providing that a representative sample of adults was tagged and that their nest site selection is unaffected by tagging. An example of this is the finding of a sample of Corncrake *Crex crex* nests by radio-tracking females. Corncrake nests are on the ground in dense vegetation and are difficult to find by cold searching. Furthermore, females are likely to desert the nest if disturbed from it. The beginning of incubation can be recognized by radio-tracking because the female remains still in one place for long periods and this can be detected from the pattern of fluctuation in the radio signal. The radio signal fluctuates when the female is walking because of changes in the orientation of the transmitter antenna. When a female was found to be inactive at the same location on several occasions, her signal was then monitored continuously. Incubating corncrakes leave the nest about once per hour for 10–15 min to feed. The start of a feeding period was identified from the change in fluctuations in the radio signal and the observer then went to the presumed nest location and searched for it (Green *et al.* 1997).

Sometimes nests are easy to find, but some are more difficult to access than others. Examples of this are nests on cliffs, in tall trees, and in deep holes. Since accessibility to researchers may well be correlated with accessibility to predators, it is important to ensure that nest checking methods are developed to the point that the sample of nests that can be observed is not markedly biased towards those that are easiest to get to. Specialized climbing techniques and devices such as endoscopes for examining the contents of hole nests may be needed.

3.3.2 Recording the stage of a breeding attempt when it is located

There are various advantages in having estimated the stage of a nesting attempt when it is first found. Counting the eggs and feeling whether they are warm or not can provide clues about whether incubation has started. The blunt end of the egg can be checked for star-shaped cracking caused by the chick preparing to hatch. These signs can sometimes be seen 2 days or more before hatching. By learning how long it takes for a typical egg of your study species to proceed from the first signs of cracking, through stages of progressive enlargement of a hole in the shell to emergence you can estimate when the egg is likely to hatch relative to the time of your observation. If there are no visible signs, the weight of the egg relative to its size can provide useful information. Eggs lose about 15% of their initial weight during incubation, mainly because of water loss (Ar and Rahn 1980). If eggs are weighed and their length and maximum breadth measured, the ratio of weight to the product of length and the square of breadth can be calculated, and from it the proportion of the incubation period elapsed. Simple home-made charts can be used to estimate this in the field (Green 1984; Galbraith and Green 1985). It can be difficult to measure small eggs safely and accurately, so an alternative method is to place the egg in a transparent vessel containing tepid water. Recently laid eggs lie on their side on the bottom of the vessel with their blunt end slightly raised. As incubation proceeds, the blunt end rises higher, though the pointed end remains on the bottom. The observer estimates the angle that the long axis of the egg makes with the horizontal floor of the vessel to quantify this change. As incubation advances the long axis of the egg eventually rises to be oriented vertically. Next the egg floats to the surface with the blunt end uppermost. The amount of shell that rises above the surface then increases as incubation progresses (van Paassen *et al.* 1984). Eggs can also be candled, that is placed on a strong light source so that the shadows cast by the developing embryo and blood vessels can be seen through the shell. However, this is difficult to do safely for small eggs and it may not be possible to see the shadows at all in species with strongly pigmented or patterned eggshells. Medical gloves can be worn to reduce the chance that handling eggs might leave scent on them that could affect the behavior of egg predators or contaminate them with pathogenic bacteria, but using gloves may make it more difficult to avoid breaking small eggs. For nests first found at the nestling stage or precocial chicks located away from the nest, age can be estimated from weights, measurements, or descriptions of plumage development. The information to do this may be found in the literature or from your own measurements and descriptions of chicks whose age you know because you observed them at hatching.

3.3.3 Precautions to take so that nests can be relocated for checking

It may seem obvious to record precisely how to relocate a nest once you have found it, but nests of many species are surprisingly easy to lose. The consequences are worse than just reducing your sample size; it can also bias estimates of nest success. This can arise because a nest being tended by adult birds at the time of a nest check is often easier to relocate than one that has failed. If you are more likely to lose nests that have failed, then nest success calculations based on the remainder will be too high. Researchers often mark nests with a visible artificial marker, such as a stick or tag, to aid relocation. This has the potential disadvantage that predators may also learn to use the markers to find nests

(Picozzi 1975). Domestic livestock and wild mammals such as Hares *Lepus europaeus* are also attracted to sticks and novelties and hence may disturb a marked nest. Using natural markers, such as stones arranged in a particular way, may reduce this problem, but any clue that is distinctive enough for you to use may also be used by a predator. Although these problems do not always occur, it seems best to avoid the risk of using visible nest markers whenever possible. A common solution is to make detailed notes and sketches of how to find the nest from an easily identifiable existing landmark, such as a bush or boulder. This has become considerably easier now that hand-held global positioning systems enable you to navigate quickly to the correct landmark. One way to find the nest location quickly from a landmark is to use a spotter, a short length of 15 mm internal diameter plastic pipe that can be attached to a branch distant from the nest with stout wire. The researcher fixes the pipe and sights through it to the nest site when it is first found and leaves it set in that orientation so that the nest can be found at subsequent checks by sighting it again from the landmark bush or tree (Simon 1998). Sometimes a nest can be very difficult to relocate even when detailed notes are available. For example, the ground nests of waders in cattle-grazed meadows are often destroyed by cows standing, lying, or even defecating on them (Beintema and Muskens 1987). Trampling by cattle can alter the appearance of the vegetation around the nest so much that it is not possible to find it to check for signs that the eggs had hatched before the cow trod on it. Green (1988) overcame this problem by cutting a slit in the turf next to the nest and burying a strip of aluminium kitchen foil a few centimeters below the surface. When nest sites were lost, a metal detector was used to find the foil and nest site.

3.3.4 Nest checking

Nest checks can affect nest success by drawing the attention of predators to the nest or by preventing parents from protecting the nest contents. Sometimes frequent nest checks have little or no effect on nest success (O'Grady *et al.* 1996), but adverse effects on nest success have been detected sufficiently often (Gotmark 1992) that precautions should usually be taken. Disturbance by observers may have particularly large effects at certain stages of breeding. For example, in Eurasian Oystercatchers *Haematopus ostralegus* the effect on the nest success of observers keeping the birds away from their nests was more severe during egg-laying than during incubation (Verboven *et al.* 2001).

It is often possible to avoid visiting a nest and disturbing the parents by viewing it from a distance to see whether a parent bird is incubating or brooding. By looking up with binoculars from the base of a nest tree, it may be possible to see the tail of an incubating or brooding bird protruding over the edge of a nest. Nests can be watched from a distance at the nestling stage and parents seen to be taking in food or removing nestling droppings or the begging calls of nestlings may be heard. If parents are seen at the nest and the stage of development of the breeding attempt was estimated when the nest was first located, then there may be no worthwhile additional information to be gained by disturbing the parents to view the nest contents. For example, if the development of a clutch of eggs was not assessed when a nest was first found, it might be considered necessary to displace a sitting parent bird at the next check to see whether hatching had occurred. However, if the stage assessment indicated that the eggs were unlikely to have hatched, it is sufficient to see the parent sitting on the nest to know that the attempt has not failed. Assessment of developmental stage can also be used to time nest checks to coincide with events of particular interest, such as hatching or the age at which young can be safely removed from the nest for ringing. This often allows the number of checks to be reduced. Other precautions to reduce the risk that nest checking will draw the attention of predators to the nest include walking to and from the nest by different routes so that the nest is not at the end of a track, restoring trampled vegetation, avoiding touching the nest, and visiting and looking at apparently suitable nest sites where no nest really exists so that predators watching or tracking the researcher are not always rewarded when they visit the same places. Martin and Geupel (1993) provide useful detail on these precautions.

It is important to record what is actually seen during nest checks, as well as what is inferred to have happened. It is good practice to prepare a list of signs that may be seen during a nest check and to know what can reasonably be inferred from them. This is especially important for the nest check at which it is discovered that there are no longer eggs or nestlings in the nest. For birds with precocial young, an empty nest that might have hatched or failed should be checked carefully for the small fragments of eggshell that fall into the nest cup when the chicks are chipping the shell open during hatching. Shell remains are taken away or eaten by the parents after hatching in some species including waders and, for these the chippings may be the only clue that hatching occurred. In other species, including many waterfowl, galliform birds, and rails, hatched shells are left in the nest and are distinguishable from damaged shells left by predators by the way the shell has opened. Hatched galliform eggs tend to have a circular cap removed at the blunt end of the egg. The texture of the shell membranes of hatched eggs is brittle and papery especially where shell fragments became detached from the membrane during hatching, whereas the membranes of depredated eggs are usually flexible and adhering to the shell. Alarm calls or distraction displays by parent birds should also be recorded as they may indicate the presence of dependent young nearby.

Empty nests of species with altricial young which contained nestlings at the last check should be examined for evidence that fledging has occurred. The nests of many species of passerine birds are altered in a distinctive way by the presence of large young about to fledge. Typically the nest rim becomes flattened and droppings accumulate in the nest cup, on the rim and outside the nest. These signs are diagnostic of young having at least reached an age close to fledging. An empty nest which lacks these signs, indicates that the nestlings would have been too young to have fledged, may be taken to have failed. However, passerine nestlings may leave the nest prematurely if disturbed and there would then be no signs of fledging. Well-grown passerine nestlings often leave fragments of the cylindrical waxy sheaths that enclosed the growing feathers behind in the nest. This is a sign that nestlings survived to an advanced age before predation or premature fledging.

The area around an empty nest that is thought to have failed should be searched for egg or shell remains apparently left by predators. These should be described carefully for damage characteristics associated with particular predators and preferably collected for later examination. Rigid cardboard or plastic cups with lids can be carried for this purpose. Subtle signs, such as the paired pinpricks left in the shell of some eggs taken by mustelids, are easier to find in a room with the help of a bright lamp and it is often useful to compare egg remains with others you have collected previously. Keeping eggshell remains also makes it possible to show egg remains to an expert on the signs left by a particular predator and compare toothmarks with the skulls of potential predators.

3.3.5 Determination of chick survival for species with precocial young

In those species whose young move away from the nest, it is usually difficult to follow the fates of chicks from a particular breeding attempt. If parents or chicks are individually marked at hatching or soon after, then mark-recapture or mark-resighting methods can be used to estimate the number of young that reach independence (Chapter 5). This can be done by making resightings or recaptures at regular intervals throughout the period when the chicks are dependent or flightless. However, the design of such studies must take into account the large movements of up to several kilometres made by broods of some galliform, anseriform, and charadriiform species. It is easy to mistake movement out of the study area for death of chicks. Having marked parents as well as chicks helps with this problem. Parent birds with precocial young often show conspicuous and characteristic behavior, such as alarm-calling and distraction displays, in the presence of the researcher. If marked parents are seen persistently not to be showing this behavior at a time when their young could not yet have become independent then it can be concluded that their chicks have died. The disappearance of marked parents from the study area can alert the researcher to search further afield for them and their brood. In some species, recently fledged juvenile birds gather at feeding or roosting areas. Searches for individually color marked birds in such areas, together with searches in subsequent years when they are adult can yield mark-resighting estimates of the proportion of marked chicks that survive to independence.

The attachment of small radio tags to chicks allows their survival and fate to be monitored (see Chapter 6). However, this method is difficult to apply to large numbers of chicks and tags and attachment methods may reduce survival. Adverse effects of tags can be checked by comparing the survival or growth of tagged and untagged chicks from the same brood, but a large study will be needed before it can be inferred that any effect is negligible. Another problem is that small tags are usually low powered and consequently have small detection distances. Tags that fall into water after becoming detached, are carried a long way or buried by predators, or are placed on chicks that move outside the study area, are frequently lost to follow up. This makes the estimation of survival rates and the unbiased assessment of causes of death problematic. Radio-tracking of parent birds can often be combined with counts of their young to measure chick survival in species where the parent and young are usually close together. For species in which young roost at night on the ground with their mother, the roost site of a radio-tagged mother can be noted at night and the researcher can locate it on the following day to look for the droppings of parent and chicks. This allows the death of all chicks to be detected from the absence of the smaller chick droppings (Green et al. 1997). The droppings can also be collected to determine diet (Chapter 10). Broods of Grey Partridge Perdix perdix chicks tracked and sampled in this way whose reconstructed diet contained a high proportion of preferred prey insects showed higher survival to independence than broods that fed mainly on less preferred prey (Potts and Aebischer 1995).

The intensive methods described above are necessary when precise estimates of early chick survival are required. However, estimates of age-specific survival of chicks can be obtained from recoveries by the public of chicks marked with numbered metal rings. This method requires that the age of each chick is estimated when it is ringed by taking a standard measurement or record of plumage development. The method relies on the fact that the proportion of ringed chicks recovered after independence is lower for young than old chicks because more of the former die before independence and are therefore not available to be recovered by the public later. If chicks are ringed at a sufficiently wide range of ages, the entire survivorship curve in relation to chick age can be reconstructed, or an index of overall chick survival can be calculated (Beintema 1995).

3.3.6 Estimation of nest success from nest check data

In most studies of nest success, nests are first located after the breeding attempt has begun. If that is the case then taking the proportion of nests found when active that escape predation or other causes of failure as a measure of nest success produces an overestimate (Green 1989). This is because nests first found at more advanced stages of development are exposed to the risk of failure for a shorter time than those found at an early stage. The Mayfield method (Mayfield 1961; 1975) enables the probability of a nest surviving from the beginning to the end of a particular stage of breeding to be estimated from nest check data without this bias. The principle is simple. The number of days after each nest was found for which it was exposed to the risk of a failure that could have been detected by the observer is tabulated together with its outcome (failed or not failed). The number of exposure days is then summed for all nests and a daily failure probability calculated as the number of observed failures divided by the total exposure days. The probability of a nest surviving the whole of the breeding stage is then given by the daily nest survival rate (1 minus the failure rate) raised to the power of the length in days of the stage. In practice nests are not usually checked every day and in calculating exposure days it is assumed that a nest that failed did so halfway through the period during which it could have failed. Suppose that a nest contained young that were estimated to be 10 days old on the penultimate check and that the last check, with signs of nest failure, was made 10 days later, 4 days after expected fledging at 16 days old. The best estimate of the time of nest failure would be midway between the 10th and 16th day of the nestling period. Information on the stage of development of eggs and nestlings and signs recorded at nest checks can often be used to define the beginning and end of the period during which failure could have occurred and improve the accuracy of Mayfield estimates.

It is important to divide the various stages of breeding appropriately before carrying out Mayfield analysis. If two stages that really have substantially different daily nest failure rates are combined and analyzed as if they were one stage, the probability of nest survival over the whole of the combined periods will be overestimated (Willis 1981). In some birds, such as waders (Charadrii), the daily failure probability during egg-laying is much higher than during incubation

(Beintema and Muskens 1987; Ens 1991), but it is often the case that relatively few nests are found during egg-laying, so the evidence for a difference is weak and the temptation is to combine the two periods.

Statistical tests of differences in Mayfield daily nest failure (or survival rates) among breeding stages, study areas or years can readily be conducted by treating each exposure day as if it was a binomial trial during which the nest can fail or not. Standard errors of the daily probability of success or failure can be calculated by the method of Johnson (1979). Daily failure probabilities can be modeled using logistic regression with the generalized linear modeling facilities available in many statistical packages (Etheridge *et al.* 1997; Aebischer 1999) and likelihood-ratio tests used to identify a minimal adequate model (Crawley 1993). This has the advantage that failure probability can be modeled as a function of several variables measured around nest sites and these can include both continuous variables (such as nest height) and categorical variables (such as nest tree species). It is often desirable to know the overall probability of nest survival over a series of successive breeding stages, such as egg-laying, incubation, and the nestling period, for which separate Mayfield estimates have been calculated. This survival rate is simply the product of the rates for the successive stages. The standard error and confidence limits of the product can be calculated using methods described by Hensler (1985). This product can be multiplied by the mean number of young fledged per successful nest to give an estimate of the mean number of young fledged per breeding attempt. Rotella *et al.* (2000) described a method that could be used to estimate both the daily nest failure rate and any adverse effect on it of nest checking. The method uses the assumption that any effect of checking would result in failure soon after the check, so longer intervals between checks yield lower apparent daily failure rates than short intervals.

3.4 Determination of the proximate causes of breeding failure

3.4.1 Signs left at the nest

The evidence left at the nest may be helpful in determining the proximate cause of failure of a breeding attempt, but assessments of this type are likely to be open to error. Consider a hypothetical case in which there are two species of egg predator that each depredate half of the egg-stage breeding failures of the study species. Both predators leave characteristic signs that allow each predator to be reliably identified, but each does so only at a proportion of nests, which differs between the two predators. At the remainder of nests, no signs or egg remains are left by either predator. The researcher, who has no information about the proportion of depredated nests at which each predator leaves signs, concludes erroneously that the predator that is less likely to leave signs is a less frequent predator, whereas in fact the two predators are equally important. Recording of signs is therefore useful in defining a list of predators, but caution is needed in assessing their relative importance.

The signs found at failed nests include intact, cold eggs, holed eggs, shell fragments, egg contents, nestlings dead in the shell, nestling body parts and intact, dead nestlings. Signs may also be found at some distance from the nest. There may also be hairs, tracks, feces or scent left by mammalian predators or feathers, down or feces left by avian predators. Hoof prints and crushed nest contents are usually found at ground nests trampled by cattle, other domestic livestock or deer. Desertion of the nest or the death of a parent or parents is indicated by a check of a previously incubated nest at which cold eggs are found and at which incubation does not resume. Close examination of remains of eggshells and egg contents may permit the identification of a predator from the type of damage to the shell and the spacing between pairs of toothmarks (Green *et al.* 1987). Shells opened by birds tend to have smaller holes with neater edges than those opened by mammals, which often crush large areas of shell. However, in many bird species there are no visible signs left at most of the nests that fail and both bird and mammal predators are capable of taking eggs long distances from nests. If knowledge of the stage of development of the breeding attempt and the absence of signs of hatching or fledging indicate that the breeding attempt cannot have ended successfully since the previous check, then it is usually reasonable to assume that an empty nest has failed because of predation.

3.4.2 Wax or plasticine eggs in the nests of wild birds

Model eggs that resemble those of the study species can be made by filling blown eggshells of similar size with wax or by moulding egg models from wax, modeling clay or plasticine and painting them. The model eggs can be added to clutches of real eggs in natural nests and may retain impressions of the bill or teeth of predators that will aid in identification of the causes of failure if they are compared with skulls, published measurements, or eggs given to captive animals. There are several reasons to be cautious about this method. The behavior of the parent birds may be affected by the addition of the model eggs. Predators may be attracted by the smell of plasticine or paint or repelled if they detect the models for what they are. The method may not work well because model eggs are removed by predators and carried away. Attaching the eggs to the nest with cord or wire to avoid this may affect the parents' behavior. Thorough pilot studies of the method are advisable if it seems to be the only way to identify the causes of nest failure. The use of egg models is considered further in Section 3.5.

3.4.3 Cameras

Video or film cameras that are either triggered by events at the nest or take a picture at fixed intervals (time-lapse) can be used to monitor nests and identify causes of failure. Triggering of the camera by events at the nest using a passive or active infra-red device or mechanical trigger is possible, but the activities of the parent birds are likely to set the camera off frequently and this may overwhelm the system's capacity to store images. Successful use of time-lapse cameras requires that the interval between frames is short enough that a predator cannot visit the nest and leave between frames. This is sometimes no more than a few seconds. Hence, time-lapse camera systems usually require considerable storage capacity, for example, on videotape, and a substantial source of electrical power to operate the recorder. Even then, it may be necessary to visit the system frequently to maintain it and change batteries, films, or tapes.

At night, nests can only be monitored using low-light cameras or flash. The use of infrared LED flash allows time-lapse cameras to be used throughout the day and night without the flash affecting the behavior of parent birds or affecting the behavior of predators (Pietz and Granfors 2000). However, checks should be made that the flash is not emitting any light visible to them, by watching to see whether they are startled when the flash fires.

The use of cameras can lead to biased identification of the causes of nest failure in several ways. The cameras themselves are necessarily visible structures fairly close to nests and may attract or repel predators and grazing animals in the way that nest markers may do (see above). This problem can be reduced by using camouflaged miniature cameras and by burying or otherwise hiding the other equipment. Herranz *et al.* (2002) compared the daily failure rates of real Woodpigeon *Columba palumbus* nests, artificial nests resembling those of this species with and without automatic cameras and nests with camouflaged cameras. The failure rate was lowest for artificial nests with uncamouflaged cameras, with rates for the other groups being higher and similar to each other. This was probably because predatory Magpies *Pica pica* were deterred from visiting nests by the conspicuous cameras. If frequent visits by the researcher to the vicinity of the nest are necessary to maintain the system, then these may attract or repel predators in the same way that visits to check nests may. However, camera systems that use a long cable or microwave link to relay image data from the camera to the video recorder may be used. This may allow the equipment that needs frequent visits to be located far enough away from the nest for disturbance to be avoided.

At present, camera systems that are most likely to allow the collection of unbiased information on the causes of nest failure are costly. Hence, it may not be feasible for a modestly funded research project to collect information at a sufficiently large sample of nests using these systems alone. Using cheaper systems runs the risk that the equipment and maintenance visits may affect predation rates and predator species. A possible compromise in some cases is to use a few cameras to validate the identification of predators based on signs left at or near nests. These signs could then be used to identify predators at large samples of nests by normal nest checking. Of course, if many nests fail without any visible signs being left, then this compromise will be unsatisfactory.

3.4.4 Temperature loggers

A sensor that responds to temperature can be placed in the nest and connected by a fine cable to a data logger, which records the temperature and time at predetermined intervals. The sensor probe can be very small (a few cubic millimetres) and the logger can have a capacity of thousands of records and can be a few cubic centimetres in volume. A small battery with an operating life of weeks or months can power the logger. By fixing the sensor probe into the nest cup and burying or otherwise hiding the connecting wires and data logger, the system can be made difficult to detect. If the logger is deployed when the nest is found, it can be recovered after the end of the breeding attempt and the temperature record can be examined to identify the time of nest failure. The sensor probe is warm when parent birds are incubating or when live nestlings are in the nest, but cools rapidly when the nest is unincubated or the nestlings have been removed or have died, so the time of failure can be discerned and any preceding absences of parent from the nest caused by disturbance may also be detected. The time of day of nest failure does not identify its cause unambiguously, but it provides useful clues. Jackson and Green (2000) used temperature loggers in the nests of waders (Charadrii) to show that there were characteristic signs left at nests that failed at night that were not present at nests that failed by day. When combined with other evidence, this indicated that predation by the nocturnally active Hedgehog *Erinaceus europaeus*, rather than by avian predators active by day, was the most frequent cause of nest failure.

3.5 Using artificial nests to measure nest success and causes of failure

In recent decades many studies have used artificial nests placed in sites that resemble the places where real birds put their nests. Frequently, nests are deployed that contain eggs from a domesticated bird species or some other readily obtained type of egg. The artificial nests are then visited at intervals and the fate of the eggs and the signs left at the nest are recorded as for the real nests of wild birds (see above). Eggs made from plasticine or modeling clay may be used to aid the identification of predators from tooth, claw, or bill marks, and cameras that automatically photograph predators can also be used (see Section 3.4). Nest failure rates can be calculated using the same methods as for real nests. The results are often interpreted as reflecting differences among years, areas, or habitats in the risk of egg predation or nest parasitism and the relative importance of different causes of nest failure of real nests. So many studies of this type have addressed similar questions that it is now possible to carry out meta-analyses of data from large numbers of independent studies (Hartley and Hunter 1998).

This approach has several advantages. The most important is that it is easier to deploy much large numbers of artificial nests than to locate similar numbers of real nests of many species. This enables powerful tests of hypotheses to be made. The method also allows some confounding variables, such as nest density and nest site type, to be manipulated or standardized. However, there are also many disadvantages. Failure rates and causes of failure of artificial nests may not resemble those of real nests for several reasons. Artificial nest sites may differ from the sites of real nests in ways that are significant to predators, such as the extent of concealment. Predators may use different cues, such as observations of parent birds, to find real nests from those they use to locate artificial nests. Parent birds may be capable of preventing predation of their nests by attacking or distracting predators, but this does not occur at artificial nests. Predators may be deterred or attracted by features of the artificial nests or their eggs that differ from those of real nests and eggs. For example, many experiments concerned with forest passerine birds in North America have used the readily available eggs of domesticated Japanese Quail *Coturnix japonica*, but these are larger than the eggs of most of the wild species that were the main focus of the studies (Haskell 1995). Quail eggs were too large for easy consumption by small rodents, which were important predators of real nests in some areas. Small rodents were more likely to damage house sparrow *Passer domesticus* eggs than quail eggs they encountered (Majer and De Graaf 2000), indicating that the eggs of sparrows or other small passerines in the experiments would make them more realistic.

Several comparisons of failures of real and artificial nests in the same areas have indicated that predation rates and predator species differ. Daily predation rates on artificial nests were usually, though not always, higher than those on real nests (Major and Kendal 1996; Davison and Bollinger 2000; Weidinger 2001). King *et al.* (1999) found that artificial nests suffered higher failure rates than real nests even when differences in concealment by vegetation were allowed for by logistic regression. Differences in failure rates might not be important if absolute rates were of less interest than differences among habitats, years, or times of year. Variation in failure rates of artificial nests might be correlated with variation in the failure rates of real nests. However, among year and within season trends of failure rates of artificial and real nests in the same areas are not always the same (Weidinger 2001).

In view of these problems, it is unsafe to assume that the rates and causes of failure of artificial nests reflect those of real nests unless studies that compare the two show that this is the case. Finding and checking enough real nests to obtain a reliable result in a comparative study of this kind might often require as much effort as would be needed to conduct the study wholly on real nests. Hence, studies of artificial nests are not recommended in most circumstances.

3.6 Measuring annual productivity

3.6.1 Why measure annual productivity?

Annual productivity or reproductive success is the number of young fledged per adult female (or adult bird or pair) per year regardless of the number of breeding attempts. In studies of population processes, it provides a measure of mean reproductive success, which can be used in simulation models of the population to estimate the population growth rate. In studies of evolutionary ecology, productivity can be assessed at the level of the individual as a component of fitness. In species in which females never lay more than one clutch of eggs per year, productivity per breeding female or pair is the same as the mean success of an individual breeding attempt and productivity per adult or pair can be calculated from this and information on the proportion of females that attempt to breed. However, most bird species make more than one breeding attempt per year, even if they only produce a replacement clutch after failure of the first clutch at the egg stage, and many species are capable of rearing two or more broods of young per year. Even so, it might seem that the success of individual breeding attempts would be a good index of productivity. The extent to which this is true depends upon how much variation among areas or years there is in the duration of the breeding season, the propensity of females to attempt to breed again after the conclusion of a previous attempt and the time required by the female to produce a new clutch after the end of the previous attempt. In several cases it is known that variation among areas or years in productivity due to these factors is considerable. The duration of the breeding season of the Common Snipe *Gallinago gallinago* and the number of breeding attempts per female were influenced by flooding and soil moisture conditions. Annual and geographical differences in the duration of the breeding season produced more variation in the number of young produced per female than variation in success per nesting attempt (Green 1988). Similarly, breeding productivity of Black-throated Blue Warblers *Dendroica caerulescens* differed among forest habitats, but the differences occurred because more breeding attempts were made per pair in the most productive habitat (Holmes *et al.* 1996). The success rate of individual breeding attempts did not differ. Variation in productivity among years in this species was also mainly accounted for by differences in the mean number of broods that each pair attempts (Holmes *et al.* 1992).

In spite of evidence from the field and from models (Ricklefs and Bloom 1977; Murray 2000) that annual productivity may not be accurately represented by the average success of individual breeding attempts, a literature review by Thompson *et al.* (2001) showed that most published papers refer only to success per attempt.

If there are adult birds that do not attempt to breed it is important to realize that measures of the number of young fledged per breeding adult or pair will overestimate productivity relative to the numbers of all adults. In such cases, it is necessary to estimate the proportion of adult birds that attempt to breed, usually separately for each age class, if the results are to be used in demographic models. Estimating the proportion of birds of a given age that attempt to breed is difficult. In some species, for example, many raptors and seabirds, some non-breeding adults occupy breeding territories or nest sites and regular checking from early in the season can indicate that they did not attempt to breed. It is sometimes difficult to exclude the possibility that a pair did make a breeding attempt, but quickly failed. However, in species in which egg-laying is always preceded by conspicuous displays or nest refurbishment, it should be possible to exclude this possibility providing that territories are visited early and often enough. Other methods for measuring the proportion of non-breeders include analysis of mark-recapture or mark-resighting data (Newton and Rothery 2001) and radio-tracking of birds of known age (Kenward *et al.* 1999). In species in which age classes can be recognized by plumage characteristics, postmortem examination of the reproductive tract of females can be used to estimate the proportion of birds in each age class that have laid eggs at any time previously (Wyllie and Newton 1999). The oviduct of females that have yet to lay eggs is thin and straight while that of females that have laid is wider and convoluted (see diagrams in Wiltschi 1961; Wyllie and Newton 1999).

3.6.2 Productivity from counts after the breeding season

Counts that yield ratios of young to adults can be used to estimate annual productivity in some species. The validity of these counts depends upon whether the survey method is equally likely to detect young and adults and whether young and adults can be distinguished reliably. These conditions are met in species, including cranes, swans, geese, and some galliforms, which are easy to observe and have full-grown young with a distinctive plumage, which remain with their parents. Surveys in autumn or winter then yield proportions of adults or pairs that have reared different numbers of young, including the proportion of females with no young, either because they were unsuccessful or did not attempt to breed. An advantage of this method is that it is relatively easy and can therefore be applied over large areas and sustained for many years. Good examples are the annual estimation of the productivity of grey partridges by counts from a vehicle in August that have continued without a break for more than 30 years on some study areas (Potts and Aebischer 1995) and series of age ratio estimates for holarctic geese and swans in their winter quarters, for example, the age ratio estimates for brent geese *Branta bernicla* extend over more than 50 years.

For most birds, the estimation of productivity is more difficult. In some species, young and adults can be distinguished by direct field observations, but families do not stay together. In such cases, differential movements of adults and young may render counts unreliable as an absolute measure, or even an index, of productivity. For example, radio-tracking showed that young Marbled Murrelets *Brachyramphus marmoratus* were more likely than adults to move away from coastal counting areas after the breeding season (Lougheed *et al.* 2002*a*). The magnitude of this differential movement might vary among study areas and years and invalidate comparisons.

3.6.3 Productivity from captures after the breeding season

Relative numbers of young and adult birds caught in nets and traps soon after the breeding season may be used to assess productivity for species that are difficult to observe or have age classes that can only be distinguished in the hand. This is likely to give an index of annual productivity, rather than an absolute measure because of age differences in susceptibility to trapping, habitat selection, and dispersal rate. If present, these sources of bias might be consistent across years and regions, so age ratios could still provide a valuable index of productivity. However, if bias varied with time or place it could invalidate the evaluation of trends or regional differences. A study of ratios of young (juvenile and first-winter) to older Blue Tits *Parus caeruleus* when first captured in mist nets by volunteer bird ringers in two winters, 1970–71 and 1971–72, showed that young to old ratios in July–September were implausibly high, presumably because of some combination of the effects listed above. However, from October to the following spring there was a consistent difference between the two winters such that sampling at any time within them would have identified that the young:old ratio was higher in 1970–71 than 1971–72 (Krebs and Perrins 1978). Similarly, the relative number of juvenile Bullfinches *Pyrrhula pyrrhula* per adult in mist net catches in October appeared to be a reliable index of productivity (Newton 1999a). The ratio of juvenile to adult Kirtland's Warblers *Dendroica kirtlandii* captured in mist nets in late summer was found to correlate well with an independent measure of annual productivity (Bart *et al.* 1999). However, juveniles were found to be about 1.7 times more susceptible to capture than adults. Age ratios were found to vary considerably among study areas, so a large, representative sample of areas is needed.

An analysis of the proportion of juveniles relative to adults in a 9-year series of season-long catches from a standardized mist-netting program carried out by volunteer ringers showed considerable homogeneity across widely distributed sites and habitats in Britain in the magnitude of year to year changes (Peach, Buckland, and Baillie 1996). This indicated that it may be feasible to use age ratios in samples of birds trapped by ringers as an annual, regional index of productivity.

3.6.4 Intensive studies of breeding

Detailed studies of individually marked birds can produce reliable estimates of annual productivity because successive breeding attempts of the same female can be identified. Such studies are most practicable where potential nest sites are restricted and can all be checked. There are several examples of this from nest box studies of hole-nesting species in which it is known that there is a negligible chance of breeding attempts occurring undetected in natural sites and where ringed individuals are identified during every breeding attempt by capturing them in the nest box. The same applies for many raptors and seabirds with restricted or highly traditional nesting sites. For species that make nests concealed in vegetation, only the most intensive studies can produce full details of every breeding attempt. Because of the early stage at which they found most nests, Roth and co-workers considered that they had located almost all breeding attempts by Wood Thrushes *Hylocichla mustelina* in their study area in a 22-year period

(Underwood and Roth 2002). However, to measure annual productivity it is not necessary to find all attempts, only the successful ones. The breeding attempts of many species become easier to detect as they proceed: for example, because adults alarm-calling or carrying food to large nestlings and especially fledglings are quite easy to detect. Hence, reliable estimates of productivity may be obtained, provided that the study area is carefully and systematically searched for evidence of this kind at intervals shorter than the duration of the period of within which the parents show conspicuous behavior. What is needed is to be sure that at most a negligible proportion of successful attempts have escaped detection.

3.6.5 Indices of productivity from surveys during the breeding season

If an index of productivity, rather than an absolute estimate, is acceptable, then it might be sufficient to map bird territories and then visit them regularly and score evidence of reproductive activity, such as the presence of nestlings or fledglings, on an ordinal scale (Vickery *et al.* 1992). Grant *et al.* (2000) showed that counts of breeding pairs of Curlews *Numenius arquata*, followed by surveys of pairs showing alarm-calling behavior characteristic of birds with chicks, could provide an index of breeding success.

3.6.6 Use of simulation models

If data are available on the success of individual breeding attempts, including those from unmarked adults, then productivity can be estimated using simulation models of breeding. To do this, additional information is needed on the propensity of females to make further breeding attempts after the success or failure of a previous attempt, the time of the season when females cease making further attempts and the interval between successive attempts. Data on nest success and replacement/multiple nesting can be combined in a simulation model to produce estimates of the number of young hatched or reared per female per season (Beintema and Muskens 1987; Green 1988; Green *et al.* 1997; Powell *et al.* 1999). Ricklefs and Bloom (1977) estimated productivity using a more general model which required information on daily nest failure rates, the distribution of nest initiation dates through the season and the durations of various stages of the breeding cycle.

3.7 Timing of breeding

Information on the timing of breeding is often an important component of a simulation model of breeding (see above) and is valuable in its own right, for example, in studies of the effects of weather on breeding biology. Studies that locate individual nests and estimate their stage of development when found yield distributions of first egg-laying dates through the season which are the most widely available data on timing of breeding. However, it should be recognized that these distributions may be biased if the effort made to search for nests, or the ease with which nests could be detected varied through the season, for example, because of the growth of concealing vegetation.

Alternative methods for studying the timing of breeding that do not involve locating nests include finding the chicks of species with precocial young and estimating their age from measurements so that dates of egg-laying or hatching can be estimated. This method is subject to the possible biases due to variation in effort and ease of chick location described above for nests. However, because the chicks of some species are much easier to find than their nests, it can be a useful method, though it only reveals the timing of those nests that hatched young, which may not be typical in their timing of all nests. Beintema et al. (1985) were able to document changes in the distribution of hatching dates of wader chicks in the Netherlands over a 50-year period using the dates on which chicks were ringed.

Data obtained when adult birds are captured can also be used to study the timing of breeding. In some species, it may be possible to identify females which have recently laid eggs by the stretched appearance of the cloacal aperture and incubating parents by the presence of a naked, oedematous brood patch. However, experience of birds at known breeding stages is usually required before assessment of these characters is reliable and in some bird species a distinctive brood patch is not developed during incubation. Assays of the level of vitellogenin-zinc in samples of blood plasma can be used as an index of vitellogenin, an egg yolk precursor, which is elevated during egg formation. Lougheed et al. (2002b) used this method to estimate the time of breeding of a Marbled Murrelet population from samples taken from adults captured at sea. All of these methods require that adults are caught and examined or sampled across a wide range of dates throughout the potential breeding season.

Examinations of mist net captured adult and juvenile Bullfinches were used by Newton (1999b) to determine the timing of the end of the breeding season, which is difficult to determine by finding nests because they become difficult to locate when vegetation is dense in late summer. Adult bullfinches begin their molt after their last breeding attempt and juveniles begin their body molt after fledging. By backdating the onset of moult from observations of juvenile and birds captured while in molt in late summer and autumn, it was possible to estimate the relative numbers of young fledged from early and late broods. This revealed that seasons with good productivity tended to be those in which breeding continued into late summer, so the duration of the breeding season was apparently an important determinant of annual productivity. This method could probably be applied to many other species, though the details of the timing and rate of moult in relation to breeding would first have to be investigated.

3.8 Measurements of eggs and chicks

Measurements of egg size and weight have already been mentioned because they permit the estimation of the stage of incubation. However, estimates of egg volume from measurements with calipers of egg length and width are valuable in their own right because female birds may lay larger eggs when food resources or foraging conditions are good and this can influence the survival of chicks. An index of egg volume can be used (the product of length and the square of maximum width) or the actual volume or fresh weight of the egg could be estimated approximately using coefficients for a species with similar egg shape from Table 1 of Hoyt (1979). More precise estimates of volume that take account of differences in shape among eggs of the same species can be made using photogrammetry (Paganelli *et al.* 1974). This method could also be used for small eggs that are difficult to measure safely with calipers. Alternatively, the fresh weight of small eggs could be determined with a portable electronic balance providing that weighing was done before significant water loss had occurred.

The thickness of bird eggshells is an important correlate of breeding success in species affected by organochlorine pesticide contamination (Newton 1979). An index of shell thickness can be calculated from the weight of the eggshells of blown eggs, usually in museum collections, after correction for the dimensions of the egg and the pieces of shell removed when the egg was blown (Nygård 1999; Green 2000). Alternatively, the thickness of blown eggshells can be measured near the equator of the egg using a specially modified micrometer in which a round-ended needle attached to the moving jaw of the micrometer is passed through the blow-hole in the side of the egg and the thickness of the shell at the other side of the egg is measured between the tip of the needle and the micrometer anvil (Green 1998). It is also possible to measure the thickness or an index of calcium content per unit shell area for living eggs non-destructively using portable X-ray or beta particle backscatter apparatus (Fox *et al.* 1975; Forberg and Odsjö 1983, 1984).

Measurements of chicks are valuable in estimating age, development, and condition. Accurate weights are required, together with measures of body dimensions such as the combined length of the head and bill, tarsus length, maximum chord wing length. For chicks of known age, the weight or some linear dimension relative to the average value from a fitted growth curve for the species gives a good indication of growth relative to the population mean. In species in which bone and feather growth are less affected by food shortage than the rate of increase in body weight, it may be possible to calculate an index of condition for chicks of unknown age by expressing weight relative to some function of linear dimensions. More details of condition indices can be found in Chapter 4.

3.9 Proximate and ultimate causes of breeding failure

It is important to recognize that observational studies lead to assessments about the relative importance of proximate factors influencing breeding success and the causes of failure. However, there may be other ultimate factors that affect breeding success indirectly. Knowledge of these ultimate factors may be useful not only in improving understanding of population processes, but also in identifying factors that may be susceptible to management for conservation purposes. An example is the finding that Cirl Bunting *Emberiza cirlus* nestlings in a threatened population frequently succumbed to predation and starvation (Evans *et al.* 1997). Analysis of the weight gain rates of nestlings showed that young in broods that were taken by predators or starved both had low growth rates and that broods that survived grew considerably faster. This suggested that habitat management to improve the food supply might reduce both causes of nestling loss.

3.10 Value of experiments to disentangle ultimate and proximate causes of breeding failure

Although the relative roles of ultimate and proximate causes may be indicated by observational studies of breeding, field experiments will often provide more robust evidence. Experimental manipulations of many aspects of bird breeding are feasible and include nest site enhancement or removal, supplementary feeding and control of predation. A detailed review is beyond the scope of this chapter and useful examples of experiments can be found in Newton (1994), Newton (1998, chapter 7), and Tapper *et al.* (1996). However, a recent development in experiments on bird breeding biology worthy of special mention is the use of manipulations of the energy requirements of parent birds and nestlings. Yom-Tov and Wright (1993) heated nest boxes of blue tits during the egg-laying period and found that interruptions of egg-laying during cold weather were reduced. Incubation requires a considerable amount of energy from parent birds and this can be reduced experimentally by slowing the rate of cooling of the eggs using a miniature electrical heating mat in the nest cup. The effects of experimental heating of clutches of Starling *Sturnus vulgaris* eggs were measured by using a datalogger which recorded temperatures sensed by a thermistor installed on top of the clutch and recording nest success for the first (manipulated) and second (unmanipulated) breeding attempts (Reid *et al.* 1999). The thermistor allowed the duration of absences of the incubating female to be recorded. Experimental females spent less time away from the nest than controls, fledged significantly more young from their first attempt, and were more likely to hatch all the eggs from their second attempt than controls. Experimental warming probably allowed the females to save resources during incubation in the first attempt and to reallocate these to chick care of the first brood and incubation of their second clutches, leading to improved breeding performance (Reid *et al.* 2000).

References

Aebischer, N.J. (1999). Multi-way comparisons and generalised linear models of nest success: extensions of the Mayfield method. *Bird Stud.*, 46 (Suppl.), S22–S31.

Ar, A. and Rahn, H. (1980). Water in the avian egg:overall budget of incubation. *Am Zool*, 20, 373–384.

Bart, J., Kepler, C., Sykes, P., and Bocetti, C. (1999). Evaluation of mist-net sampling as an index to productivity in Kirtland's warblers. *Auk*, 116, 1147–1151.

Beintema, A.J. (1995). Fledging success of wader chicks, estimated from ringing data. *Ringing and Migration*, 16, 129–139.

Beintema, A.J. and Muskens, G.J.D.M. (1987). Nesting success of birds breeding in Dutch agricultural grasslands. *J. Appl. Ecol.*, 24, 743–758.

Beintema, A.J., Beintema-Hietbrink, R.J., and Muskens, G.D.M. (1985). A shift in the timing of breeding in meadow birds. *Ardea*, 73, 83–89.

Crawley, M.J. (1993). GLIM for Ecologists. Blackwell Scientific Publications, London.

Davison, W.B. and Bollinger, E. (2000). Predation rates on real and artificial nests of grassland birds. *Auk*, 117,147–153.

Ens, B.J. (1991). Guarding your mate and losing the egg: an oystercatcher's dilemma. *Wader Stud. Group Bull.*, 61(Suppl.) 69–70.

Etheridge, B., Summers, R.W., and Green, R.E. (1997). The effects of illegal killing and destruction of nests by humans on the population dynamics of the Hen Harrier *Circus cyaneus* in Scotland. *J. Appl. Ecol.*, 34, 1081–1105.

Evans, A.D., Smith, K.W., Buckingham, D.L., and Evans, J. (1997). Seasonal variation in breeding performance and nestling diet of Cirl Buntings *Emberiza cirlus* in England. *Bird Stud.*, 44, 66–79.

Forberg, S. and Odsjö, T. (1983). An X-ray back-scatter method for field measurements of the quality of bird eggshells during incubation. *Ambio*, 12, 267–270.

Forberg, S. and Odsjö, T. (1984). Influence from the embryonic development of hen eggs on the calcium index measured by an X-ray back-scatter method. *Ambio*, 13, 40–42.

Fox, G.A., Anderka, F.W., Lewin, V., and MacKay, W.C. (1975). Field assessment of eggshell quality by beta-backscatter. *J. Wildlife Manage.*, 39, 528–534.

Galbraith, H. and Green, R.E. (1985). The prediction of hatching dates of Lapwing clutches. *Wader Stud. Group Bull.*, 43, 16–18.

Gotmark, F. (1992). The effects of investigator disturbance on nesting birds. In D.M. Power (ed.) *Curr. Ornithol.*, 9, 63–104. Plenum Press, New York.

Grant, M.C., Lodge, C., Moore, N., Easton, J., Orsmann, C., and Smith, M. (2000). Estimating the abundance and hatching success of Curlew *Numenius arquata* using survey data. *Bird Stud.*, 47, 41–51.

Green, R.E. (1984). Nomograms for estimating the stage of incubation of wader eggs in the field. *Wader Stud. Group Bull.*, 42, 36–39.

Green, R.E. (1988). Effects of environmental factors on the timing and success of breeding of Common Snipe *Gallinago gallinago* (Aves: Scolopacidae). *J. Appl. Ecol.*, 25, 79–93.

Green, R.E. (1989). Transformation of crude proportions of nests that are successful for comparison with Mayfield estimates of nest success. *Ibis*, 131, 305–306.

Green, R.E. (1998). Long-term decline in the thickness of eggshells of thrushes, *Turdus* spp., in Britain. *Proc. R. Soc. B*, 265, 679–684.

Green, R.E. (2000). An evaluation of three indices of eggshell thickness. *Ibis*, 142, 676–679.

Green, R.E., Tyler, G.A., Stowe, T.J., and Newton, A. V. (1997). A simulation model of the effect of mowing of agricultural grassland on the breeding success of the Corncrake (*Crex crex*) *J. Zool., Lond.*, 243, 81–115.

Hartley, M.J. and Hunter, M.L. (1998). A meta-analysis of forest cover, edge effects and artificial nest predation rates. *Conserv. Biol.*, 12, 465–469.

Haskell, D.G. (1995). Forest fragmentation and nest predation: are experiments with Japanese Quail eggs misleading? *Auk*, 112, 767–770.

Hensler, G.L. (1985). Estimation and comparison of functions of daily nest survival probabilities using the Mayfield method. *Statistics in Ornithology*, eds. P.M. North and B.J.T. Morgan, pp. 289–302. Springer-Verlag, Berlin.

Herranz, J., Yanes, M., and Suarez, F. (2002). Does photo-monitoring affect nest predation? *J. Field Ornithol.*, 73, 97–101.

Holmes, R.T., Sherry, T.W., Marra, P.P., and Petit, K.E. (1992). Multiple brooding and productivity of a neotropical migrant, the Black-throated Blue Warbler (*Dendroica caerulescens*) in an unfragmented temperate forest. *Auk*, 109, 321–333.

Holmes, R.T., Marra, P.P., and Sherry, T.W. (1996). Habitat-specific demography of breeding Black-throated Blue Warblers (*Dendroica caerulescens*): implications for population dynamics. *J. Anim. Ecol.*, 65, 183–195.

Hoyt, D.F. (1979). Practical methods of estimating volume and fresh weight of bird eggs. *Auk*, 96, 73–79.

Jackson, D.B. and Green, R.E. (2000). The importance of the introduced hedgehog (*Erinaceus europaeus*) as a predator of the eggs of waders (Charadrii) on machair in South Uist, Scotland. *Biol. Conserv.*, 93, 333–348.

Johnson, D.H. (1979). Estimating nest success: the Mayfield method and an alternative. *Auk*, 96, 651–661.

Kenward, R.E., Marcstrom, V., and Karlblom, M. (1999). Demographic estimates from radio-tagging models of age-specific survival and breeding in the Goshawk. *J. Anim. Ecol.*, 68, 1020–1033.

King, D.I., DeGraaf, R.M., Griffin, C.R., and Maier, T.J. (1999). Do predation rates on artificial nests accurately reflect predation rates on natural bird nests. *J. Field Ornithol.*, 70, 257–262.

Krebs, J.R. and Perrins, C.M. (1978). Behavior and population regulation in the Great Tit (*Parus major*). In *Population Control by Social behavior*, ed. F.J. Ebling and M.J. Stoddart. Institute of Biology, London.

Lougheed, C., Lougheed, L.W., Cooke, F., and Boyd, S. (2002*a*). Local survival of adult and juvenile Marbled Murrelets and their importance for estimating reproductive success. *Condor*, 104, 309–318.

Lougheed, C., Vanderkist, B.A., Lougheed, L.W., and Cooke, F. (2002*b*) Techniques for investigating breeding chronology in Marbled Murrelets, Desolation Sound, British Columbia. *Condor*, 104, 319–330.

Majer, T.J. and DeGraaf, R.M. (2000). Predation on Japanese Quail vs. House Sparrow eggs in artificial nests: small eggs reveal small predators. *Condor*, 102, 325–332.

Major, R.E. and Kendal, C.E. (1996). The contribution of artificial nest experiments to understanding avian reproductive success: a review of methods and conclusions. *Ibis*, 138, 298–307.

Martin, T.E. and Geupel, G.R. (1993). Nest-monitoring plots: methods for locating nests and monitoring success. *J. Field Ornithol.*, 64, 507–519.

Mayfield H.F. (1961). Nesting success calculated from exposure. *Wilson Bull.*, 73, 255–261.

Mayfield, H.F. (1975). Suggestions for calculating nest success. *Wilson Bull.*, 87, 456–466.

Murray, B.G. (2000). Measuring annual reproductive success in birds. *Condor*, 102, 470–473.

Newton I. (1979). *Population Ecology of Raptors*. Berkhamsted, Poyser.

Newton, I. (1994). Experiments on the limitation of bird breeding densities: a review. *Ibis*, 136, 397–411.

Newton, I. (1998). *Population Limitation in Birds*. Academic Press, London.

Newton, I. (1999*a*). Age ratios in a Bullfinch *Pyrrhula pyrrhula* population over six years. *Bird Stud.*, 46, 330–335.

Newton, I. (1999*b*). An alternative approach to the measurement of seasonal trends in bird breeding success: a case study of the Bullfinch *Pyrrhula pyrrhula. J. Anim. Ecol.* 68, 698–707.

Newton, I. and Rothery, P. (2001). Estimation and limitation of numbers of floaters in a Eurasian Sparrowhawk population. *Ibis*, 143, 442–449.

Nygård, T. (1999). Correcting eggshell indices of raptor eggs for hole size and eccentricity. *Ibis*, 141, 85–90.

O'Grady, D.R., Hill, D.P., and Barclay, R.M.R. (1996). Nest visitation by humans does not increase predation on Chestnut-collared Longspur eggs and young. *J. Field Ornithol.*, 67, 275–280.

Paganelli, C.V., Olszowska, A., and Ar, A. (1974). The avian egg: surface area, volume and density. *Condor*, 76, 319–325.

Peach, W.J., Buckland, S.T., and Baillie, S.R. (1996). The use of constant effort mist-netting to measure between-year changes in abundance and productivity of common passerines. *Bird Stud.*, 43, 142–156.

Picozzi, N. (1975). Crow predation on marked nests. *J. Wildlife Manage.*, 39, 151–155.

Pietz, P.J. and Granfors, D.A. (2000). Identifying predators and fates of grassland passerine nests using miniature video camers. *J. Wildlife Manage.*, 64, 71–87.

Potts, G.R. and Aebischer, N.J. (1995). Population dynamics of the Grey Partridge *Perdix perdix* 1793–1993: monitoring, modelling and management. *Ibis*, 137 (Suppl.), S29–S37.

Powell, L.A. ,Conroy, M.J., Krementz, D.G., and Lang, J.D. (1999). A model to predict breeding season productivity for multibrooded songbirds. *Auk*, 116, 1001–1008.

Reid, J.M., Monaghan, P., and Ruxton, G.D. (1999). The effect of clutch cooling rate on Starling, *Sturnus vulgaris*, incubation strategy. *Anim. Behavi.*, 58, 1161–1167.

Reid, J.M., Monaghan, P., and Ruxton, G.D. (2000). Resource allocation between reproductive phases: the importance of thermal conditions in determining the cost of incubation. *Proc. R. Soc. B*, 267, 37–41.

Ricklefs, R.E. and Bloom, G. (1977). Components of avian breeding productivity. *Auk*, 94, 86–89.

Rotella, J.J., Taper, M.L., and Hansen, A.J. (2000). Correcting nesting success estimates for observer effects: maximum-likelihood estimates of daily survival rates with reduced bias. *Auk*, 117, 92–109.

Simon, J.C. (1998). Nest relocation using PVC "spotters". *J. Field Ornithol.*, 69, 644–646.

Tapper, S.C., Potts, G.R., and Brockless, M.H. (1996). The effect of an experimental reduction in predation pressure on the breeding success and population density of Grey Partridges *Perdix perdix. J. Appl. Ecol.* 33, 965–978.

Thompson, B.C, Knadle, G.E.,Brubaker, D.L., and Brubaker, K.S. (2001). Nest success is not an adequate comparative estimate of avian reproduction. *J. Field Ornithol.*, 72, 527–536.

Underwood, T.J. and Roth, R.R. (2002). Demographic variables are poor indicators of Wood Thrush productivity. *Condor*, 104, 92–102.

van Paassen, A.G., Veldman, D.H., and Beintema, A.J. (1984). A simple device for determination of incubation stages in eggs. *Wildfowl*, 35, 173–178.

Verboven, N., Ens, B.J., and Dechesne, S. (2001). Effect of investigator disturbance on nest attendance and egg predation in Eurasian Oystercatchers. *Auk*, 118, 503–508.

Vickery, P.D., Hunter, M.L., and Wells, J.V. (1992). Use of a new reproductive index to evaluate the relationship between habitat quality and breeding success. *Auk*, 102, 697–705.

Weidinger, K. (2001). How well do predation rates on artificial nests estimate predation on natural passerine nests? *Ibis*, 143, 632–641.

Willis, E.O. (1981). Precautions in calculating nest success. *Ibis*, 123, 204–207.

Wiltschi, E. (1961). Sexual and secondary sexual characteristics. In *Comparative Physiology of Birds*, Vol 2, ed. A.J. Marshall. Academic Press, London.

Wyllie, I. and Newton, I. (1999). Use of carcasses to estimate the proportions of female Sparrowhawks and Kestrels which bred in their first year of life. *Ibis*, 141, 504–506.

Yom-Tov, Y. and Wright, J. (1993). Effect of heating nest boxes on egg-laying in the Blue Tit (*Parus caeruleus*). *Auk*, 110, 95–99.

4

Birds in the hand

Andrew Gosler

4.1 Introduction

This chapter is concerned with the trapping of birds for research with a view to releasing them as soon as possible, after having "processed" them as necessary. The goal is to release the bird, none the worse for its experience, back into the wild, where it will behave normally.

However much we might discover about the lives of birds through simple observation, there is a point beyond which we cannot do more without direct physical contact. Thus the skills needed to catch birds with minimal disturbance are central to ornithological research, but this also raises welfare, ethical, and legislative issues that do not apply in observational studies (but see Chapters 9 and 10). Nevertheless, the bird in the hand is a mine of useful information that cannot be obtained otherwise, and my main focus is on how to obtain the maximum information in the minimum time (thus reducing disturbance and stress to the birds). Developing the necessary skills requires experience, which cannot be obtained without training. Without these skills, you would endanger the birds, reduce the quality of data obtained, and in many countries would be acting illegally. In my experience, however, few people lack the aptitude to become safe and competent bird-handlers, and a conscientious approach at the outset will bring rewards long term.

There are broadly three reasons why researchers might need, temporarily, to take birds from the wild: (a) for individual marking or attachment of tags, (b) to observe details at close-quarters that cannot be recorded otherwise, (c) to conduct some scientific procedure or obtain material (biopsy) that cannot otherwise be obtained. Accomplishing such procedures (a–c) is known as *processing*. I shall consider each in some detail, but first we should consider some broad ethical and legislative issues, and then how to catch the birds.

4.2 Welfare, ethical, and legislative issues

The intervention caused by trapping and processing may have very minor consequences, such as the loss of a few minutes feeding, or major ones, such as exposure to a predator. Furthermore, the same procedure might have different consequences under different conditions; losing a few minutes feeding in summer when food is abundant and days are long might be quite different from midwinter when both food and time to find it might be in short supply. Indeed the balance of costs to a bird from being trapped might change through the day (Gosler 2001).

Disregard for the welfare of birds may prejudice their lives and their contribution to the population (with consequences for conservation status or population studies); it is also likely to reduce the quality and reliability of data, and it may be a legal necessity to take heed of such issues. The degree to which these considerations affect a particular study will vary across taxa, seasons, and countries, but the best advice must be first to weigh up the necessity of the research itself against any cost to the birds, then the necessity of the methods employed (in case a non-interventionist approach might yield equally valid results), and finally, having accepted their necessity, to aim for the highest possible standards within the constraints of the research needs themselves. Current thinking on the use of animals in science is to consider what is known as the three "R"s. These are "Replacement," for example, use a non-sentient rather than sentient species (not applicable here) or an abundant species rather than a rare one, "Refinement" of technique to minimize stress, etc. and "Reduction" of numbers to the minimum necessary for the work to produce valid results (for further information on the three "R"s, see Salem and Rowan 2001 and Hawkins *et al.* 2001; and for a modern discussion of animal stress, see Moberg and Mench 2000).

Outside the laboratory, the legislation most relevant to field ornithologists regulates the capture and marking (ringing or banding) of birds. Legislative constraints naturally vary by countries, and it is essential that researchers obtain whatever permits are necessary to operate legitimately within the country concerned. Where specific legislation exists (e.g. most European states, USA, Canada, and Australia) researchers must obtain a Government permit issued under the appropriate legislation. Also, where a national ringing scheme or banding program is in place, you should use rings or bands supplied by that scheme. This is not just a courtesy and an established convention; it also prevents confusion and unnecessary administration for the ringing schemes concerned, and for you. The administration of permits is also often undertaken on behalf of the governments by the ornithological institutions that run the ringing schemes.

In the first instance, the Worldwide Web gives basic information about many national ringing schemes and banding programs.[1]

In countries where there is no bird-ringing scheme and no specific legislation covering ringing, it will usually also be necessary to obtain some form of state permission to trap birds in the country. If difficulty is found in obtaining relevant contact details for a country, the national ringing schemes of other countries (such as the British Trust for Ornithology in the United Kingdom[2] should be able to advise). In countries with no ringing/banding scheme of their own, international agreement dictates which nationality's rings may be used there (typically a former colonial nation), and it will be necessary to operate through the ringing scheme of that country. So, for example, while British rings should be used in the West African state of Gambia, in neighboring Senegal, French rings must be used. Further information on legislation and ethical matters relating to animals (including bird ringing) is given in Brooman and Legge (1997), while the North American banding manual (NABM), which can be viewed online, is useful in North America.[3]

The requirement to capture and mark, in the most benign way, the minimum number of birds necessary for the research involved does not simply mean that fewer is more ethical. If too few birds have been processed to yield statistically meaningful conclusions, then the captures made will have been wasted, and that is, arguably, unethical too. Much depends on the questions asked and on statistical power. For example, to determine the timing of molt might take just a few dozen captures; to map migration may require many thousands of birds to be ringed in order to provide a relatively small number of recoveries. The costs to the birds should be small in either case.

4.3 Catching the birds

A wide range of techniques is available for live-trapping birds, many of them developed long ago. The diversity of capture techniques reflects both local independent developments for similar purposes (e.g. the wide variety of walk-in traps for shorebirds), and the diverse requirements for trapping different taxa (wildfowl, raptors, passerines). In any case, some trapping methods favor naïve or vulnerable birds. For example, cage traps may be avoided by all but starving

[1] At the time of writing, the following URLs are useful sources of contact email addresses for national ringing and banding programs: www.birdsinthe.net, www.wildlifeweb.f9.co.uk/intro.html/content.htm, www.euring.org/ContactSchemes.html, http://web.tiscali.it/sv2001/cormo_centre.htm#Australia, www.aves.net/the-owl/blnkobsv.htm, www.uct.ac.za/depts/stats/adu/safring-index.htm   [2] www.bto.org
[3] www.pwrc.usgs.gov/bbl/manual/manual.htm

individuals, and this sort of trapping bias may seriously compromise the scientific value of the results obtained. In general, methods in which the bird is unaware of any risk are preferable to those in which a bird's fear of novel situations (*neophobia*) might come into play.

Whatever method is used, some general guidelines apply. First, remember that a trapped bird is vulnerable: it may be stressed, unable to escape from predators and exposed to weather effects such as heat, wind, or rain. Operations must be planned to reduce this stress, and the associated risks. This means not trapping in wind, rain, or direct sun unless shelter is provided within the trap, or you can guarantee the rapid removal (*extraction*) of captured birds. Even in the tropics, hypothermia is a more universal risk to birds than hyperthermia, but note that some deep-forest birds (e.g. Ant-thrushes, Formicaridae) can die of shock if brought into direct sun. Predators sometimes learn to associate traps with food and, if you are not alert to this risk, will incorporate your trap sites into their daily foraging rounds (trap-lining), if allowed to. This is especially a problem in the tropics where a great range of predators, including ants, snakes, birds including coucals and raptors, carnivores, primates and megadermatid bats, have been known to take birds from mist-nets. Alternating sessions between different net lanes or trap series can reduce this risk.

Once removed from the trap or net, birds must be kept individually (birds of some species will kill each other if kept together) in bags or larger keeping cages. These must be kept clean and dry, and placed out of direct sun. Speed of operation means having enough helpers to cover all procedures adequately so that no bird is in a net for more than about 20–30 min from capture (10 min if birds are exposed in nets to tropical sun). In allowing time, remember that there might be unforeseen problems arising from difficult extractions, unusually large catches, snakes etc. All people involved must be adequately trained or supervised, and must know about any local hazards; remember also that the temperament of some people changes in pressurized situations.

It is advisable to map net and trap sites with a GPS, and adopt some system (e.g. removable reflective labels) to prevent traps/nets from being lost or forgotten in a net round; count traps when set out and when collected back. Knowledge of the area in which you are trapping, and of the species likely to be encountered, is invaluable if available. For example: (a) hummingbirds are hardy, but need to feed frequently, so you should have some fresh sugar-water to hand to give them on arrival at the ringing station or before release (protrusible-tongued birds such as spider-hunters, Nectariniidae, are usually happy to drink while being held in the hand). (b) Long-legged waders (shore-birds) can suffer from leg-cramp (*capture myopathy*) if kept from standing for any length of time. They must therefore be transferred from nets or bags to keeping cages in which they can stand as quickly as possible. In many areas, but especially in the tropics, bird activity tends to peak early in the day and may peak again just before rain (requiring extra care). Trapping later in the day may be unproductive, but the profitability of netting in the evening varies, and extra vigilance is necessary if trapping at night.

Apart from direct means of catching birds (e.g. pulli—see below, swan-hooks and batfowling), a useful distinction to make is between traps, which tend to be set for individual birds, and nets, which typically catch many birds simultaneously. Some traps must be baited with some sort of lure such as food, and nets too may be more effective if bait is provided nearby. Both traps and nets present risks to birds if operated by unskilled staff. The seminal work of Bub (1991) on bird trapping carries 456 text figures, illustrating more than 150 different trapping methods. Given this, then, it is remarkable that today most birds are trapped for research using just a handful of methods, described below.

4.3.1 The breeding season

The breeding season presents a period of intense enquiry for the researcher, and intense activity and sensitivity to disturbance for most bird species. Nests, eggs, chicks, and adults at the nest, are vulnerable to predators. This means that the number of visits should be minimized, and in general trapping of adults and marking of young should be left as late as is consistent with preventing the brood leaving the nest prematurely, and with logistic considerations. For many passerines there is a risk of desertion if the incubating females are lifted off the nest. However, for some it is safe to do so, in which case this may offer a valuable means of sampling. Advice should be sought for particular species before undertaking such a procedure.

Young birds prior to fledging are known collectively as *pulli* (singular pullus). For the researcher, there are some obvious advantages to ringing birds as pulli, since their exact origin and age are known, and it might be possible to trap the parents also. The disadvantages concern welfare, since pulli are vulnerable to rough handling, predators and the elements. Nidicolous young (which stay in the nest after hatch) should not be ringed until flight feathers have emerged by at least a quarter from the pin; nidifugous young (which leave the nest shortly after hatching) of many species can be ringed from hatching but always check the fit of the ring to be sure that it cannot slip down over the foot, and that it will be large enough to take the full-size leg. Some researchers have tried inserting plasticine inside the fitted ring, which gets squeezed out as the leg grows. However, there are several reasons why this might be unsafe (e.g. plasticine might go hard in water) and the method is not generally recommended. Do not mark nest positions conspicuously as crows can learn to associate the marks with food (see Chapter 3).

4.3.2 Cage traps

These range from the simple cage propped up on a broken stick (drop trap), which is knocked out the way by a bird, through a variety of cages which the bird must enter for food, thereby tripping a door-release mechanism (either on the floor, for example Potter trap, or roof, for example Chardonneret trap), to large permanent cage traps such as the crow trap (an aviary-sized cage with a mesh funnel set downwards in the roof) or Helgoland trap—a huge horizontal mesh funnel set in an area with little vegetation or cover and planted inside (typically with some hardy berry-bearing shrubs to provide food as well as shelter) to attract birds, which must be driven by the trapper toward a clear-fronted catching box (which the bird sees as a way out).

In the case of wader traps (e.g. wader funnel trap, Ottenby trap, wader nest traps), rail traps, and Helgoland traps, birds simply walk or fly through the trap entrance and can continue to feed within the trap. Walk-in traps, which can be used to trap waders at the nest (but replace the clutch with dummy eggs, so keeping the real eggs safe and warm), are typically less than about 1 m³, while Helgoland traps, which are typically built at coastal observatory sites, can be up to a hectare in extent. In some of these traps, the birds might not realize that they are trapped until the researcher arrives. Such "*passive*" traps are benign for this reason. Large traps such as Helgolands also have a side door through which the trapper leaves the trap after the birds are caught. This is also useful to deactivate the trap when nobody is available to work it. Trap deactivation is valuable also for *pre-baiting*. With any form of trapping (or netting) where birds must be concentrated near the trap site, or where neophobia would prevent the bird from entering the trap (*trap-shyness*), it is essential to set the trap for a few days with food inside so that the bird can enter *and leave* at will, to gain confidence. Larger cage traps offer the opportunity to include shelter within the trap so that operation may be continued in poor weather.

4.3.3 Spring traps

Unlike cage traps, which can be active or passive, spring traps are always active, and consist of some sort of spring-loaded mechanism to close off an entrance, through which the bird must pass to reach bait. Because of the spring loading, great care must be exercised in their setting, and rapid extraction of the bird is important. Also, unlike some cage traps, spring traps catch only one of several birds at a time, and many traps might be set at once if capture-frequency and processing-time permit. A very effective spring-trap is the bow-net, which consists of a circle of netting (e.g. 30–50 cm radius) held flat open on the ground by a wire mechanism. A bird (ground foragers, robins, chats, wheatears) is attracted by a bait such as a live mealworm tethered at the center of the circle of net. To take the bait, the bird must be well within the circle of net, and on taking the bait the mechanism is closed by a mousetrap spring so that the bird is caught within a semi-circular sandwich of netting. The trap must also be pegged down so that it cannot be moved (e.g. by the bird inside).

4.3.4 A couple of nestbox traps

A useful passive trap for catching parent birds (tits, flycatchers etc.) in their nestbox consists simply of an inverted "U"-shaped piece of wire pinned lightly (so that it is free to move) to the inside of the box just above the entrance hole so that the "arms" of the "U" hang down over the inside of the hole. Thus the birds can enter by pushing the wire out the way (which they will do) but not escape when the wire falls back down behind them. Note that unless the bird shows at the entrance, there is nothing to alert the researcher to the bird having been caught, and both parents can enter the box without the trap having to be reset. Because this trap is difficult to use in woodcrete boxes, researchers at the Edward Grey Institute (EGI) designed a spring trap to close a light metal door behind a bird after it had entered. It consists of a metal plate with a hole larger than, but corresponding to, the entrance hole on front of the box. The sliding door holds a flat steel spring in compression when open and the door is held open by one end of a pivoted perch that the bird depresses on entering the box, so releasing the mechanism and closing the door. Such traps must not be used until the brood is at least half-grown because disturbance during laying or incubation may cause many species to desert.

4.3.5 Noose-carpet traps

In circumstances where a bird returns to a particular spot, such as its nest, a carpet of monofilament nooses can be used to trap adult birds. A large number (50+) of loose nooses are tied to a circular wire frame large enough to encircle the nest cup and its contents. The frame must be secured so that the bird cannot fly off with it. The nooses are arranged to sit over the nest-contents so that when the bird settles to incubate, it unknowingly catches its feet in the nooses. Upon the researcher's approach the bird leaves hastily, and, if set properly, will be caught by a monofilament noose holding the leg. Since capture occurs on your approach, the birds should not be caught for long. Nevertheless, birds can be highly agitated by this and so, as with any trap set over the nest during incubation, it is essential that eggs are replaced by dummies prior to using this trap, and be sure not to allow the real eggs to chill while in your care. Noose carpets work well with larger non-passerines such as raptors and gulls but must not be used for species that are known to be sensitive at the nest during incubation.

4.3.6 Mist-nets

One capture method outstrips all others as the option preferred by researchers today: the mist-net. However, they are of little or no use for larger species (e.g. pigeons, wildfowl, gulls, and raptors). A mist-net is a vertically erected fine (almost invisible) net, typically of terylene (beware nylon, which tends to be too inelastic for bird safety) supported horizontally as net panels or *shelves* between a series of strings (shelf-strings) set about 50 cm apart, which are themselves attached to vertical poles via string loops. The netting forming the shelf must have sufficient vertical slack in it to form a pocket between the self-strings. Nets are available in a variety of lengths (e.g. 6-, 9-, 12-, or 18-m) and mesh-gauges (e.g. *c.*32 or 60 mm stretched knot-to-knot– larger meshes allow larger species to be taken) or can be made up from loose material to almost any specification. Poles can be in the form of aluminum sections (sectioned tent poles can be ideal) that slot together, or 4-m lengths of bamboo or similar material. The poles, and net strung between them, are held under tension by guy cords either tied to vegetation, or to pegs or poles (on mud-flats) in the ground. Birds are caught when they fly or walk into the net, becoming entangled, and ending up in a pocket of netting supported from a shelf-string. On entry to a net, birds can spin into a pocket and become tangled by wings and feet in addition to their heads passing through the netting.

The mist-net's ubiquity reflects its flexibility (it can be used in almost any terrestrial habitat for a wide range of species) and its portability (an 18-m long net setting 2.5-m top to bottom, folds into a bag weighing less than 0.5 kg), but not its ease of use because more than any other trapping method (apart from cannon-netting), mist-netting requires patience, dexterity, and experience if birds are to be extracted unharmed. However, the ubiquity of the mist-net is also testimony to the fact that most people can acquire these skills rapidly, and world-wide many millions of birds have been trapped safely by mist-net for ringing and release.

To extract a bird from a mist-net, identify the side on which the bird entered, and then put your hands in the pocket to untangle it until it can be taken by the base of the legs and carefully drawn away from the net, clearing feet, wings, and head in sequence. Care must always be taken not to damage flight or tail feathers on extraction. A very handy tool for the netter, supplied by the BTO, is the *Quickunpick*, which is actually a stitch cutter supplied with sewing machines, and which can be used as a probe to loosen tight netting, or if needs be, to cut a mesh of the net.

However, mist-net extraction can only be learned from practice, in the presence of experienced colleagues, so I shall go into no further detail, except to offer the following tips:

1. Mist-nets are almost invariably more effective if set against a dark background rather than the open sky, and be sure that guys are strong enough to hold the length of net set. When setting up, it helps to lay the poles out along (not across) the net lane so that you have some leeway if you misjudge where the net will come to when you start running it out. If netting in forest, setting nets diagonally along the net lane (ride or transect) will catch birds traveling along them as well as across. Do not set nets across paths, as mammals (including bats) tend to use these. Net-damage can best be reduced by setting nets angled in net-lanes off paths and cut long enough to contain the whole net off the main path, in a fish-bone pattern.

2. The height of the bottom shelf is critical because of predators. Avoid the temptation in tropical forest to set this on the ground to catch terrestrial species as this leads to real tangles of birds and litter and exposes birds to ants (which can kill birds), and other predators. Setting the net 30–40 cm above ground is ideal when the ground is bare or above the level of the longest grass to prevent ants entering nets from the vegetation. Also, check for ant nests near the ride. If it is essential to catch terrestrial species, set the net at 15–20 cm and clean the trails of leaves and twigs to make army ants easier to locate (but beware snakes, which like to bask on cleared net lanes adjacent to thick cover). It is of course essential to set nets higher if over water.

3. If you are unfortunate enough to catch a snake in a net or trap do not try to remove it by hand but encourage it out from a distance. On expedition, nets are likely to suffer damage from a range of nontarget animals (always be wary of livestock, buffalo, big cats etc.). So learn how to mend nets, and never store them wet or in plastic bags.

4. Large insects, including hymenoptera, can be trapped in mist-nets, and for a variety of reasons including bird welfare, these are best removed. Bamboo poles left along net lanes may be occupied by insects, in particular ants, wasps, and bees. Ants may inflict a painful bite, and are best knocked out of the end of the pole before setting the net. Wasps and bees bore holes into the poles, so be sure not to cover the hole with your hand when setting the net up as you will get stung. If you are allergic to these stings, it is better not to leave poles out.

5. In some areas the theft and illegal use of mist-nets for hunting is a problem. Be aware of this when exposing netting activities to the public, for whatever reason (including education).

4.3.7 Clap-nets and whoosh-nets

Although in special circumstances mist-nets can be used horizontally (e.g. manually to catch or "flick" swifts), they are generally not so used. Clap-nets and whoosh-nets, in contrast, are always set furled horizontally on the ground, and whereas the mist-net is a passive capture method (the net is static), the clap-net or whoosh net is thrown rapidly over the birds by powerful elastic bands. The net is also "fired" actively by the trapper who decides when birds are safe (i.e. away from the elastics and leading edge) within the catching area; they can then be removed immediately after capture. Although clap-net material is heavier than mist-netting, care must be taken to check that it will not become snagged in vegetation during the net's release. Failure to deploy it properly will reduce the catching area, and so also the size of the catch.

Clap nets can be 2–10 m² in area, and more than one can be operated simultaneously. They are extremely effective for trapping any flocking species that feeds or roosts on the ground, including finches, buntings, starlings, and shorebirds. Clap-net sites are best if pre-baited for a few days, but remember that when clap-net elastics are set under tension, they can be dangerous to the trapper as well as the birds.

4.3.8 Cannon-nets

A development of the woosh-net, used to trap waders, gulls, and wildfowl in prodigious numbers, is the cannon-net; basically a large woosh-net powered by gunpowder. Four projectiles fired simultaneously from small cannons are attached by ropes to the leading edge of the net. When fired, the net is carried out over the heads of the birds, which are subsequently trapped beneath the net. Hence, the angle of the cannons is critical: it must be set higher for long-necked species (e.g. geese) than for small waders. While all the other traps and nets covered in this chapter can be operated by a single worker, a cannon-net needs a team of people. The weight of four cannons, the net, associated wiring, charges, firing box, and all the keeping bags (or sacking cages) for the birds, is considerable; and the need to remove large numbers of birds and dismantle the net quickly in advance of the tide means that experience is, again, essential. There is also the legal issue of holding the gunpowder charges.

4.3.9 Capture by hand

If accessible, the adults of many species can simply be captured at night while at roost; small passerines in nestboxes offer an example. If this is done, processing is best done in silence, with minimal use of artificial light and the bird should be returned to its roost site as soon as possible in darkness. Roost sites should be monitored subsequently (visual inspection during the day for droppings may be sufficient) to assess the roost-desertion rates from this method, as a shortage of suitable roost sites might affect the survival prospects of small birds in winter. Other direct capture methods have been devised for a range of species. For example, batfowling is a traditional method of catching seabirds on land by using a long-handled hand net held above the head, and if approachable, swans can be caught with a swan-hook: basically a pole with a metal hook like a narrow shepherd's crook.

4.4 Individual marking

Birds are marked whenever it is necessary to identify them either as individuals, or as members of a class such as a particular cohort. In general, individual identification is to be preferred (see Chapter 5). Always fit any ring or mark before taking any measurements, in case the bird escapes, but only after it has been identified to species (or subspecies if relevant).

As implied above, the universal marking method for birds is the leg ring (ring or band) fitted to the tarsometatarsus (below tarsal joint, not tibiotarsus above joint), but it is not the only kind of mark available. Metal or plastic rings are available in a variety of sizes (e.g. 27 in the United States and the United Kingdom) covering everything from hummingbirds (United States), whose bands must be trimmed to fit (see NABM), to swans. Metal rings carry a unique number and usually (depending on size) an address to which a finder can report the ring (e.g. Zool. Mus. Denmark). They are also available in a variety of designs and metals (from pure aluminum to stainless steel) depending on the biology of the birds concerned and the risk to the ring of wear or corrosion. These rings are obtained in series or strings of, for example, 100 rings numbered in sequence (e.g. VA32101–100).

For recovery, metal rings generally require recapture of the bird for the inscription to be read. Plastic rings are generally for individual identification in the field without recapture. For smaller species, they are typically available as plain (i.e. no inscription) celluloid split rings in a variety of colors, which are applied singly or in combination. These rings are manufactured commercially for the pet trade. For larger species (not feasible if internal diameter less than 8.0 mm), rings can be engraved and made to measure by the researcher from plastic laminate (DARVIC), which hot water softens enough to shape in a simple mould.

Metal rings are fitted with specially designed pliers. In some schemes, the rings have to be opened (e.g. with circlip pliers) before they can be closed on the bird's leg (Figure 4.1). Plastic color-rings are fitted with a tool like a shoehorn. Large

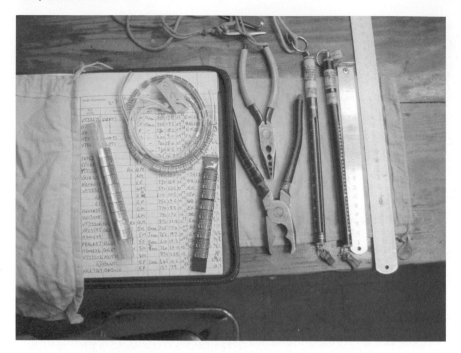

Fig. 4.1 Standard ringing kit consisting of (left to right) bird bag, metal rings of various sizes (the plastic "string" labelled 2.3 carries 100 rings of 2.3 mm internal diameter), field logbook, two sizes of ringing pliers, balances (the bird is placed in a container, which is then clipped to the balance), and stopped rules for measuring wing and tarsus lengths. (Photo: Andrew G. Gosler)

DARVIC rings must be opened by hand or with strong circlip pliers before being closed on the leg. Note that some countries will not have as large a range of ring sizes available as in the United States and the United Kingdom so that either rings will have to be overlapped (without covering the inscription), cut down in the field or special arrangements made to permit the use of rings from another country.

Color-rings are often used on their own for behavioral–ecological studies but have limitations. Not all colors are identified with equal facility under field conditions (at range, on short tarsus etc.), and some colors change (light blue, light green, yellow, and mauve) over time, especially under tropical conditions. For small species, bands with combinations of stripes with different widths may be better than letters or combinations of different-colored rings. Celluloid split rings are also prone to cracking and falling off, in which case small split DARVIC rings are better. It is generally wise to glue plastic rings, especially in hot countries where temperatures may be high enough to open the rings. It should also be noted that color-rings are thought to cause deformity in some short-legged species such as Tyrannid flycatchers (especially noted in *Mionectes oleagineus*).

Fig. 4.2 Colour ringing is the most standard means of following individuals to study their behavior, demography or migration. This Black-tailed Godwit *Limosa limosa* chick, was ringed as part of a study determining the migration decisions of individuals and their demographic consequences (Gill *et al.* 2001). It was ringed in July 2001 in eastern Iceland and subsequently seen in September that year on South Uist, Scotland, presumably having just arrived from Iceland, in November 2002 on the Dee estuary, England and then in January 2003 in Wexford Slobs, SE Ireland. (Photo: Peter Potts)

Finally, while generally ignored in field studies, some species react behaviorally to color-rings in unexpected ways, such as incorporating them into their mate-selection criteria—the so-called "Burley effect" (Burley 1986, 1988).

The standard way to record color rings is top to bottom giving the bird's left leg first and then its right. A common abbreviations are R: Red, W: White, Y: Yellow, O: Orange, G: dark Green, L: Lime (or light) green, B: dark Blue, P: Pale blue, N: Black. The color ringed bird in the photo in Figure 4.2 is dark green over orange on the left leg and red over red on the right. It is thus GO-RR. The tibio-tarsal joint is usually shown by "//". Thus if the orange had been below the joint this would have been G//O-RR.

Where the species concerned moves over long distances and/or many ornitho-logists may be color-ringing the same species in different places, it is advisable to coordinate color-ring combinations among workers. In Europe this is done especially for gulls and waders.

The following are a few of the principal alternatives to rings for specialized use.

1. Patagial tags (wing tags) are attached to the wing by a nylon rivet through the skin (patagium) of the leading edge of the "forearm," and are useful for identifying individuals in flight. They are usually made of soft plastic in a conspicuous color, can carry a number, and are typically used on larger birds such as corvids and raptors. However, numbered metal patagial tags, which are supplied for game managers, are useful for marking some galliforms, which have tarsal spurs or knobs that preclude the use of rings. Patagial tags can suffer from the problem of being "preened" into the plumage by the bird and so rendered invisible. Also, as on plastic rings, some colors fade.

2. Flipper bands are flattened metal clip-type bands carrying a number for marking penguins. The band is fitted at the base of the flipper by the body and great care must be taken over the fitting as injury can result if the fit is poor (subcutaneous PIT tags might be preferred for this reason—see 6 and Chapter 7 below).

3. Leg-flags are simple plastic flags made by the researcher from colored plastic for fitting to shorebird (wader) legs. The bird's identity can be indicated by the color of flag and/or a number printed on it.

4. Neck-collars of DARVIC can be used on geese and swans. The bird's identity is indicated by the DARVIC color and inscription. Care over the fit is essential for obvious reasons.

5. Dyes can also be used to mark plumage. In his studies of the energetics of hirundines, Bryant (1984, 1997) used tippex® as a short-term dye on the remiges to identify the birds in flight. A more frequently used permanent plumage dye, however, is picric acid (2,4,6-trinitrophenol), which stains feathers yellow. It is therefore typically used on white birds (swans, seabirds). Care should be exercised in its use because, although it is safe once applied to plumage, under certain conditions it is explosive. While picric acid lasts with little fading until the next molt, human hair-colorants can be used to dye plumage in the short term. Such dyes can be used to paint numbers on plumage, but subsequent preening may distort their appearance; the dyes often wash out within a few weeks, and some containing bleach can damage feathers.

6. As described in Chapter 7, birds can also be fitted with remotely sensed devices, such as radio transmitters (which can indicate activity as well as location), satellite transponders and PIT tags, and with a variety of telemetry devices, such as depth gauges, that can be interrogated on recapture of the bird.

4.5 Notes on bird handling

The adaptations of birds for flight, including pneumatized (hollow) bones, and features of plumage, mean that birds are light in weight and must be handled with care. Two methods for holding small (small enough to be held in one hand) birds have become standard. In the head forward grip, the bird is "caged" by the fingers and the bird's head protrudes between the folded index- and fore-fingers (Figure 4.3). This method is standard in North America and the United Kingdom, where it is known as the "*ringer's grip*" (Pyle *et al.* 1987; Redfern and Clark 2001). In this position the bird is restrained from biting the handler, and the ring, tarsus, and wing can be manipulated as required by the index finger and thumb, leaving the other hand free to handle pliers and other instruments. The bird should be held in the left hand by a right-handed observer and *vice versa*. In the second method, which is favored in parts of continental Europe, birds are held in a reverse grip in which restraint is exercised by the little-finger and ring-finger, while the index finger and thumb can manipulate the tarsus for ringing (Svensson 1994, p.21).

What feels most natural varies between people, but if several measurements are to be taken, less manipulation is required if the ringer's grip is adopted. In either position, it is important that minimal pressure is applied, so that the bird is restrained comfortably without restricting respiration at either the trachea or

Fig. 4.3 This Rose-ringed Parakeet is being held in the standard (head-up) ringer's grip in which its body is supported by the thumb, ring, and little fingers, and the head restrained by the index and middle fingers. This individual also has a large number "4" painted on the breast with hair dye. This mark lasted for several weeks during which it was not recaptured, but was identifiable within the flock. (Photo: Chris Butler)

rib cage. Feathers can be bent to some extent, but shafts of remiges and rectrices should not be broken, as this will affect flight. Remember that many heavy-billed herbivores can inflict a nasty bite, as can most piscivores, and that raptors strike with the feet as well as the bill. Especially in the tropics, liquid iodine is considered vital for scratches and cuts inflicted by birds.

4.6 The bird at close quarters

Once the bird has been identified and ringed, various other attributes can be recorded, and samples taken, before release. The measurements listed below are described in greater detail in Pyle *et al.* (1987), Baker (1993), Jenni and Winkler (1994), Svensson (1994), and Redfern and Clark (2001).

4.6.1 Age and molt

Birds are aged in years relative to the year of hatch, and apart from nestlings and nidifugous precocial young, whose hatch-year is obvious, age is best determined from plumage and soft-part details. Guides to ageing are available for European and North American birds, but knowledge of the molt pattern of the species is also useful. In strongly seasonal temperate latitudes, this poses relatively little problem since most species breed at a specific time of year and the molts fit into a well-defined cycle (**breed–molt**–{migrate}–{molt}–{migrate}–**breed** etc.), and in Europe and North America excellent handbooks containing such details are available (e.g. Pyle *et al.* 1987; Baker 1993; Jenni and Winkler 1994; Svensson 1994). But in areas where the timing of breeding is less well defined, for example "during the rains," these cycles may be less clear and, coupled with or general ignorance of the species concerned, this means that ageing the birds can be difficult or impossible. In these situations, it becomes important routinely to record soft-part details such as the color of iris, bill, mouth-lining, feet, and bare facial skin routinely since these may enable you to age the birds retrospectively when you have worked out how. For example, some nestling passerines (e.g. Sylvine warblers [Svensson 1994], and Indigobirds [Payne 2002]) have tongue-spots, which remain until after fledging, and these can aid ageing.

Here also, skull ossification may be valuable. In passerines the cranium is not fully ossified at fledging. This means that there is effectively a "window" in the bone which is visible beneath the skin (without surgery). However, "skulling" must be performed with the feathers parted in very good direct light and is difficult to learn without guidance. Also, the rate of ossification differs between species so that, while a bird with incomplete ossification of the cranium is undoubtedly juvenile, one with an ossified cranium is not necessarily adult; it depends on the month of observation.

But we should come back to considering the typical pattern of molt, how it indicates age, and how the age should be recorded. Apart from some long-lived non-passerines (e.g. gulls), most species can only be aged as first-year (hatch-year or within a year of hatch), or older. However, once ringed, a bird marked in its first year is effectively ageable thereafter. Many species have a distinct juvenile plumage, grown in the nest. These plumages are typically more cryptically colored than those of the adult, but even in species in which juveniles have an adult-type plumage, the feathers are typically downier in texture, especially under-tail coverts and under-wings, and the underparts and under-wing covert plumage is sparse, leaving bare areas of skin (novices beware brood patches of adult females, see below). In some taxa, juvenile remiges (e.g. galliforms) and/or rectrices (e.g. ducks, some passerines) are more pointed, and because the strength and patterning of feathers differs with age, as does the time since the last molt, the amount of wear on the tips of primaries and other feathers may be much greater on juveniles (e.g. shorebirds).

This juvenile plumage is typically replaced during a partial molt (post-juvenile or first pre-basic molt) a few months after fledging, but still broadly in the natal area. This molt affects all body plumage, and may affect all or some tail feathers, but in most species does not affect the primary and secondary remiges or the greater primary coverts. Molt of the alula, carpal covert and greater coverts may vary between individuals. All this means that birds with the juvenile-type of primary coverts and perhaps contrast in greater coverts between old, unmolted juvenile feathers, and the fresh, molted feathers, may be identified as birds of the year. Many species will retain these juvenile characters for the next year until they undergo their first complete molt as a post-breeding adult (post-breeding or adult prebasic molt). This complete molt includes all feathers and tracts so that the bird is now indistinguishable from older individuals, but first-years in such species can be aged while breeding when 1 year old.

This pattern is typical of resident passerines at temperate latitudes (but beware some species of sparrows, larks, starlings, and a few others that have a complete post-juvenile molt), but variations abound. Many migrants defer molt until they reach the wintering grounds, and some species also have a partial pre-breeding body-molt. These factors can obscure or eliminate the contrast between juvenile and adult feathers. Again, the details for particular species are available in standard handbooks.

In recognition of these patterns of molt, ringing schemes have devised systems of age codes that make allowance for deficiencies in available information. The European (EURING) scheme is widely used and well-known. This is a numerical code based on calendar years whereby the number increases with advancing age. Odd-number codes (1, 3, 5, etc.) describe birds whose hatch-year is known, while even-numbered ages (2, 4, 6, etc.) are not precisely known. Thus a pullus is coded "1," it becomes "3" between fledging and the year's end, and turns "5" on 1 January. A bird of completely unknown age is coded "2." The same bird becomes 4 on 1 January (as it could not be in its hatch-year), and "6" the following year. Hence, using this system a resident temperate-latitude passerine with "typical" molt cycle (e.g. a Common Blackbird *Turdus merula*, Great Tit *Parus major*, American Robin *Turdus migratorius* or a Golden-crowned Kinglet *Regulus satrapa*) would be aged 1 in the nest, 3 or 5 as a first-year (before or after 1 January respectively), and adults would be aged 4 or 6 (before or after 1 January respectively). The scheme differs somewhat in North America, where 3 = HY (hatch year), 5 = SY (second year), 4 = AHY (after hatch year), and 6 = ASY (after second year). In the tropics and Southern Hemisphere, these codes can make little sense because of the different patterns of seasonality which lead many birds to breed at the turn of the year.

4.6.2 Sex

Most species show some degree of sex difference, either in color (sexual dichromatism) or size (sexual dimorphism). However, in most sexually dimorphic species (males typically larger except in raptors and some waders), the statistical separation on size is only partial, that is, there is some overlap between the sexes. Thus the ideal species in which to distinguish sex are dichromatic (e.g. ducks, gamebirds, some passerines). Although many passerines and other species are technically monochromatic, there is nevertheless a sufficient sex-difference in the plumage through, for example, its reflectivity (plumage typically more glossy in males) to "sex" a significant proportion of the birds unequivocally.

For many sexually monochromatic species, it has been possible to produce an index for sexing based on a Discriminant Function Analysis (DFA) of a combination of size measures (see below), so that one might be able to say, for example, that a bird with a wing-length greater than 87 mm, bill greater than 14.0 mm and tarsus length greater than 35.0 mm has a 0.90 probability of being male. While this method has better discriminating power than any single measurement, it is rarely foolproof or applicable to all populations. Thus in such cases, if definite sex identity is essential, molecular methods (based on DNA) must be used. Thanks to the Polymerase Chain Reaction (Chapter 9) and the fact that avian erythrocytes are nucleated, birds can now be sexed from minute quantities of blood (Griffiths and Tiwari 1993, 1995).

In theory it should be possible to determine the sex of birds through examination of the cloaca since, for example, males may have a pronounced cloacal protuberance, at least when breeding (Chiba and Nakamura 2002, 2003). However, in practice this is difficult for many species, or when not breeding, although wildfowl, cranes and some others can be sexed by eversion of the cloaca since the male has a distinct phallus (Baker 1993, p29). Breeding females (e.g. passerines) develop a distinctive brood-patch in which the feathers of the belly are lost, and the skin becomes hot, highly vascularized and oedematous.

4.6.3 Weight

The weight of a bird is a basic measurement that can be used as a measure of size in cross-species comparisons. However, as it incorporates variation in both size and condition (e.g. fat reserves), it should not be used alone as either a measure of size or condition in intra-specific studies. Weight is readily determined either with a lightweight precision spring-balance (e.g. manufactured in Switzerland by Pesola®), or on an electronic, digital, pan balance. Again, these are now available as small, precise, highly portable battery or solar-operated units (e.g. Tanita® TPK100). Birds of up to 50 g should typically be weighed to a precision of 0.1 g, though greater resolution is desirable for the smaller hummingbirds. For birds between 50 and 300 g weight to the nearest gramme should suffice, and for even heavier birds to the nearest 10 or 100 g as appropriate. The bird should be immobilized in some form of restraint (a polythene funnel or cone weighing c.0.2 g is ideal for smaller birds), and this should be clipped onto the spring balance or, for digital balances, closed with a bulldog clip of known weight and laid on the pan of the digital balance. When weighing in the field, windy conditions can distort readings, but one way to avoid this problem is to suspend the bird in its cone within a larger windproof container such as a large jar (or a plastic or cardboard tube).

4.6.4 Color, for example, UV reflectance

It is often desirable for work on sexual selection and systematics, to record variation in color traits. The colors of plumage areas can be recorded visually by reference to a standard color chart such as Küppers (1978). However, this has several problems due to variation in illumination, observer eyesight, print quality, and fading. Furthermore, recent work has indicated that some plumage colors are reflected in the ultraviolet range (<400 nm), and so cannot be detected by the human eye. If you are studying such colors, you can use an electronic color detector (such as the PS1000 diode array spectrometer available from Ocean Optics, Dunedin USA), which reflects a standard light source (DS2000 deuterium-halogen) onto a small area of feather (or any other surface) and analyses the wavelength profile reflected back in terms of spectral location (color or hue), purity (chroma), and intensity (brightness). The reflected spectrum can be visualized and analyzed on a microcomputer (e.g. see Ornborg *et al.* 2002). The drawback of this equipment is that it is expensive, cumbersome, and requires a power source.

4.7 Size

4.7.1 Body size

There are many reasons in ecological and evolutionary research for measuring the size of individual birds. For comparative studies, where most variance under consideration is among (rather than within) taxa, weight gives a reasonable approximation. However, because this reflects condition (e.g. weight changes through the day) while size cannot, weight alone is an inappropriate measure to study intraspecific variation. Thus some measure of overall body size is required. No single measure is ideal for what is really a surrogate for the overall skeletal size.

Traditionally, total bird length measured by straightening the bird out laid on its back on a rule (see Svensson 1994), is frequently given in field-guides. However, in practice this is difficult to do with a live bird, and is neither highly repeatable nor a reliable indicator of body size since it includes independently variable features such as bill- and tail-length. The most direct measure of body size is the length of the sternal keel, as taken by Bryant and Jones (1995) to measure Sand Martins *Riparia riparia*. Again however, this measure is difficult to take and not very repeatable.

A more meaningful, reliable, and now widespread method, suggested by Rising and Somers (1989), is to take several measures such as wing, tail, tarsus, and bill, and to use the first Principal Component from a Principal Components Analysis (PCA) since this represents the component of variance in each which correlates with the other measures largely through the shared correlate of overall size. Although PCA is the ideal, it may not be practicable. If, for example, time constraints mean that only one measure of size can be taken, then wing-length is recommended because it gives the best approximation to PC1 in full-grown individuals of most species studied (Gosler *et al.* 1998). If chicks are to be included in sampling, tarsus-length might be preferred because in many species it is full-grown earlier than the wing.

4.7.2 Wing

Wing-length, defined as the distance from the carpal joint to the tip of the longest primary on the closed wing, is the single measure of size most commonly recorded in birds. It is measured on a stopped stainless-steel rule (e.g. BTO wing-rule) with the wing held in its natural closed position or as near to it as possible (i.e. do not pull the wing out) (Figure 4.4). With the bird in the ringer's grip (left hand if right-handed) with its back adjacent to the palm, gently slide the rule under the wing and draw it down until the carpal joint abuts the end stop. With the left thumb securing the carpal at this point, draw down the primaries with the right thumb to straighten them and read off the rule (proximal side to the

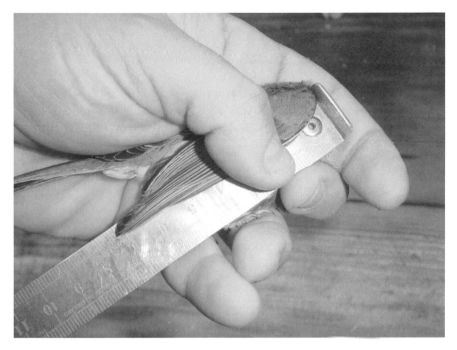

Fig. 4.4 Measuring wing length of a small bird using a stopped rule. Note that the bird's back is in the palm of the hand, the carpal joint abuts the end-stop, that the wing is closed and the leading edge of the wing forms a line, as near as possible, parallel with the edge of the wing rule, and the wing as gently but firmly flattened and straightened onto the rule. This procedure gives a highly repeatable, and reliable measure of general body size for many species. This Great Tit has a wing length of 73 mm. (Photo: Andrew G. Gosler)

bird) at the wing tip to the nearest 0.5 or 1 mm. The pressures required should be adequate to flatten the wing onto the rule and straighten the leading edge of the wing without either pulling the carpal away from the end-stop, or damaging the wing. Do not try to over-straighten the rounded wings of certain non-passerine groups, and beware of birds in molt, whose wing tip might not be fully grown. With a little practice, this measurement becomes highly repeatable (Gosler *et al.* 1995*a*). In multiple-observer studies, it is wise to have all observers measure a sample of birds to standardize their methods.

An alternative to the "flattened-straightened wing" measure described here, is to measure just the third outermost primary. This should be done using a standard rule, available from the Swiss ringing scheme (Vogelwarte Sempach). The rule has a vertical pin at the tip, which is inserted between the second and third primaries while the third primary is straightened on the rule and its length read off (Jenni and Winkler 1989). This has advantages when skins and live-bird measurements must be compared, it may be easier to use on larger birds, and may reduce inter-observer error (although this is not proven—Gosler *et al.* 1995*a*), but is best not used on very small birds.

Length is not the only measure that can be taken of the wing. The relative lengths and pattern of notching and emargination of the primaries (wing formula) may be of use in specific identification and taxonomic studies (see Pyle *et al.* 1987; Svensson 1994). Wingspan and wing-area can be taken also. These may be necessary for flight-performance calculations but are not taken as routine measurements for practical and welfare reasons (but see Pennycuick 1989 if required). Also they are less repeatable than the more usual measures.

4.7.3 Tail

In general, within species (at least in passerines), tail-length and wing-length are highly correlated (Gosler *et al.* 1998). However, especially in studies of sexual selection, it may be desirable to measure the length and shape of the tail. For overall tail-length, slide an unstopped steel rule between the tail feathers (rectrices) until it stops at the feather bases, and read off the longest feather length from the rule. It is helpful to hold the bird so that the underside of the tail can be viewed easily. The difference in length between the longest and shortest tail feathers is also best measured from under the tail (depth of notch of fork—for example, Barn Swallows *Hirundo rustica*).

4.7.4 Tarsus

Another metric that gives a good indication of overall size (and excellent when combined with wing-length in PCA) is the tarsus-length, strictly the length of the tarsometatarsus. Earlier methods described in the literature proved to be highly unrepeatable for live birds but were necessary to measure the "set" legs of skins. However, the method described here is highly reliable. With the bird in the ringer's grip, the tarsus should be held between the thumb and ring-finger (left hand) so that the tibiotarsus and tarsometatarsus form an acute angle. The foot is held at right angles to the tarsus and the measurement from these two jointed right-angle bends taken either with OD (outside diameter) dial or Vernier callipers, or on a stopped rule. In either case, the measurement (maximum tarsus) can be read to 0.1 mm in small birds (under 100 g) and to 1 mm in larger birds. An alternative—the minimum tarsus—is to measure in the same way with callipers, but into the notch at the tarsal joint (Figure 4.5).

(a)

(b)

Fig. 4.5 Measuring "maximum" tarsus using the OD calliper of a Vernier. This measurement is typically taken to 0.1 mm in species of the size of this European (Wood) Nuthatch. Note that (a) the foot forms a right angle to the tarsus, and that it is essential that the tarsus is held parallel to the calliper (above); (b) this measurement can be taken on a wing rule (below), and that the two methods should give identical measurements. (Photo: Andrew G. Gosler)

4.7.5 Tarsus-and-toe

In some non-passerine groups (e.g. waders, rails) a useful additional size measure is the tarsus-and-toe, taken by holding tarsal joint at right angles (see tarsus), and placing the tibiotarsus up against the stop of a wing-rule. The foot and toes are then stroked out flat onto the rule and measurement taken at the tip (see Baker 1993, p.14).

4.7.6 Bill

Bill-size (length, depth, width) can contribute information toward a size PCA (see above) but expresses much ecological information in its own right, since variation in bill-size and shape constrains the diet that the bird can take. Because of its ecological importance, it is often found to correlate poorly with overall body size within species. The exact measurement to take depends on the type of bird because bill-morphology varies too much between taxa to give a single recommendation. In general, bill-length is taken with ID (inside diameter) dial or Vernier from the bill's base at the skull (*naso-frontal hinge*) to the tip (*dertrum*) (Figure 4.6). With practice, this can be taken repeatably in small birds to 0.05 or 0.1 mm. In waders and similar taxa, the measurement should be taken to the feathering, which is clearly demarcated, rather than to the skull, and in species with a cere (e.g. raptors) the measurement is taken to this rather than the skull. Special care must be taken in placing the calliper at the proximal end of the bill due to its proximity to the bird's eye. For taxa with complex (e.g. highly curved) bills it may be necessary to use or devise other measures such as from the tip to the nostril, and close-up photography may be helpful for taking complex measurements such as curvature of the culmen (e.g. Gosler 1999).

Bill-depth is ecologically a highly significant trait. It is taken at right-angles to the cutting edges (*tomia*) of the mandibles, specifying where along the bill it is taken. The typical reference point is at the deepest point of the gonys (the distal portion of the bill where the two lateral arms (*rami*) of the lower mandible are fused). In gulls and some seabirds, the bill-depth at the gonys is sexually dimorphic and a good indicator of body-size, but it is a useful and repeatable measure in many other taxa. It should be taken with the OD-calliper of a Vernier to 0.05 or 0.1 mm in small birds. Do not apply excessive pressure but take care that the bill is closed (and the upper mandible not retracted) when taking the metric. When taking this measurement on waders, remember that the bill surface is sensitive to touch.

The ratio of bill-depth/bill-length is known as the bill-index. This gives a useful measure of bill shape (relative bill-depth), which is ecologically relevant in

(a)

(h)

Fig. 4.6 Measuring bill-length (a) and depth (b) of a Great Tit. The bill-length is taken with the "ID" calliper, the bill-depth with the "OD" calliper. Both metrics can be taken reliably (i.e. repeatably) to 0.1 or even 0.05 mm with Vernier or dial callipers. Variation within and between species in bill-size and shape is ecologically important. (Photo: Andrew G. Gosler)

some species. Bill-width is more rarely used than the above metrics because of the difficulty in defining where, along a tapering object, to take the measurement: it is usual to do so at the gape (but beware pulli and recently-fledged juveniles with a gape flange), and at the gonys.

4.7.7 Total-head

The distance from the bill-tip to the center of the back of the skull (known as total-head or head-and-bill measurement) has been found to give a reliable measure of overall body size in certain non-passerine taxa (especially waders, some seabirds, especially gulls, and wildfowl, especially swans). The metric should be taken with the OD-calliper of a Vernier large enough to accommodate the whole head comfortably, and apply minimal force when taking the metric. Remember that some of these birds (waders, wildfowl) have sensitive, highly innervated, bills. It can be useful to attach a metal plate to the "jaw" of the calliper that goes at the back of the head, as this helps to standardize its position.

4.7.8 Claw, eye-ring, and other measures

If it is necessary to measure claws, this should be done with the ID-calliper of a Vernier and the measurement taken from the base of the upper surface to the claw tip. Older works advise the use of a pair of dividers to take precise measurements, and then reading the distance off from the dividers with a rule. While fine for skins, it is inadvisable to use such sharp implements near sensitive areas of a live bird, so ID-callipers are preferable for any other measures (e.g. eye-ring width etc.).

4.8 Condition

In its present sense, condition means some qualitative assessment (which may be determined quantitatively) of the bird that has a direct bearing on its fitness. Although it might be affected by the size of the bird, it is defined such that size is *not*, per se, a measure of condition. Condition measures typically reflect the size of nutrient reserves or the ability to resist parasites or disease, and might be reflected in attractiveness to a potential mate. Hence color measures (above) might act as indicators of condition.

4.8.1 Asymmetry

The symmetry of an animal (measured as asymmetry) itself gives a measure of the animal's condition during development of the trait (fluctuating asymmetry). Any trait that can be taken on both sides of the bird (wing, tarsus, tail tips, etc.) can be assessed for symmetry, and used as an indicator of condition and fitness

(Møller 1998; Shykoff and Møller 1999). However, observers are themselves asymmetrical, and typically one side will be preferred as feeling more "natural" because of the observer's own left or right-handedness. Normally this does not matter, but when determining a bird's symmetry, the fact that we are more proficient in measuring one wing, tarsus etc. than the other can become a problem, because one side is measured more reliably than the other. This problem needs to be addressed more frequently than it has in the past (Helm and Albrecht 2001).

4.8.2 Relative mass

If it is not possible to measure nutrient reserves directly on the live bird, a reasonable estimate can be obtained by considering the mass (or weight) relative to the size of the bird. This is typically done by regressing mass on a size measure (PC1 or wing-length—see above) and using the residual scores (i.e. a bird above the regression line is heavy for its size etc.) as a measure of condition. This gives only an approximation to fat reserves (Gosler et al. 1998) because a bird may be relatively heavy for various reasons (fat mass, pectoral-muscle mass, fullness of gut, etc.). Relative mass is widely used and often simply called "condition" or "condition index," but for some kinds of analysis it can give misleading results because body size is measured imperfectly (see e.g. Gosler and Harper 2000). A useful extension of this method, however (mass/bill-length), using the fact that the bill grows more or less linearly, can be used to assess the condition of partially-grown wader chicks (see Beintema 1994).

4.8.3 Fat reserves

The characteristic most frequently associated with individual condition is the quantity of fat carried by the bird. However, the assumption that high and low fat loads indicate birds in "good" and "bad" condition respectively comes from our anthropocentric viewpoint as a terrestrial mammal, our knowledge of the energy requirements of long-distance migrants, and the fact that birds picked up dead under extreme cold conditions (waders and wildfowl) are usually seen to have starved. We must be careful how we regard fat loads because they carry a cost (Witter and Cuthill 1993; Gosler 2001). Nevertheless, the observation of visible subcutaneous fat in many species (especially passerines) is quick and reliable, and can give insight into the birds' biology (Gosler et al. 1995b; Gosler 1996, 2002; Carrascal et al. 1998).

Most birds deposit fat in discrete depots, which can be assessed by a standard scoring system (Figure 4.7). One such scale for this was developed by McCabe (1943) for assessing the fat on museum skins. It subsequently proved reliable for use on live birds and was adopted by the BTO's Biometrics Working Group

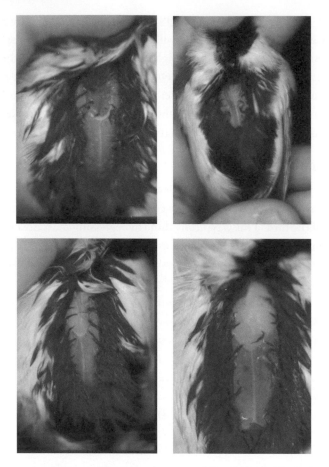

Fig. 4.7 In many species, such as passerines at middle and higher latitudes, the level of fat reserves can be assessed visually from live birds because the amount of fat accumulated beneath the skin within the tracheal pit and on the abdomen are directly proportional to the total body fat. In this Great Tit, the feathers have been parted; the fat is then visible as a yellow–pink mass beneath the skin; its quantity can then be scored by reference to the fixed points formed by the two arms of the furculum. Top left, score 1, fat just visible in the bottom of the pit. Top right, score 2, pit one third full. Below left, score 4, the pit is just full. Below right, score 5, fat bulges beyond the level of the furculum and onto the pectoral muscle. (Photo: Andrew G. Gosler)

(Gosler *et al.* 1998). In this system, visible fat is assessed in the tracheal pit (between the halves of the wishbone) on a scale (0–5) where zero represents no fat, and "5" represents fat filling and bulging out of the pit. It has been shown in several species that fat in these discrete deposits reflects the total fat load in the body. Moreover, it can be seen simply by blowing the feathers aside.

McCabe's system is inadequate for recording the high fat loads of long-distance migrants. Kaiser's (1993) system based on both the tracheal pit and abdominal fat bodies consists of eight main score categories with a further four subcategories in each giving a total resolution of 32 score classes. An advantage of these scoring systems is that they are inherently scaled relative to the size of the bird (e.g. a fat score of 5 means the same whether it is a kinglet or a thrush). A disadvantage is that, although they tend to be normally distributed, fat scores are bounded between limits and this can have implications for statistical analysis (Greenwood 1992).

The total fat content of a bird can also be measured by determining the bird's total body electrical conductivity (TOBEC). This method uses the fact that the electrical conductivity of tissues varies systematically with its fat content. TOBEC machines can be used to assess the fat content of live birds. However, they are expensive, not readily portable, the bird must be immobilized within the chamber of the machine during use and, for each size class (e.g. *Regulus*, *Parus*, *Turdus*, etc.) if not every species, its readings must be calibrated against a series of birds on which total lipid extractions have been undertaken. Studies indicate that its precise results are little better than those achieved by non-destructive fat-scoring methods (Brown 1996; Speakman 2001).

4.8.4 Muscle protein

Another useful measure of condition concerns muscle mass, especially of the pectoral muscles, which constitute 15–21% of total body mass in birds generally. Muscle mass varies through changes in labile protein reserves and wastage. It can be assessed from the cross-sectional shape of the pectoral muscles (visually or by feeling the shape) and recorded on a 4- or 5-point scale (Gosler 1991; Gosler *et al.* 1998; Redfern and Clark 2001), or by recording its shape by using fine wire or dental alginate (Bolton *et al.* 1991).

4.8.5 Physiological measures

Various measures of physiological stress can be obtained from microscopic and/or chemical analysis of blood components (erythrocytes, plasma etc.) (see Brown (1996) and Chapter 10 for further details).

4.8.6 Molt and plumage

The study of molt is a major line of enquiry in its own right, which in general must be assessed in the hand (some details of molt of some tracts in some birds can be determined by observation in the field) or by collecting shed feathers from roosting sites. Although a detailed assessment of molt would require each feather tract to be assessed in turn, much information can be gleaned from a quick assessment of wing and body molts. The simplest scheme is to record the state of progress of molt through the primaries by scoring each feather 0 (old) to 5 (new full-grown) and summing across all primaries (thus a passerine with 10 primaries scores 50 when molt is completed). Molt of other tracts can be assessed similarly as 0 (old), 1 (in molt), or 2 (molt complete), or just recorded as in molt. The extent of post-juvenile molt (e.g. greater coverts) varies between individuals in many species (Jenni and Winkler 1994), and may indicate the bird's age and condition at molt (Gosler 1991).

The condition of the plumage itself, and especially the presence of fault bars in remiges and rectrices, indicate inadequate nutrition during feather growth, and are thus worth recording (Murphy *et al.* 1989). Furthermore, there is evidence that the growth rates of feathers during molt indicates the bird's nutritional status during its growth. Feather structure differs depending on whether growing in the day or night; a fact that leaves growth-bands across the feather. Thus by measuring the widths of these bands, the feather's growth rate can be determined. This method is called *ptilochronology* (Grubb 1989, 1995). Ideally, assessment is made by comparing bars in a tail feather induced by plucking to grow, with those in the plucked feather. This allows condition to be assessed outside the molt period; although the width of bars on feathers grown at the usual time has also been used as a measure of condition during molt (Carlson 1998). Although the value of ptilochronology has been well-demonstrated (Brodin 1993), its use has been criticized (Murphy 1992).

4.8.7 Parasites

Birds have a number of parasites, which may be important with reference to the condition of the bird (but be wary of attributing poor condition to high parasite infestation, since the birds condition may be the cause of the infestation not the response to it), or of interest in their own right. Ectoparasites include fleas, feather lice, feather mites, ticks, and parasitic flies such as hippoboscids. Most avian ectoparasites pose little risk to humans, although some bird fleas will sample human blood before giving up on it. The presence, distribution, density, and specific identities of ectoparasites can be assessed visually by searching through plumage under good light parting the feathers with a paintbrush (rather ineffective), or by putting the bird into a jar containing ether vapor (e.g. on a cotton-wool swab), and with its head exposed (Fair-Isle Apparatus), so that the parasites fall off. Feather mites, which line up between the barbs of the remex vanes, can be scored (0–5) on each feather by holding the wing up to the light. For further details and methods see Loye and Zuk (1991).

Birds provide a habitat for endoparasites, chiefly in the blood and gut. Gut parasites may be obtained from feces collected during ringing but this is likely to give a poor indication of gut parasite load. Blood parasite assessment requires blood biopsy (see below and Chapter 10).

4.9 Biopsy

Recent developments allow more precise measurements of condition, metabolic rate, and life history to be made, from blood and feather samples, than have ever previously been possible. The technical details need not concern us here (see Chapter 9) but a short list may be helpful. In some countries special legislation may apply to these procedures (e.g. ASPA in the United Kingdom) so a specific licence may be required. From small blood samples, typically taken by venipuncture of the ulnar wing vein (Hawkins *et al.* 2001), genetic, condition, and stress analyses (fat, protein, and corticosterone) can be undertaken (Brown 1996). It is well established that such sampling does not cause lasting harm to the birds (Stangel 1986). The ratio of isotopes (e.g. of Carbon and Hydrogen) in feather keratins can be measured from small feather samples, and these can be used (with caution because C isotopes can also vary with diet and habitat) to determine, within broad regional limits, where the bird molted because these ratios differ geographically (Hobson 1999; Bearhop *et al.* 2000; Wassenaar and Hobson 2000). If the bird can be trapped and retrapped within a short time (e.g. 36 h), a small injection of doubly-labelled water can be used to assess metabolic rate, which is reflected in the rate at which the ratio of labelled to unlabelled water changes over time, after equilibration (Tatner and Bryant 1986; Bryant 1997). Feces for analysis (Chapter 10) can easily be collected during handling, with the advantage that details of the individual will be known. Finally, for further observations that can be made of trapped migrants, such as preferred migratory orientation (e.g. Busse 1995), see Chapter 7.

Acknowledgements

I am indebted to Peter Kennerley, David Wells, Paul Leader, David Melville (all Far East) Jeremy Lindsell (Africa), Jose Arevalo (South America), Paul Salaman (South America), Leon Bennun/Colin Jackson (Africa and tropical ringing generally) and Peter Potts for their advice on tropical issues, and to the editors for their constructive comments on an earlier draft.

References

Baker, J.K. (1993). *Identification Guide to European Non-Passerines*. BTO Guide 24, Thetford.
Bearhop, S., Phillips, R.A., Thompson, D.R., Waldron, S., and Furness, R.W. (2000). Variability in mercury concentrations of great skuas Catharacta skua: the influence of

colony, diet and trophic status inferred from stable isotope signatures. *Mar. Ecol.—Prog. Series*, 195, 261–268.

Beintema, A.J. (1994). Condition indexes for wader chicks derived from body-weight and bill-length. *Bird Stud.*, 41, 68–75.

Bolton, M., Monaghan, P., and Houston, D.C. (1991). An Improved Technique for Estimating Pectoral Muscle Protein Condition from Body Measurements of Live Gulls. *Ibis*, 133, 264–270.

Brodin, A. (1993). Radio-ptilochronology—tracing radioactively labelled food in feathers. *Ornis Scand.*, 24, 167–173.

Brooman, S. and Legge, D. (1997). *The Law relating to Animals*. Cavendish Publishing, London.

Brown, M. (1996). Assessing body condition in birds. In Nolan, V. and Ketterson, E.D. (eds.) *Curr. Ornithol.*, 13, 67–135.

Bryant, D.M. (1984). Foraging energetics of swallows and other birds. *Ibis*, 126, 456.

Bryant, D.M. (1997). Energy expenditure in wild birds. *Proc Nutr. Soc.*, 56, 1025–1039.

Bryant, D.M. and Jones, G. (1995). Morphological changes in a population of Sand Martins *Riparia riparia* associated with fluctuations in population size. *Bird Stud.*, 42, 57–65.

Bub, H. (1991). *Bird Trapping and Bird Banding*. Cornell University Press, New York (first published 1978 as *Vogelfang und Vogelberingung* A. Ziemsen Verlag, DDR).

Burley, N. (1986). Comparison of the band colour preferences of two estrildid finches. *Anim. Behav.*, 34, 1732–1741.

Burley, N. (1988). Wild zebra finches have band colour preferences. *Anim. Behav.*, 36, 1235–1237.

Busse, P. (1995). New technique of a field study of directional preferences of night passerine migrants. *Ring*, 17, 97–116.

Carrascal, L.M., Senar, J.C., Mozetich, I., Uribe, F., and Domenech, J. (1998). Interactions among environmental stress, body condition, nutritional status, and dominance in great tits. *Auk*, 115, 727–738.

Carlson, A. (1998). Territory quality and feather growth in the White-backed Woodpecker *Dendrocopus leucotos*. *J. Avian Biol.*, 29, 205–207.

Chiba, A. and Nakamura, M. (2002). Female cloacal protuberance of the polygynandrous Alpine Accentor *Prunella collaris*: histological features and possible functional significance. *Ibis*, 144, E96–E104.

Chiba, A. and Nakamura, M. (2003). Anatomical and histophysiological characterization of the male cloacal protuberance of the polygynandrous Alpine Accentor *Prunella collaris*. *Ibis*, 145, E83–E93.

Gill, J.A., Norris, K., Potts, P., Gunnarsson, T., Atkinson, P.W., and Sutherland, W.J. (2001). The Buffer effect and large-scale population regulation in migratory birds. *Nature*, 412, 436–438.

Gosler, A.G. (1991). On the use of greater covert moult and pectoral muscle as measures of condition in passerines with data for the Great Tit *Parus major*. *Bird Stud.*, 38, 1–9.

Gosler, A.G. (1996). Environmental and social determinants of winter fat storage in the Great Tit *Parus major*. *J. Anim. Ecol.*, 65, 1–17.

Gosler, A.G. (1999). A comment on the validity of *Parus major newtoni* (Prazak 1894).

Gosler, A.G. (2001). The effects of trapping on the perception, and trade-off, of risks in the Great Tit *Parus major*. *Ardea* 89 (special issue), 75–84. *Bull. Brit. Orn. Club.*, 119, 47–55.

Gosler, A.G. (2002). Strategy and constraint in the winter fattening of the great tit *Parus major. J. Anim. Ecol.*, 71, 771–779.

Gosler, A.G. and Harper, D.G.C. (2000). Assessing the heritability of body condition in birds: a challenge exemplified by the great tit *Parus major* L. (Aves). *Biol. J. Linnean Soc.*, 71, 103–117.

Gosler, A.G., Greenwood, J.J.D., Baker, J.K., and King, J.R. (1995*a*). A comparison of wing length and primary length as size measures for small passerines—a report to the British Ringing Committee. *Ring. and Migr.*, 16, 65–78.

Gosler, A.G., Greenwood, J.D.D., and Perrins, C. (1995*b*). Predation risk and the cost of being fat. *Nature*, 377, 621–623.

Gosler, A.G., Greenwood, J.J.D., Baker, J.K., and Davidson, N.C. (1998). The field determination of body size and condition in passerines: a report to the British Ringing Committee. *Bird Stud.*, 45, 92–103.

Greenwood, J.J.D. (1992). Fat scores: a statistical observation. *Ring. and Migr.*, 13, 59–60.

Griffiths, R. and Tiwari, B. (1993). The isolation of molecular genetic markers for the identification of sex. *Proc. Natl. Acad. Sci. USA*, 90, 8324–8326.

Griffiths, R. and Tiwari, B. (1995). Sex of the last wild Spix's macaw. *Nature*, 375, 454.

Grubb Jr., T.C. (1989). Ptilochronology: feather growth bars as indicators of nutritional status. *Auk*, 106, 314–320.

Grubb Jr., T.C. (1995). Ptilochronology: a review and prospectus. *Curr. Ornithol.*, 12, 89–114.

Hawkins, P., Morton, D.B., Cameron, D., Cuthill, I., Francis, R., Freire, R., Gosler, A., Healy, S., Hudson, A., Inglis, I., Jones, A., Kirkwood, J., Lawton, M., Monaghan, P., Sherwin, C., and Townsend, P. (2001). Laboratory birds: refinements in husbandry and procedures. Fifth report of the BVAAWF/FRAME/ RSPCS/UFAW Joint Working Group on Refinement. *Lab. Anim.*, 35(Suppl. 1), 162.

Helm, B. and Albrecht, H. (2000). Human handedness causes directional asymmetry in avian wing length measurements. *Anim. Behav.*, 60, 899–902.

Hobson, K.A. (1999). Stable-carbon and nitrogen isotope ratios of songbird feathers grown in two terrestrial biomes: implications for evaluating trophic relationships and breeding origins. *Condor*, 101, 799–805.

Jenni, L. and Winkler, R. (1989). The feather-length of small passerines: a measurement for wing-length in live birds and museum skins. *Bird Stud.*, 36, 1–15.

Jenni, L. and Winkler, R. (1994). *Moult and ageing of European passerines*. Academic Press, London.

Kaiser, A. (1993). A new multicategory classification of subcutaneous fat deposits of songbirds. *J. Field Orithol.*, 64, 246–255.

Küppers, H. (1978). *DuMont's farben atlas*. DuMont Buchverlag, Köln, Germany.

Loye, J.E. and Zuk, M. (eds.) (1991). *Bird-parasite interactions*. Oxford University Press, Oxford.

McCabe, T.T. (1943). An aspect of the collector's technique. *Auk*, 60, 550–558.

Moberg, G.P. and Mench, J.A. (eds.) (2000). *The Biology of Animal Stress—basic principles and implications for animal welfare*. CABI, New York.

Møller, A.P. (1998). Developmental instability as a general measure of stress. *Stress Behav.*, 27, 181–213.

Murphy, M.E. (1992). Ptilochronology: accuracy and reliability of the technique. *Auk*, 109, 676–680.

Murphy, M.E., Miller, B.T., and King, J.R. (1989). A structural comparison of fault bars with feather defects known to be nutritionally induced. *Can. J. Zool.*, 67, 1311–1317.

Ornborg, J., Andersson, S., Griffith, S.C., and Sheldon, B.C. (2002). Seasonal changes in a ultraviolet structural colour signal in blue tits, *Parus caeruleus. Biol. J. Linnean Soc.*, 76, 237–245.

Payne, R.B., Hustler, K., and Stjernstedt, R., *et al.* (2002). Behavioural and genetic evidence of a recent population switch to a novel host species in brood-parasitic indigobirds *Vidua chalybeata. Ibis*, 144, 373–383.

Pennycuick, C.J. (1989). *Bird Flight Performance*. Oxford University Press, Oxford.

Pyle, P., Howell, S.N.G., Yunick, R.P., and DeSante, D.F. (1987). *Identification Guide to North American Passerines*. Slate Creek Press, Bolinas California.

Redfern, C.P.F. and Clark, J.A. (2001). *The Ringer's Manual* (5th edn.). BTO, Thetford.

Rising, J.D. and Somers, K.M. (1989). The measurement of overall body size in birds. *Auk*, 106, 666–674.

Salem, D.J. and Rowan, A.N. (eds.) (2001). *The State of the Animals 2001*. Humane Society Press, Washington.

Shykoff, J.A. and Møller, A.P. (1999). Fitness and asymmetry under different environmental conditions in the barn swallow. *Oikos*, 86, 152–158.

Speakman, J.R. (ed.) (2001). *Body composition analysis of animals: a handbook of non-destructive methods*. Cambridge University Press, Cambridge.

Stangel, P.W. (1986). Lack of effects from sampling blood from small birds. *Condor*, 88, 244–245.

Svensson, L. (1994). *Identification guide to European passerines* (4th edn.). Stockholm.

Tatner, P. and Bryant, D.M. (1986). Flight Cost of a Small Passerine Measured Using Doubly Labeled Water—Implications for Energetics Studies. *Auk*, 103, 169–180.

Wassenaar, L.I. and Hobson, K.A. (2000). Stable-carbon and hydrogen isotope ratios reveal breeding origins of red-winged blackbirds. *Ecol. Appl.*, 10, 911–916.

Witter, M.S. and Cuthill, I.C. (1993). The ecological costs of avian fat storage. *Phil. Trans. Roy. Soc. London B*, 256, 299–303.

5

Estimating survival and movement

James D. Nichols, William L. Kendall, and Michael C. Runge

5.1 Introduction

Goals of bird conservation programs typically are expressed in terms of either abundance or quantities such as extinction probability that are strongly influenced by abundance. Abundance is accordingly the state variable used in most models of bird populations, and its estimation is therefore important (Chapters 1, 2). Changes in abundance over time are functions of four fundamental demographic parameters: reproduction, survival, emigration, and immigration. Conservation programs that seek to bring about changes in abundance must do so via management actions that influence one or more of these four parameters (see Chapters 12, 13, 14). Estimation of these quantities and, more importantly, the relationship between these quantities and environmental variables, bird density and conservation actions, forms a central methodological topic in bird conservation. Methods for studying reproduction have been presented in Chapter 3, and this chapter deals with methods for estimating survival and movement in and out of populations. More detailed treatment of the material presented here can be found in Seber (1982) and Williams *et al.* (2002), and for birds, in particular, in Clobert and Lebreton (1991).

Sometimes it is possible to draw inferences about survival and movement based on counts of birds. For example, estimation of survival is sometimes based on counts of birds in different age classes. Although appropriate methods exist for such estimation (e.g. Udevitz and Ballachey 1998; Williams *et al.* 2002), they require restrictive assumptions about time- and age-specific sampling probabilities and population growth that are often difficult to meet. We thus tend to agree with Clobert and Lebreton (1991) that such methods have not been generally useful for birds. However, a recent Bayesian analysis successfully used counts of first-year and older Whooping Cranes (*Grus americana*) to estimate age-specific survival and recruitment rates (Link *et al.* 2003), and this approach holds promise for similar sampling situations. Inferences about movements can also be based on counts of birds. For example, Johnson and Grier (1988) drew inferences about duck movements based on year-to-year changes in abundance estimates in different regions. Although reasonable inference is sometimes possible, confidence in the findings is limited because of the influence of other demographic variables. It is difficult to attribute changes in abundance to the action of a single demographic parameter, when all four parameters act in concert to determine abundance. In this chapter, we therefore consider only methods based on marked individuals, because such methods are well developed and permit separate estimation of both survival and movement rates.

5.2 Tag type and subsequent encounters

Robust estimation of rates of survival and movement usually entails capturing and marking birds with individual marks so that they can be recognized at subsequent encounters (see Chapter 4 for methods of capture and marking). The kind of mark applied determines the appropriate method for re-encountering marked birds. For example, if birds are tagged with satellite transmitters then re-encounter data are downloaded from satellites. If birds are tagged with standard radio-transmitters (Chapter 6), then re-encounters are obtained via receivers that can be handheld or mounted on vehicles or fixed structures. If bird tags are visible from a distance (e.g. color rings or legbands, patagial tags, neck collars; see Chapter 4) then re-encounters occur as repeat observations by investigators and, in some cases, members of the public. If bird tags are not visible from a distance but can only be read from a bird in hand (e.g. standard metal rings or legbands, passive integrated transponder [PIT] tags; see Chapter 4), then re-encounters occur via recaptures by investigators or by recoveries of dead birds by members of the public. Some investigators use so-called "batch marks" (e.g. dyes or marks of a single color) to identify birds caught at a particular time or location, or hatched in a particular year. Such batch marks are much less useful than individual marks for the purpose of estimating rates of survival or movement. We do not discuss their use here.

5.3 Survival rates

5.3.1 Radio-telemetry

Field sampling for survival studies using radiotelemetry usually involves a single study area that is small enough to be traversed within a few days at most. Thus, investigators with radio receivers try to cover the area at specified sample periods (e.g. once each week), listening for radio signals and identifying birds as alive or dead. Such status identification sometimes requires actually locating and observing the bird, although some transmitters are equipped with "mortality sensors," based on either temperature or motion, that indicate whether the bird is alive or dead and that so do not necessarily require location of the bird.

If all radioed birds are detected when alive and are also detected as dead during the first sampling period following death, then the data needed for each bird are simply the sampling occasion of initial capture and release with a radio, the last sample period of detection as a live bird, and, in the case of death, the sample period during which the bird was encountered dead. We can also summarize the data for each marked bird as an encounter history, using codes 1 = marked and alive, 2 = marked and newly dead (i.e. the bird died following the previous sample period and before the current period) and 0 = not yet marked or died during a previous period. The encounter history is a row of these codes, with an entry for each sample period. Thus the encounter history 0 1 1 1 2 0 would denote a bird marked in period 2, detected alive in periods 3 and 4, found dead in period 5, and so not detected in period 6 of a 6-period study. These data can then be modeled in either of two basically equivalent ways, using either binomial survival models or models based on time at death (Williams *et al.* 2002).

We will illustrate the binomial survival modeling approach (also see Heisey and Fuller 1985) and define s_i as the probability that any bird alive at sampling period i is still alive at sample period $i + 1$. We would model the above capture history, 0 1 1 1 2 0, as:

$$P(0\ 1\ 1\ 1\ 2\ 0 \mid \text{release in 2}) = s_2 s_3 (1 - s_4).$$

Thus, s_2 denotes the probability associated with the bird surviving from week 2 until week 3, and s_3 denotes the probability that the bird survives from week 3 to week 4. The $(1 - s_4)$ term indicates the probability that the bird did not survive the interval between weeks 4 and 5 (we found the bird dead in week 5). We would have a similar probability for each observed encounter history. The product of these probabilities over all birds in the study would constitute the model for the entire data set and could be used to estimate the model parameters, the s_i. Nesting studies described in Chapter 3 use similar encounter histories and similar survival models to estimate daily nest survival probabilities and success.

In general, we could obtain estimates under various models of this sort using a software package such as MARK (White and Burnham 1999). Program MARK can also be used to fit competing binomial models (e.g. interval survival varies over time and sample period or is instead constant; survival differs for two groups of birds such as males and females or is instead the same for both sexes) and to discriminate among them based on model selection procedures or likelihood ratio tests (e.g. Lebreton *et al.* 1992; Burnham and Anderson 2002). Goodness-of-fit tests should also be conducted as part of the testing or selection procedure (Pollock *et al.* 1985, 1990; Burnham *et al.* 1987), as both likelihood ratio tests and model selection procedures assess relative model fit and are therefore strictly appropriate for inference only when the most general model in the pair or model set fits the data adequately. When the general model does not fit well, quasilikelihood methods based on the goodness-of-fit statistic can be used to adjust model test and selection results for lack of fit (e.g. Burnham *et al.* 1987; Lebreton *et al.* 1992; Burnham and Anderson 2002). Time at death models (not described here) and associated estimators, such as Kaplan–Meier, can frequently be implemented using comprehensive statistical software packages such as SAS (see Pollock *et al.* 1989*a,b*).

In most studies, point estimates themselves are not of primary interest, even if these estimates are of fundamental parameters such as survival probability. Instead, biologists are interested in the relationship between these parameters and such quantities as environmental covariates and management actions. One approach to covariate modeling is to write survival probability for a specific time period as a linear-logistic function of time-specific environmental or management covariates. For example, if s_i is daily survival probability (probability of surviving from day i to day $i + 1$) and x_i is a minimum temperature over the interval i to $i + 1$, then survival can be modeled as a linear-logistic function of temperature using the following expression:

$$s_i = \frac{e^{\beta_0 + \beta_1 x_i}}{1 + e^{\beta_0 + \beta_1 x_i}}. \tag{5.1}$$

where β_0 and β_1 are model parameters to be estimated, with β_1 reflecting the nature of the relationship between temperature and survival. If the relationship between survival and temperature is hypothesized not to be monotonic, but to instead involve higher survival at intermediate temperatures, then an additional quadratic term (e.g. $\beta_2 x_i^2$) can be added to the model. This flexible modeling approach can be implemented using MARK (White and Burnham 1999).

If the linear-logistic model does not provide an adequate parametric structure for the problem of interest, then another approach models the hazard or instantaneous risk of mortality over the period i to $i + i$, where i is a short time interval (e.g. 1 day). This hazard, h, is related to the daily survival probability as: $h = -\ln(s)$. Proportional hazard models (Cox 1972; Cox and Oakes 1984) provide an alternative approach to equation (5.1) for covariate modeling of survival data. Under this approach, the time-specific hazard is modeled as the product of a baseline hazard and an exponential term reflecting the level of the covariate.

Proportional hazards modeling can be implemented in many biomedical statistical packages and in program MARK (White and Burnham 1999). Although the above discussion is focused on time-specific covariates, individual covariates may also be of interest. For example, Pollock *et al.* (1989*b*) modeled survival of wintering Black Ducks *Anas rubripes* as a function of individual body mass at the time of radio attachment.

The questions about survival that are of most interest to scientists and managers require discriminating among competing models. For example, we might model weekly survival as a function of a weekly management action (e.g. different levels of food provisioning), under the hypothesis that increased food improves survival. A competing model is that natural foods are sufficient and that the amount of food provided by managers is not relevant to survival. Under this hypothesis, we might specify a statistical model in which survival varied over time but independently of food (i.e. there would be no food covariate or associated parameter in this model). We would fit both models to the data and compute either likelihood ratio tests (under a hypothesis testing approach) or Akaike's Information Criterion (AIC; under a model selection approach) to decide which model is most appropriate for the data and, hence, which hypothesis is supported by the data (e.g. Lebreton *et al.* 1992; Burnham and Anderson 2002; Williams *et al.* 2002). Under some study designs (e.g. random selection each week from a small number of management treatments), we could fit models that include both time effects and management effects and thus consider the possibility that management is relevant to weekly survival, but that additional time effects are important as well. The important point is that this sort of modeling, with databased model selection, is a key component of science and science-based management (also see Hilborn and Mangel 1997; Nichols 2001; Williams *et al.* 2002).

The above discussion of estimation and modeling has been based on an ideal field situation in which birds are always detected with probability 1 for the duration of the study. Detection of radioed animals is seldom perfect in actual field studies. In reality, some radios typically fail during the study, birds sometimes leave the study area either temporarily or permanently with respect to the study duration, and birds with functioning radios are sometimes missed despite searches. These kinds of problems must be dealt with in the modeling of observation histories (e.g. see White and Garrott 1990; Pollock *et al.* 1995). A potentially severe problem occurs when radio signals are lost for many birds that die (e.g. because predators or scavengers destroy the radio when handling the dead bird). If all such losses of radio contact can be assumed to reflect dead birds, then there is no problem, but in the usual case of some undetected temporary or permanent emigration, it is not clear how to model a bird with which radio contact is simply lost during the study. In some cases, inferences about this problem are possible when the study also includes birds marked with standard tags, permitting estimation of survival with capture–recapture/resighting data as well as with telemetry (see Bennetts *et al.* 1999). As in any effort to estimate parameters of natural animal populations, the primary recommendation is to tailor the estimation model to the realities of the sampling process as much as possible.

5.3.2 Capture–recapture/resighting

Capture–recapture and resighting studies are often carried out in single, local study areas, and data collection is usually by investigators and not by the general public. The duration of the sampling period and the length of the interval between sampling occasions can vary substantially and depend on study objectives. Sampling 1 day each week has been used to estimate postfledging survival (e.g. Krementz *et al.* 1989), whereas most studies involve longer sampling occasions and intervals. Many studies of bird populations are carried out during the breeding season, producing multi-year data sets with sampling periods of 6–8 weeks each year. If standard metal legbands are used to mark birds, then sampling typically involves recapturing birds each year (e.g. with mist nets, nest traps, rocket nets, etc.). If colored legbands, neck collars (e.g. for geese and swans), nasal discs (e.g. for ducks) or patagial tags are applied, then resampling may involve observations of individual birds with spotting scopes or binoculars. In the remainder of this section, we will refer to capture–recapture models with the understanding that this is a general descriptor that pertains also to resighting studies.

Data resulting from a capture–recapture study differ from radio-telemetry data in that deaths are not typically observed in the former type of study. Instead, it is possible to estimate a parameter frequently termed "apparent survival" or "local survival" to emphasize the fact that its complement includes both death and permanent emigration from the study area. The data are typically summarized as individual capture histories, which are simply rows of 1's and 0's indicating whether each bird is (1) or is not (0) captured at each sampling period of the study. For example, a 5-year study of a breeding population might yield one or more birds with the following capture history: 0 1 0 1 0. The five entries in the row represent the 5 years of the study. A bird with this history was first captured during sampling in year 2 of the study. It was marked and released following capture, was not detected in year 3, was recaptured or resighted in year 4 and not detected in the final year of the study, 5.

As in the modeling of radio-telemetry data, the key to estimation of survival parameters from capture–recapture data is to develop a reasonable model of the processes that give rise to the data. Following the original work of Cormack (1964), Jolly (1965), and Seber (1965), the Cormack–Jolly–Seber (CJS) model requires two sets of parameters. Capture (or resighting) probability, p_i, is the probability that a bird present in the study area during sampling occasion i is recaptured (or resighted) during that period. Survival probability, ϕ_i, is the probability that a bird alive and in the study area at sample occasion i is still alive and in the local population at sampling occasion $i + 1$.

Consider a 2-period study in which a bird is captured and released in period 1 and is either recaptured in period 2 (capture history 1 1) or not (capture history 1 0). Figure 5.1 presents a tree diagram of the possible events and their associated parameters. The probability for a particular capture history is obtained by working backwards from the history and by multiplying the parameters associated with each branch of the tree that led to the history. Thus, given release in period 1, the probability associated with capture history 1 1 is simply $\phi_1 p_2$. There are 2 different paths leading to capture history 1 0, and when we sum the two products of parameters along these paths we obtain: $\phi_1(1 - p_2) + (1 - \phi_1) = 1 - \phi_1 p_2$ as the probability associated with a bird released in period 1 and not recaptured in period 2. The two summed components reflect the two possible sets of events producing this capture history; the bird could have survived until period 2 and not been caught then, or it could have died or permanently emigrated between periods 1 and 2.

Now consider a longer capture history. Given that a bird is first caught and released in period 2, the probability associated with the capture history 0 1 0 1 0 can be written in terms of these parameters as:

$$P(0\ 1\ 0\ 1\ 0\,|\,\text{release in 2}) = \phi_2(1 - p_3)\phi_3 p_4(1 - \phi_4 p_5).$$

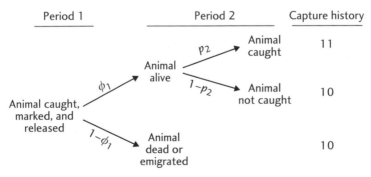

Fig. 5.1 Tree diagram of events and their probabilities for a bird released in period 1 of a 2-period capture–recapture study under the CJS model.

The initial ϕ_2 is the probability associated with the bird surviving from period 2 to 3, $(1 - p_3)$ corresponds to the bird not being captured in period 3, ϕ_3 corresponds to the probability of surviving from period 3 to 4, p_4 denotes capture probability in period 4, and $(1 - \phi_4 p_5)$ is the probability that the bird either does not survive until period 5 or survives but is not captured then (i.e. both possibilities are included in this term). Note that a key difference between the modeling of capture–recapture data and radio-telemetry data concerns the modeling of trailing 0's (0 entries that occur following the last 1). In the modeling of telemetry data for which detection probability is 1, 0's that occur following death are not modeled, as there is no uncertainty associated with them. However, trailing 0's in capture–recapture studies must be modeled, as their meaning is ambiguous.

The capture–recapture data (consisting of a capture history for every bird marked during the study) and the probability model (each history has an associated probability, constructed as in the above example) are then combined into a likelihood function, and the parameters of the model are estimated. Computation of estimates, their variances, and covariances, under different models and computation of test statistics and model selection criteria are usually accomplished using computer software such as MARK (White and Burnham 1999). The CJS model outlined above can be modified in numerous ways for various reasons. For example, reduced-parameter models in which parameters are assumed constant over time provide estimates with smaller variances than those produced by time-specific models. The logistics of capture–recapture sampling, combined with animal behavior, may result in the need to model parameters as a function of the previous capture history in order to deal with such phenomena as trap response in capture or survival probabilities. A special kind of capture–history dependence, especially useful in avian capture–recapture studies conducted at certain times of the year, involves incorporation of a transient parameter, reflecting the possibility that an unknown number of unmarked birds are transients with no chance of returning to the study area (Pradel *et al.* 1997). Parameters are frequently thought to be age-specific for birds, and such variation can be included in the modeling. As with the analysis of telemetry data, interest will often be focused on covariate relationships in which survival is modeled as a function of environmental or management covariates (e.g. using the linear-logistic relationship in equation (5.1)) or even individual bird covariates. A key step in these analyses again involves the selection of the most appropriate model from a set of competitors. Recent descriptions of capture–recapture modeling are provided by Burnham *et al.* (1987), Lebreton *et al.* (1992) and Williams *et al.* (2002).

Design issues relevant to capture–recapture/resighting studies include timing of sample periods and spatial coverage of sampled areas. Sample periods themselves are the periods during which birds are captured, recaptured, and re-observed, and may range from durations of 1 day (e.g. for estimation of weekly postfledging survival rates on local areas, Krementz et al. 1989) to 2–3 months (e.g. for estimation of annual survival rates). Sample periods typically should be short relative to the intervals that separate them and should occur during seasons when birds are relatively stationary and not migrating. Specifically, the sample periods should be sufficiently short that they include little mortality, but sufficiently long that the entire study area can be searched for birds or subjected to capture efforts. The spatial sampling of the study area should be such that all birds on the area should have similar, a priori probabilities of being caught or observed each sample period. If the entire area cannot be covered in each sampling period, then the sections to be covered can be randomly selected at each sample period. Investigators should ensure that certain portions of the study area are not consistently avoided or poorly sampled.

Temporary emigration from the study area can cause problems with the estimation of survival rate. Kendall et al. (1997) found that these problems can be largely remedied by collecting data under Pollock's (1982) robust design. Here sampling for each primary period of interest (e.g. the sample periods discussed thus far) consists of at least two distinct secondary capture sessions that encompass the entire study area and are closely spaced in time. For example, it might take the investigator 4 days to sample all portions of a local study area, so the investigator might conduct such 4-day sampling during each of three consecutive weeks in a breeding season. Each 4-day sample period would be considered a secondary sampling period, whereas the three sets of such periods combined would constitute the primary sample period for the year. Study design for survival rate estimation with open-population and robust design models is discussed in more detail by Williams et al. (2002), and figures for use in selecting needed sampling intensities are provided by Pollock et al. (1990).

5.3.3 Band recovery

Band recovery studies generally involve the application of metal legbands to birds at one or more study locations where recaptures of previously banded birds are either infrequent or ignored. Estimation is instead based on bands recovered by members of the public, often throughout the range of the bird. Typical cases in which recoveries far outnumber recaptures include banding of waterfowl on breeding grounds or at molting sites, with subsequent recoveries coming from hunters who report bands from birds they shoot on migration and wintering areas. Data from such studies thus consist of the number of birds banded each year and the number of these that are recovered (dead bird encountered by a member of the public, and the band reported to the investigators, usually via a central bird-banding repository such as the US Bird Banding Laboratory or the British Trust for Ornithology).

Such band recovery data are modeled using survival parameters, corresponding to the probability of a bird surviving from the time of banding in 1 year to the time of banding in the following year. Sampling parameters analogous to capture probabilities are also needed in the modeling. In the special case where recoveries are restricted to reports from hunters of birds that have been shot, recovery rates (one kind of sampling parameter) are of interest themselves as indices of hunting intensity (e.g. Anderson 1975; Brownie *et al.* 1985). Band recovery models are simply a special case of capture–recapture models. Band recovery models do not assume that all dead birds are encountered and reported, but instead view the number of birds reported as some unknown fraction of the total number dying in a year. A key difference between capture–recapture and band recovery studies involves the interpretation of the estimated survival parameter. As noted above, the complement of apparent or local survival in capture–recapture studies includes both death and permanent emigration. In most band recovery studies based on recoveries of dead birds by members of the public, most or all dead birds have some non zero probability of being recovered regardless of where death occurs. Thus, permanent emigration is not possible and resulting survival estimates can be viewed as estimates of true survival (complement includes only mortality).

Estimation models for band recovery data were initially developed by Haldane (1955), Seber (1970), and Robson and Youngs (1971). An excellent synthetic treatment containing models with different underlying assumptions about time-specificity of parameters was provided by Brownie *et al.* (1985). Survival can be modeled as a function of covariates in band recovery models (North and Morgan 1979; Conroy *et al.* 1989). Modeling, model selection, and estimation are now in most cases best conducted using program MARK (White and Burnham 1999). Recent discussion of modeling and estimation using band recovery data is provided by Williams *et al.* (2002). Capture–recapture data and band recovery data can be combined for the purpose of estimation. Because of the different interpretations of survival parameters in the two classes of models, the combination of methods permits separate estimation of true survival and fidelity, the probability of returning to the study area (banding location) conditional on survival (Burnham 1993; Williams *et al.* 2002).

Study designs for band recovery studies typically involve banding at one of more local areas at a specific time each year (e.g. at the end of each breeding season). As with capture–recapture designs, the banding period should be relatively short, should include relatively little mortality, and should occur at a time of the year when birds are relatively stationary. The recovery period need not be short and may include either the hunting season or the entire year. Because birds are encountered twice at most (initial banding and possibly recovery as a dead bird), it is not possible to estimate survival rates based on banding of young (first-year) birds only, unless fairly restrictive modeling assumptions are made (Brownie *et al.* 1985). Banding studies relying on recoveries, rather than recapture or re-observation data, should thus include banding of both adults and young birds. Additional information on study design and sample sizes is presented by Brownie *et al.* (1985) and Williams *et al.* (2002).

5.4 Movement

5.4.1 Radio-telemetry

Although satellite telemetry studies can effectively record bird locations anywhere on earth, the majority of radio-tracking studies of bird movement still involve one or more local study areas. We consider both the field situation and statistical modeling for one and multiple areas separately. If detection probability for radioed birds is 1, then studies of a single area can be used to address questions about the probability of a bird departing the study area either permanently or temporarily. As described for radio-telemetry studies of survival, field sampling requires periodic sampling of the study area. At each period, instrumented birds are located and designated as either alive on the study area (denote as 1), dead (reflecting death following the previous sample period) on the area (denote as 2), or not present on the area (denote as 3), indicating movement off the area following the previous sample period. A 0 is then used to denote the sample periods before an animal is tagged and after an animal has died or departed the study area. Thus, encounter history 0 1 1 1 3 0 would indicate a bird radio-tagged during sample period 2, relocated on the study area in periods 3 and 4, and not present on the area in period 5 (hence moved off the study area).

As with survival estimation, we must develop a probabilistic model to describe the sequence of events depicted by the encounter history. In survival estimation, the event of interest is bird death, whereas in this type of movement modeling, the event of interest is departure from the study area. Also as with survival modeling, two basically equivalent ways exist for conducting this modeling, one based on the time elapsed until movement and the other based on a binomial model of movement between each pair of sampling periods (see Bennetts *et al.* 2001; Williams *et al.* 2002). Here we outline the binomial modeling approach. Define s_i as the probability of survival from sample period i to $i + 1$ for a bird that remains on the study area and f_i as fidelity, or the probability that a bird alive and on the study area in sample period i does not permanently emigrate from the study area between sample periods i and $i + 1$. Given these two sets of parameters, we would model the above encounter history as:

$$P(0\ 1\ 1\ 1\ 3\ 0\ |\ \text{release in period 2}) = s_2 f_2 s_3 f_3 (1 - f_4).$$

The s_2 parameter denotes the probability of survival between sample periods 2 and 3, f_2 corresponds to fidelity during the time interval from period 2 to 3, s_3 and f_3 correspond to survival and fidelity, respectively, between 3 and 4, and $(1 - f_4)$ is the probability that a bird alive in the study area in period 4 permanently emigrates before period 5.

As noted above, some field studies involve sampling multiple locations at each sample period. Birds are caught, radio-tagged, and released at one or more of the locations, and each location is searched with a radio receiver for marked birds at each sample period. Such studies permit estimation of the probabilities of moving among the different study locations. Encounter data from such a study must specify not only fate with respect to presence in the study system (consisting of all sampled locations), death, and emigration from the study system, but also location within the study system. Assume a simple system in which birds are radio-tagged and sampled at two locations, A and B. Encounter histories can be written with numbers reflecting fate and letters reflecting location. So encounter history 0 A1 A1 B1 B3 indicates a bird marked in location A at period 2, located again in location A at period 3, located in location B at period 4 and not located in either A or B (departed the study system) in period 5 (the notation B3 simply indicates that the bird was last detected at location B). If we again assume that detection probability for radio-marked birds within the study system is 1, then the following parameters are needed to model encounter history data:

s_i^R = probability that a bird in location R at period i that does not permanently emigrate the study system survives until period $i+1$;

ψ_i^{RS} = probability that a bird in location R at period i and alive in the study system at period $i+1$ is in location S in period $i+1$;

f_i^R = probability that a bird in location R at period i does not permanently emigrate from the study system between periods i and $i+1$.

Given these definitions, we can write the probability for the above example capture history as:

$$P(0\ A1\ A1\ B1\ B3\ |\ \text{release in location A, period 2})$$
$$= s_2^A f_2^A (1 - \psi_2^{AB}) s_3^A f_3^A \psi_3^{AB} (1 - f_4^B).$$

The bird was located in location A at period 2 and survived until period 3 (associated probability, s_2^A), it did not permanently emigrate the study system (associated probability f_2^A), it stayed in location A (associated probability, $1-\psi_2^{AB}$), it survived again until period 4 (associated probability, s_3^A) and again did not permanently emigrate (f_3^A), it moved to location B (associated probability, ψ_3^{AB}), and it then departed the 2-location study system (associated probability $1-f_4^B$).

Estimation under such movement models can be obtained using the multistate modeling structures of program MARK (White and Burnham 1999) and then modifying them to reflect the specifics of radio-telemetry data. An alternative approach would be to write the model directly into the flexible software SURVIV (White 1983). As noted in previous sections, point estimates themselves are not typically of primary interest. Instead, competing models of biological interest are developed and model testing or selection procedures (e.g. see Lebreton *et al.* 1992; Burnham and Anderson 2002; Williams *et al.* 2002) are used to discriminate among the competitors. As with the survival models, movement parameters can be modeled as functions of other quantities (e.g. distance between locations, ratio of fitness indicators in the two locations, difference in management actions between two locations; see Nichols and Kendall 1995). Surprisingly, the sort of probabilistic movement modeling described here has not been implemented frequently, and reports of results from previous telemetry studies directed at movement have tended to be descriptive. As noted in the discussion of radio-telemetry survival studies, the assumption of detection probability equal to 1 is not always justified in telemetry studies. In such cases, the data can be modeled using capture-resighting models (e.g. Bechet *et al.* 2003).

5.4.2 Capture–recapture/resighting

Various methods exist for drawing inferences about movement based on capture–recapture data. Limited inferences are possible using capture–recapture from a single study site (e.g. see Nichols 1996; Nichols and Kaiser 1999; Bennetts *et al.* 2001). For example, temporary emigration can be estimated using Pollock's (1982) robust design (Kendall *et al.* 1997). The proportion of newly caught birds that are transients can be estimated (Pradel *et al.* 1997), as can departure probabilities and lengths of stay on migration stopover areas (e.g. Schaub *et al.* 2001).

Here we focus on the use of multistate capture–recapture models for studies of multiple locations (Arnason 1973; Brownie *et al.* 1993; Williams *et al.* 2002). Birds are marked and released on all study locations during sampling periods and recaptured or resighted in subsequent sampling periods, either at the location of release or another location. Study duration will depend on objectives. Some of the early uses of multistate models to estimate movement were based on annual sampling periods, but movement studies with shorter time intervals are also possible. Data from such a study can again be summarized as capture histories, with one history for each bird in the study. If A and B are two study locations, then we can denote capture in each location by the location letter and noncapture by 0. Thus, we can write an example capture history as 0 A 0 A B, denoting a bird that was caught and released in location A at period 2, not captured at period 3, caught again in location A at period 4 and caught in location B at period 5 of a 5-period study.

We define the following parameters to model multistate capture–recapture data:

S_i^R = probability that a bird alive in location R at sample period i is still alive and in the study system (consisting of the set of sampled locations) in period $i+1$ (note that this survival is again an apparent or local survival in the sense that its complement includes both death and permanent emigration from the study system);

ψ_i^{RS} = probability that a bird alive in location R at sample period i that survives in the study system until period $i+1$ is located in location S at $i+1$

p_i^R = probability that a bird in location R at period i is recaptured or resighted during sample period i.

Using these parameters, we can write the probability associated with the example capture history as:

$$P(0 \text{ A } 0 \text{ A B} \mid \text{release in location A, period 2})$$
$$= S_2^A[(1-\psi_2^{AB})(1-p_3^A)S_3^A(1-\psi_3^{AB})+\psi_2^{AB}(1-p_3^B)S_3^B\psi_3^{BA}]p_4^A S_4^A\psi_4^{AB}p_5^B$$

The bird is initially caught in location A and released at period 2. The bird is known to survive until period 3 (because it was seen after that period), and the probability associated with this event is S_2^A. The bird is not caught in period 3, hence its location in period 3 is unknown. The probability model for events occurring between periods 2 and 4 is thus written as a sum of two probabilities (in brackets) corresponding to the two alternative locations where the bird could have been in period 3. The bird could have remained in location A the entire time, $(1-\psi_2^{AB})(1-p_3^A)S_3^A(1-\psi_3^{AB})$, or it could have moved to location B at period 3 and then back to location A at period 4, $\psi_2^{AB}(1-p_3^B)S_3^B\psi_3^{BA}$. Regardless of which of the two paths was taken, the bird was caught in location A at period 4 (associated probability p_4^A), survived until period 5 (probability S_4^A), moved from A to B (probability ψ_4^{AB}), and was caught in location B at period 5 (probability p_5^B).

As was the case for the single-location model in which the focus was on survival estimation, the capture history data and the corresponding probability model (each history has an associated probability as above) are combined to form a likelihood function, and estimates are then obtained using software such as MARK (White and Burnham 1999). The general model with time- and location-specific parameters can be constrained in various ways. For example, it may be that movement between pairs of locations is expected to be symmetric ($\psi_i^{RS}=\psi_i^{SR}$). Covariate modeling can be used to investigate biologically interesting hypotheses, as movement between two locations can be modeled as a function of such factors as distance between the locations, the ratio of fitness indicators between the locations, and density at the locations (Nichols and Kendall 1995). Although the focus of this section is on movement, we note that location-specific survival probabilities can also be estimated using multistate modeling. These provide survival estimates in situations where animals move among locations and where survival may vary over locations. Constraints involving survival (e.g. $S_i^R=S_i^S$) and covariate modeling of survival are also frequently of biological interest. Again, competing models expressing different hypotheses of biological interest about movement or survival can either be tested using likelihood ratio tests or evaluated using a model selection approach (Burnham and Anderson 2002).

The parameterization described above is most useful when bird movement between locations occurs near the ends of the interval separating sampling periods. Although this approach seems reasonable for migratory birds (e.g. Hestbeck et al. 1991; Spendelow et al. 1995) returning to breeding or wintering locations at the end of each sample year, there are other situations where movement may occur at any time during the interval. In such cases, it is possible to parameterize with transition parameters that combine survival and movement, $\phi_i^{RS}=S_i^R\psi_i^{RS}$, thus requiring no assumption about the timing of movement. Another modeling approach is to view time of movement as a random variable with known distribution (Joe and Pollock 2002), although user-friendly software for implementing this approach is not yet available.

5.4.3 Band recovery

Band recovery data can also be used to draw inferences about bird movement. We generally envisage two sampling situations. In one, banding and recovery occur at different times of the year and at different locations. For example, in North America, it is common for banding of ducks to occur on the breeding grounds, whereas the hunting season recoveries occur during the fall and winter. In this situation, inference is sometimes possible about movement from a particular banding area to two or more recovery areas, although the nature of the inferences depend on assumptions about the constancy of survival rates among locations and the permanence of migration "decisions" (Schwarz *et al.* 1988; Schwarz and Arnason 1990). For example, do birds decide where to winter during their first migration and thereafter consistently return to that same locality, or can birds visit different wintering locations in different winters?

In the other sampling situation, banding and recovery occur on the same areas. Winter banding of North American waterfowl occurs following the hunting season in late winter, whereas early winter recoveries occur in the same general locations. Under this sampling situation, multistate band recovery models may be useful (Schwarz 1993; Schwarz *et al.* 1993). These models may be viewed as special cases of multistate capture–recapture models and involve the same kind of thinking and modeling as described in the previous section.

Covariate modeling and constrained models incorporating interesting biological hypotheses can be developed using these band recovery models as with the previous models. Tests and model selection criteria can again be used to discriminate among competing models. Band recovery models have not seen much use in estimating bird movement parameters, probably because large numbers of recoveries are needed to obtain reasonably precise estimates.

5.5 Summary and general recommendations

Methods based on sampling marked birds exist for estimating parameters associated with survival and movement. Utility of these methods depends on consideration of the type of data resulting from sampling and the subsequent modeling of these data in terms of parameters of interest that describe the processes underlying data generation. When detection probabilities of marked birds are 1 (when birds are detected at will on study locations), the modeling of bird encounter histories can be based on the biological parameters of interest (e.g. survival, fidelity, movement). When not all birds on sample areas are detected during sampling, modeling is still possible, but the models become more complicated as they must also include parameters corresponding to recapture and resighting probabilities.

This distinction between data types and their associated models leads to the simple observation that for studies of equal sample size (equal numbers of marked animals), mark types (e.g. radios) for which detection probability is 1 will yield more precise estimates than studies using mark types (standard tags and bands) with variable and unknown detection probabilities. However, radios are much more expensive than conventional tags and bands, so in cases where it is possible to mark large numbers of birds, the following type of question is likely to arise: should I mark 30 animals with radios or 300 animals with conventional tags? Informed answers to such questions will require pilot data or guesses about capture-resighting probabilities and development of simulation- or approximation-based sample size figures such as those of Pollock *et al.* (1990) that can be used to compare estimator precision under different scenarios. However, estimator precision is not the only quantity of relevance to such study design decisions, as potential for estimator bias (e.g. via radio effects on survival, or dependence of fate and censoring) will also be relevant. As noted previously, mechanical problems and difficulty in detecting signals in some habitats and sampling situations can also lead to radio-telemetry detection probabilities <1.

When field methods do not permit all marked birds to be detected, then study designs should seek to minimize variation in detection probabilities among marked birds and to identify and record important sources of variation (e.g. bird location, bird sex) that still exist. The use of Pollock's (1982) robust design offers several advantages for capture–recapture studies, including the ability to account for movement to areas outside the study area(s). Band recovery models tend to be most useful when large numbers of birds (e.g. 100s to 1000s) can be banded and when recovery rates (probability that a banded bird alive at the time of banding dies, is found and has its band reported) are relatively high (e.g. >0.04). Such high recovery rates typically occur only for hunted species. Recapture and resighting probabilities in many intensive capture–recapture studies of birds are fairly high (e.g. >0.2 and sometimes >0.5) and permit precise estimation of apparent survival. Once again, however, precision is not the only relevant quantity. The complement of apparent survival from capture–recapture studies includes both death and permanent emigration. The permanent emigration component may be small for adult birds but is frequently very large for young birds. The complement of survival estimates resulting from most band recovery studies includes only death, so there is an advantage to the use of such estimators when recovery rates are sufficiently large.

In general, studies directed at questions about survival or movement should be designed in a manner that exploits available field methods and their respective analytic and modeling approaches. Rather than focusing exclusively on the selection of single data types and associated designs, it is becoming increasingly clear that hybrid designs offer many advantages. Combining sources of information, such as capture/resighting, radio-telemetry, recoveries, or sightings between formal sampling periods, can be used to address questions about bird survival and movement, sometimes permitting estimation of otherwise inaccessible parameters, as well as increasing the precision of estimates (e.g. see Barker 1997;

Powell *et al.* 2000; Kendall and Bjorkland 2001; Lindberg *et al.* 2001). We believe that such designs, that exploit the advantages of different data types, will be an important research focus for the next decade.

Finally, although this chapter has dealt with methodological considerations, we urge the reader to retain focus on the population-dynamic questions that motivate the use of these methods. Conditional on the study design and component field methods, questions of interest can be addressed by incorporating interesting biological hypotheses into models and then using tests or model selection criteria to discriminate among competing models. This discrimination then forms the basis for the conduct of science and its application to management and conservation.

References

Anderson, D.R. (1975). Optimal exploitation strategies for an animal population in a Markovian environment: a theory and an example. *Ecology*, 56, 1281–1297.

Arnason, A.N. (1973). The estimation of population size, migration rates, and survival in a stratified population. *Res. Population Ecol.*, 15, 1–8.

Barker, R.J. (1997). Joint modeling of live-recapture, tag-resight, and tag-recovery data. *Biometrics*, 53, 666–677.

Bechet, A., Giroux, J.-F., Gauthier, G., Nichols, J.D., and Hines, J.E. (2003). Spring hunting changes the regional movements of migrating greater snow geese. *J. Appl. Ecol.*, 40, 553–564.

Bennetts, R.E., Dreitz, V.J., Kitchens, W.M., Hines, J.E., and Nichols, J.D. (1999). Annual survival of snail kites in Florida: radio telemetry versus capture-resighting data. *Auk*, 116, 435–447.

Bennetts, R.E., Nichols, J.D., Lebreton, J.D., Pradel, R., Hines, J.E., and Kitchens, W.M. (2001). Methods for estimating dispersal probabilities and related parameters using marked animals. In *Dispersal*, eds., J. Clobert, E. Danchin, A.A. Dhondt, and J.D. Nichols, pp. 3–17. Oxford University Press, Oxford, UK.

Brownie, C., Anderson, D.R., Burnham, K.P., and Robson, D.R. (1985). Statistical inference from band recovery data—a handbook, 2nd ed. U.S. Fish and Wildlife Service Resource Publication 156.

Brownie, C., Hines, J.E., Nichols, J.D., Pollock, K.H., and Hestbeck, J.B. (1993). Capture–recapture studies for multiple strata including non-Markovian transition probabilities. *Biometrics*, 49, 1173–1187.

Burnham, K.P. (1993). A theory for combined analysis of ring recovery and recapture data. In *Marked Individuals in the Study of Bird Population*, eds., J.D. Lebreton and P.M. North, pp. 199–213. Birkhauser Verlag, Berlin.

Burnham, K.P. and Anderson, D.R. (2002). *Model Selection and Inference: A Practical Information-Theoretic Approach*, 2nd ed. Springer-Verlag, New York.

Burnham, K.P., Anderson, D.R., White, G.C., Brownie, C., and Pollock, K.P. (1987). Design and analysis of methods for fish survival experiments based on release-recapture. *Am. Fish. Soc. Monogr.* 5.

Clobert, J. and Lebreton, J.-D. (1991). Estimation of demographic parameters in bird Populations. In *Bird Population Studies: Relevance to Conservation and Management*, eds. C.M. Perrins, J.-D. Lebreton, and G.J.M. Hirons, pp. 75–104. Oxford University Press, Oxford.

Conroy, M.J., Hines, J.E., and Williams, B.K. (1989). Procedures for the analysis of band-recovery data and user instructions for program MULT. U.S. Fish and Wildlife Service Resource Publication 175.

Cormack, R.M. (1964). Estimates of survival from the sighting of marked animals. *Biometrika*, 51, 429–438.

Cox, D.R. (1972). Regression models and life tables. *J. Roy. Statist. Soc., Series B*, 34, 187–220.

Cox, D.R. and Oakes, D. (1984). *Analysis of Survival Data*. Chapman and Hall, London.

Haldane, J.B.S. (1955). The calculation of mortality rates from ringing data. *Proceedings of the International Congress of Ornithology*, 9, 454–458.

Heisey, D.M. and Fuller, T.K. (1985). Evaluation of survival and cause-specific mortality rates using telemetry data. *J. of Wildlife Manage.*, 49, 668–674.

Hestbeck, J.B., Nichols, J.D., and Malecki, R.A. (1991). Estimates of movement and site fidelity using mark-resight data of wintering Canada geese. *Ecology*, 72, 523–533.

Hilborn, R. and Mangel, M. (1997). *The Ecological Detective. Confronting Models with Data*. Princeton University Press, Princeton, New Jersey.

Joe, M. and Pollock, K.H. (2002). Separation of survival and movement rates in multi-state tag-return and capture–recapture models. *J. Appl. Statist.*, 29, 373–384.

Johnson, D.H. and Grier, J.W. (1988). Determinants of breeding distributions of ducks. *Wildlife Monog.*, 100, 37pp.

Jolly, G.M. (1965). Explicit estimates from capture–recapture data with both death and immigration—stochastic model. *Biometrika*, 52, 225–247.

Kendall, W.L. and Bjorkland, R. (2001). Using open robust design models to estimate temporary emigration from capture–recapture data. *Biometrics*, 57, 1113–1122.

Kendall, W.L., Nichols, J.D., and Hines, J.E. (1997). Estimating temporary emigration using capture–recapture data with Pollock's robust design. *Ecology*, 78, 563–578.

Krementz, D.G., Nichols, J.D., and Hines, J.E.. (1989). Postfledging survival of European starlings. *Ecology*, 70, 646–655.

Lebreton, J.D., Burnham, K.P., Clobert, J., and Anderson, D.R. (1992). Modelling survival and testing biological hypotheses using marked animals: a unified approach with case studies. *Ecol. Monogr.*, 62, 67–118.

Lindberg, M.A., Kendall, W.L., Hines, J.E., and Anderson, M.G. (2001). Combining band recovery data and Pollock's robust design to model temporary and permanent emigration. *Biometrics*, 57, 273–282.

Link, W.A., Royle, J.A., and Hatfield, J.S. Demographic analysis from summaries of an age-structured population. *Biometrics*, 59, 778–785.

Nichols, J.D. (1996). Sources of variation in migratory movements of animal populations: statistical inference and a selective review of empirical results. In *Population Dynamics in Ecological Space and Time*, eds. O.E. Rhodes, R.K. Chesser, and M.H. Smith, pp. 147–197. University of Chicago Press, Chicago.

Nichols, J.D. (2001). Using models in the conduct of science and management of natural resources. In *Modelling in Natural Resource Management*, eds. T.M. Shenk and A.B. Franklin, pp. 11–34. Island Press, Washington, DC.

Nichols, J.D. and Kaiser, A. (1999). Quantitative studies of bird movement: a methodological review. *Bird Stud.*, 46(Suppl.), S289–S298.

Nichols, J.D. and Kendall, W.L. (1995). The use of multistate capture–recapture models to address questions in evolutionary ecology. *J. Appl. Statist.*, 22, 835–846.

North, P.M. and Morgan, B.J.T. (1979). Modeling heron survival using weather data. *Biometrics*, 35, 667–681.

Pollock, K.H. (1982). A capture–recapture design robust to unequal probability of capture. *J. Wildlife Manage.*, 46, 757–760.

Pollock, K.H., Hines, J.E., and Nichols, J.D. (1985). Goodness-of-fit tests for open capture–recapture models. *Biometrics*, 41, 399–410.

Pollock, K.H., Winterstein, S.R., Bunck, C.M., and Curtis, P.D. (1989*a*). Survival analysis in telemetry studies: the staggered entry design. *J. Wildlife Manage.*, 53, 7–15.

Pollock, K.H., Winterstein, S.R., and Conroy, M.J. (1989*b*). Estimation and analysis of survival distributions for radio-tagged animals. *Biometrics*, 45, 99–109.

Pollock, K.H., Nichols, J.D., Brownie, C., and Hines, J.E. (1990). Statistical inference for capture–recapture experiments. *Wildlife Monogr.*, 107, 97pp.

Pollock, K.H., Bunck, C.M., Winterstein, S.R., and Chen, C.-L. (1995). A capture–recapture survival analysis model for radio-tagged animals. *J. Appl. Statist.*, 22, 661–672.

Powell, L.A., Conroy, M.J., Hines, J.E., Nichols, J.D., and Krementz, D.G. (2000). Simultaneous use of mark-recapture and radio telemetry to estimate survival, movement, and capture rates. *J. Wildlife Manage.*, 64, 302–313.

Pradel, R., Hines, J.E., Lebreton, J.-D., and Nichols, J.D. (1997). Capture–recapture survival models taking account of transients. *Biometrics*, 53, 60–72.

Robson, D.S. and Youngs, W.D. (1971). Statistical analysis of reported tag-recaptures in the harvest from an exploited population. *Biometrics Unit*, Cornell University, Ithaca, NY. BU-369-M.

Schaub, M., Pradel, R., Jenni, L., and Lebreton, J.-D. (2001). Migrating birds stop over longer than usually thought: an improved capture–recapture analysis. *Ecology*, 82, 852–859.

Schwarz, C.J. (1993). Estimating migration rates using tag-recovery data. In *Marked individuals in the study of bird population*. eds. J.D. Lebreton and P. M. North, pp. 81–104. Birkhauser Verlag, Berlin.

Schwarz, C.J. and Arnason, A.N. (1990). Use of tag-recovery information in migration and movement studies. *Am. Fish. Soc. Symp.*, 7, 588–603.

Schwarz, C.J., Burnham, K.P., and Arnason, A.N. (1988). Post-release stratification in band-recovery models. *Biometrics*, 44, 765–785.

Schwarz, C.J., Schweigert, J.F., and Arnason, A.N. (1993). Estimating migration rates using tag recovery data. *Biometrics*, 49, 177–193.

Seber, G.A.F. (1965). A note on the multiple-recapture census. *Biometrika*, 52, 249–259.

Seber, G.A.F. (1970). Estimating time-specific survival and reporting rates for adult birds from band returns. *Biometrika*, 57, 313–318.

Seber, G.A.F. (1982). *The Estimation of Animal Abundance and Related Parameters.* MacMillan, New York.

Spendelow, J.A., Nichols, J.D., Nisbet, I.C.T., Hays, H., Cormons, G.D., Burger, J., Safina, C., Hines, J.E., and Gochfeld, M. (1995). Estimating annual survival and movement rates within a metapopulation of roseate terns. *Ecology*, 76, 2415–2428.

Udevitz, M.S. and Ballachey, B.E. (1998). Estimating survival rates with age-structure data. *J. Wildlife Manage.*, 62, 779–792.

White, G.C. (1983). Numerical estimation of survival rates from band recovery and biotelemetry data. *J. Wildlife Manage.*, 47, 716–728.

White, G.C. and Burnham, K.P. (1999). Program MARK: survival rate estimation from both live and dead encounters. *Bird Stud.*, 46, 120–139.

White, G.C. and Garrott, R.A. (1990). *Analysis of Wildlife Radio-tracking Data*. Academic Press, San Diego, CA.

Williams, B.K., Nichols, J.D., and Conroy, M.J. (2002). *Analysis and Management of Animal Populations*. Academic Press, New York.

References | 175

6

Radio-tagging

Robert Kenward

6.1 Introduction

Radio-tagging has been used for more than 40 years to reveal where animals are (location), how they are (physiology, alive or dead), and what they are doing (behavior). Whereas academic studies can be based on choosing a common and conspicuous species that is easy to mark and watch, conservation is usually focused on species that are rare, elusive, or living in remote areas. Radio tags are often the only practical way to record the basic requirements of such species, such as the areas and habitats they use for foraging and sheltering, or how individuals interact for mating or when transmitting disease. Radio-tags can reveal the fate of every valuable animal in a release scheme, and small samples of rare animals provide details of conditions along migration routes. Radio-tagging is therefore an essential tool in conservation ecology.

The development of radio-tags was made possible by invention of the transistor, leading to microelectronics and tiny power supplies that now permit tags of 300 mg. These Very High Frequency (VHF) tags can be detected at hundreds of meters, or a few kilometers if antennas are long enough, by converting brief signal pulses to audible beeps in very sensitive receivers. By tuning to different tag frequencies, individual animals can be tracked on foot or from vehicles (including aircraft) and located by triangulation. Accuracy is typically 10–100 m. Although the smallest VHF tags function for only a week or so, 2 g tags can transmit signals for months. At 20 g, tags can transmit VHF signals detectable at 1–100 km for 2–3 years.

A second type of radio-tag can record locations automatically anywhere on the globe. Although the ultra high frequencies (UHF) of these tags penetrate vegetation poorly, high power is used to communicate with satellites. In the ARGOS system, Doppler principles are applied to locate tags that transmit one very stable

frequency, with signal coding to identify individuals. The high power requirement raises tag mass to at least 15 g, and tracking accuracy is (at best) several hundred meters. However, locations can be recorded automatically at 10-m accuracy
by tags that receive UHF signals from the Navstar global positioning system
(GPS). Minimal tags of 30 g can store the locations, but must be recovered to
extract the data. Tags that can relay GPS data to satellites or mobile phone systems
are too large for most birds.

Although animals have been radio-tagged for more than 40 years, the techniques are seldom used to their full potential. One problem is the considerable
knowledge, planning, and skill required to make best use of the tagging. Another
is the need to avoid adverse effects of tagging on the welfare of animals and the
quality of information. Great care is needed in obtaining suitable equipment, field
skills, and data. This chapter gives pointers for success, but wider reading is essential. Recent reviews are Fuller *et al.* (in press), or more extensively Kenward
(2001), with Millspaugh and Marzluff (2001) for analysis techniques.

6.2 Choice of techniques

6.2.1 Constraints on radio tagging

Radio tagging is practical only if animals can carry large enough tags for long
enough to give the required data. Tags at 10% of body-mass have been used, but
any above 2–3% are liable to reduce survival, especially on birds. Except in the
smallest tags, mass is constrained mainly by the power supply, which limits the life
and power of the tag. Figure 6.1 shows typical transmission lives available for
VHF tags of increasing mass. The smallest tags rely on single silver-oxide cells,

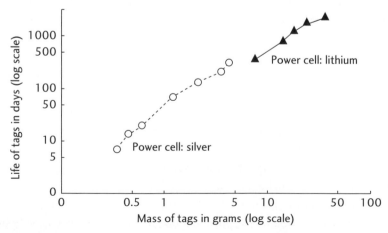

Fig. 6.1 Tag transmission life expected from bird tags of different mass.

which give only 1.5 V and therefore require two cells in series to reach the same power as tags with lithium cells, at 3.0–3.7 V. Although solar cells can be used to boost energy of rechargeable cells, the tags must be exposed to light (which may result in increased drag) and eventually lose cell efficiency. Tags can also be switched by microprocessors, to save power by transmitting only at desired seasons or times of day. However, recent increases in tag efficiency can give similar lives from good quality primary cells more reliably than with either micro-processors, which can be vulnerable to static, or solar cells.

An important constraint on detection range is antenna efficiency. With a wave-length of 2 m at 150 MHz (a common frequency for radio tags in Europe), an "ideal length" quarter-wave antenna is 500 mm. This length decreases in pro-portion to increasing frequency. A quarter wavelength is thus 434 mm at 173 MHz, as permitted in the United Kingdom, and 347 mm at 216 MHz, as used in some other countries. In practise, antennas are usually shorter than the ideal, with ground-plane or antenna loading systems to compensate. Efficiency reduces as an inverse power function of length (though not appreciably down to about 70% of the ideal) and also falls slightly with decreased width. As a result, a transmitter that can be detected in line of sight at 40 km when coupled to an antenna with efficient length and robust width (e.g. on a large bird) may not be detectable much beyond 1 km with a 100 mm antenna on a small bird. Extra life can be secured by decreasing the rate and duration of signal pulses, without appre-ciable reduction in detection range from typical receivers until pulses are below 10 ms. See www.biotrack.co.uk for software to help select tags with optimal mass, life, and range.

Another important consideration is cost. Automated tracking is most expen-sive. A budget of US$5000 buys 1–2 tags for tracking by satellites with Doppler or GPS-relay systems, about five tags for recovery with GPS data, or 20 of the VHF tags with a receiver for manual tracking. Doppler-system tags are at their best for migration studies (Chapter 7), for example, to identify important staging or wintering sites. Although the tags are expensive, a small number on a rare species can provide basic data with more detail and immediacy than ring (band) returns. GPS tags are more accurate than Doppler-system tags for studies of habi-tat and other resources (Chapter 11) and can be more cost-effective than VHF tags for birds in dangerous or remote areas. However, GPS-relay tags may remain suitable only for species with body-mass well above 1 kg. Ease of recapture will continue to constrain the use of storage-only GPS tags.

6.2.2 Applications and advantages

Despite some exciting tracking by satellite that has revealed unanticipated bird movements (Jouventin and Weimerskirch 1990) and migration events

(Fuller *et al.* 1995; Higuchi *et al.* 1996; Meyburg and Meyburg, 1998), including a pesticide hotspot (Woodbridge *et al.* in press), most avian radio-tagging has involved VHF systems. Thus, VHF tagging has revealed the need to create reserves that conserve unexpectedly large home range areas and to take account of unanticipated behavior in conspicuous species, for example, when they forage by night (Evans *et al.* 1985). Tagging females is ideal for finding well-concealed nests or recording survival of precocial broods (Chapter 3), while tiny tags in eggs or on the young can indicate causes of loss (Willebrand and Marcström 1988). Tags are also invaluable for recording survival (Chapter 5) and causes of death (Chapter 8) for species that provide few ring returns due to rarity or remoteness, and especially to monitor every individual in experiments or reintroduction programs (Chapter 12).

These are all applications where it is important to record the location and status of animals at particular times. However, if it is simply a matter of recognizing individuals, cheaper techniques can be appropriate. Conspicuous birds can be marked with wing-tags or color patterns or even recognized from photographs or DNA in their young (Evans *et al.* 1999; Wink *et al.* 1999) and minimal survival estimated. In line of sight, even insects can be tracked using tiny harmonic radar transponders (Riley *et al.* 1996). Long-distance movement routes can be recorded without radio-tags, provided that animals can be recaptured to remove tags that store data. Tags with photo-sensors record the time of dusk and dawn for each date, which gives an approximation of latitude from the interval between them and of longitude from the absolute times (Wilson *et al.* 1992). These tags have cheaper and lighter components than GPS tags, but are also much less accurate.

Nevertheless, all these other methods are vulnerable to bias from differential detection or recovery, whereas data from radio-tagged animals can be recorded systematically to a specified level of precision. This is important, because biased information can be seriously misleading. An example is a case where humans were thought to be slowing re-colonization by raptors. The majority of deaths recorded with ring returns were from deliberate killing or impact with human artifacts. However, radio-tracking in the same areas showed that deaths were much less frequently caused by human activities, survival was better than estimated by ringing and many adults were not breeding (Kenward *et al.* 2000). A focus on human impacts, probably because dead birds are found most easily where human activities cause deaths (Newton 1979), had diverted attention from other factors that constrain recolonization.

Radio-tagging can also be used to improve other techniques that are less costly, and therefore often best for the volunteer effort that is growing so important in conservation. If a bias can be quantified, correction factors may be applied, for example, for birds missed during transect surveys that are recorded by their radios

(Brittas and Karlbom 1990) or nests that go undetected during visual searches (Hill 1998). There is much scope for improvements of this type in census work (Chapter 2).

Beyond the speed and precision with which radio-tagging can supply data is the huge amount of information on individual life-histories that can come from long-life tags. Tags that last for months or years from fledging can show how performance (e.g. survival, dispersal, productivity) relates to individual use of habitats and other resources as well as to age. Such data are important components for individual-based modeling. Although data for individual-based demographic modeling has mainly come from visual observations (Goss-Custard 1996; Sutherland 1996), the first demographic modeling based on functional responses originated from radio-tagging (Kenward and Marcström 1988). Radio-tagging also has huge potential in field experiments, not only by providing detailed information, but also by enabling minimal samples (because there is minimal unexplained variance in tests if all outcomes are recorded).

However, the benefits of radio tagging may be lost if data are collected and analyzed in ways that are biased, or the sample of tagged animals is biased (e.g. by differential capture) or there are adverse impacts of tagging. It can be hard to obtain control data to demonstrate absence of bias from tagging, especially with the low statistical power that results from small samples. Tag attachment must therefore be planned with great care.

6.2.3 Considerations for tag attachment

Radio tags have been attached externally to bird beaks, necks, backs, legs, tail feathers, and patagia, or implanted, depending on species and study requirements (Table 6.1). Implanting usually requires veterinary supervision and licensing and may put bird health at risk. Implants also have low antenna efficiency (unless a transcutaneous antenna further increases the health risk) and are therefore best avoided unless required for measuring physiological parameters.

The choice of attachment also depends on requirements for tag mass, sensors, and detachment. Tags on 1–2 tail feathers are shed at the molt, but tend to be molted prematurely if a feather carries more than 1% of bird mass (i.e. more than 2% on 2 feathers). Tail-mounts (Figure 6.2(a)) are convenient for mounting sensors that indicate behavior from tail-posture (Kenward et al. 1982) or, as they are relatively remote from the skin, for measuring ambient temperature. Nasal saddles have been used to indicate feeding activity from head posture (Swanson and Keuchle 1976), but must be very small if they are not to affect behavior. Sensors on necklaces (Figure 6.2(b)) and leg tags can indicate general activity or mortality. However, antennas are apt to break on leg tags and both methods are

Table 6.1 *Limitations of tag attachment methods for birds, from low (+) to high (+ + +) in each category. Risk-assessment risk is from tests in >10 (2), <10 (1), or no (0) published studies*

Technique	Handling time	Skill requirement	Rate of loss	Risk: impact on wearer	Other limitations
Leg-mount	+	+	+ + +	+[1]	Not for very small birds
Tail-mount	+ +	+ +	+ +	+[1]	Not for very small birds
Necklace	+	+	+	+ +[1]	Behavior, icing, crop-shape
Backpack: glue	+	+ +	+ + +	+[1]	Detachment rate varies greatly
body-loops	+ +	+ + +	+	+ + +[2]	Needs very careful fitting
wing-loops	+	+	+	+ +[1]	Needs very careful design
thigh-loops	+	+ +	+	+[1]	Based on 2 assessments
Patagial	+	+ +	+ +	+ +[1]	Only for large, slow fliers
Nasal-saddle	+	+ +	+ +	+ +[0]	Only for medium to large birds
Implant	+ + +	+ + +	+	+ +[1]	Poor signal range, invasive

best avoided on water birds in cold regions where they can load with ice. Patagial tags are only for species that flap their wings slowly, and have been successful on Californian Condors (Wallace *et al.* 1994).

Tag sensors typically convey information by modulating the length of signal pulses or the interval between them. Pulse interval modulation is most common, because the resulting change in pulse rate is easy to detect by ear. Sensors are used in many ways and can be coded with complex information. Thus, death can be indicated by a sensor that is simply modulated by body temperature, or by a change in signal rate that is triggered by absence of motion in an activity sensor during a pre-set time period. Tags with microprocessors can be coded to insert extra signal pulses. For example, they can vary the repetition rate of all signals to indicate temperature and have an extra signal in every tenth interval if there has been no activity.

Tags close to mass limits must be close to the center of lift, and are therefore mounted with backpack or lumbar harnesses. Backpack harnesses, in which a neck loop ahead of the wings is joined under the breast to a body loop behind the wings (Figure 6.2(c)), are the most commonly used attachment on birds and have thus provided many records of impacts on behavior or performance. Assuming the use of safe harness material, such as Teflon ribbon, problems may largely be due to poor fitting. Problems have been recorded mainly with precocial birds and raptors, in which growth conditions and sexual dimorphism

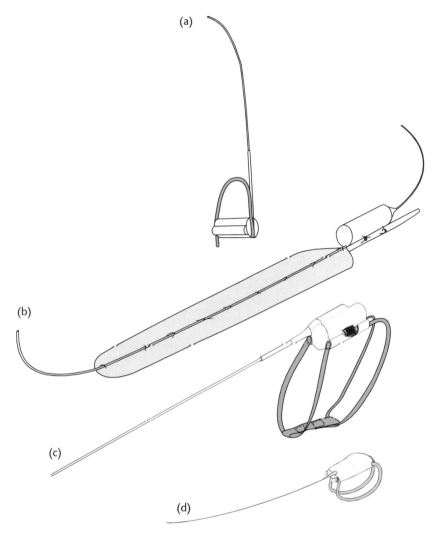

Fig. 6.2 Radio-tags used on birds, for attachment as (a) necklace; (b) tail mount; (c) back-pack with body loops and breast-strap; (d) backpack with under-wing loops.

can greatly affect body size, and have even been found to vary between field workers in the same project (Patton *et al.* 1991). When birds vary little in body-size, harness loops can be a standard size. When species must be marked while still growing, under-wing loops (Figure 6.2(d)) can be satisfactory for backpacks, but tags must be very carefully designed (Hill *et al.* 1999). Lumbar harnesses, with a loop round each leg (Rappole and Tipton 1991), are also a relatively recent development. Neither wing-loop nor lumbar designs have yet been tested on many species.

For short-term studies, small tags glued to skin or feathers on bird backs have proved satisfactory in several projects (e.g. Graber and Wunderle 1966; Raim 1978; Green *et al.* 1990). Surgical glues are now the preferred option, as other adhesives (e.g. epoxy-based) can cause inflammation. The most durable attachments, which can last several months, occur after use of biologically compatible cleansing solvents and by gluing to the relatively immobile skin over the synsacrum rather than to the thorax (R.E. Green, personal communication).

In summary, tag mass, recovery requirements, and animal welfare considerations influence the choice of attachment techniques. Whereas some techniques are quite straightforward, application of a precautionary principle indicates that, whenever possible, the most risky methods (Table 6.1) should be avoided. People should certainly not fit backpack harnesses without proper training. Moreover, there will always be some impact of putting extra weight on an animal, no matter how carefully a tag is attached. There should therefore always be tests for adverse impacts, even if only by comparing results with different tag attachments and mass. There is also the issue of tag detachment, which may be simple to arrange with a weak-link system, but which is hard to arrange reliably for long-life necklaces and harnesses, and impossible for implants. In cases of doubt, an ethics committee may be required to assess the trade-off between necessity of data for species conservation and impact on individual animals.

6.3 Forward planning

6.3.1 Equipment

It is important to order the most suitable equipment, and in good time. Manufacturers often make a wide range of equipment, but may favor particular niches, such as ARGOS tags (e.g. www.microwt@aol.com, www.northstarst.com), GPS tags and logging equipment (e.g. www.lotek.com, www.televilt.se), or general VHF equipment (e.g. www.biotrack.co.uk, www.holohil.com, www.sirtrack. landcareresearch.co.nz, www.titley.com.au and US companies listed at www. bio telem.org/manufact.htm). Some firms deliver more reliably than others and have tight schedules booked months in advance, so be sure to order in good time. Helpful firms will schedule orders provisionally, before funding is assured, to avoid a last-minute rush. Manufacturers will require an identity code for each ARGOS tag, so contact the ARGOS system (www.argosinc.com) as early as possible.

Long-established firms have wide experience with many species, and some have biologists to plan the optimal tag (and project) design for each biological question. Therefore, choosing the right firm is in many ways more important than choosing the details of the tag. Read web-sites carefully and ask other researchers in the same field for advice. Many prefer, where practical, to stay with proven tag types until new designs are well tested. Conscientious manufacturers provide details of their quality controls, such as temperature-cycling systems.

Receivers for VHF projects last a long time, so it can be wise to obtain capabilities beyond the immediate needs, if the budget permits. Tags are identified by separate frequencies, typically 10 kHz apart. A pilot project with low budget may have to manage with a receiver for 10–20 tag frequencies. However, subsequent quantitative work will need to distinguish many more tag frequencies. Most receivers cover a band of only 1–2 MHz, but one new model covers most of the VHF band and is therefore suitable for projects in many different countries. It is convenient to store frequencies and, if you need to search for tags on dispersing or wide-ranging animals from moving vehicles, to scan automatically through them at chosen time intervals.

As an alternative to these specialized receivers, commercial "scanner" receivers are available at low cost to cover the same frequency range. However, their scanning is for seeking signals, not to dwell on pre-set frequencies. Moreover, these receivers are not designed with gain controls suitable for close-range tracking, or to be robust, waterproof, and easy to use with gloves. They are therefore best for undemanding tracking and to keep in reserve for when the mainstay receiver is unavailable. Servicing requirements of specialized receivers should be infrequent, but check when buying for availability of prompt servicing or loan in case of emergencies.

Any receiver can be used for logging data. The simplest logging involves recording the signal from a headphone socket onto audio or paper tapes. This approach typically records for one tag frequency at a time, for instance to record presence on a nest. Several tagged birds can be recorded visiting a feed site if the receiver will step through frequencies, provided that one channel records an identifier signal. Alternatively, a receiver may have an interface for data transfer and for control by auxiliary hardware. Some tracking receivers that lack an interface in the basic models can have it added as an option. Very sophisticated logging is possible with such systems, for instance to check some frequencies more often than others or even to provide paging or other alerts for rare events (e.g. probable death or dispersal).

6.3.2 Mobile tracking

When tracking on foot, it is important to have lightweight, robust receivers with simple controls. Many receivers are not waterproof, in which case it must be possible to operate them through plastic bags during rain. A light source is useful at night, and headphones or an ear-plug make it easier to detect faint signals and changes in signal strength. For directional tracking at short range, it should be easy to reduce the signal reception gain to a level at which signals from nearby tags are only just detected. Power supplies should last for at least the length of a tracking day, with reserve power sources available (e.g. plug in battery packs in case an overnight recharge has failed). Antennas should be light and not cumbersome. Yagi antennas with 3–4 elements are very suitable, to give optimal gain and directionality, and can now be obtained with flexible elements that do not impede passage through brush.

When tracking from vehicles, it is important that other equipment does not interfere with the very sensitive receiver reception. Diesel or thoroughly suppressed engines are important, and the effect of communications equipment, GPS receivers and computers should be checked before starting a journey. The plasma in strip lighting can give a lot of interference. On boats and road vehicles, a pneumatic mast is ideal for raising antennas (www.clarkmasts.co.uk, www.aoaqps.com/hilomast.htm), which can be longer than for hand-held work. A 5–6 element Yagi on a 3–5-m mast more than doubles the reception range compared with a 3-element hand-held antenna. A compass repeater from the mast to the cabin is often indispensable (e.g. at night), and accurate GPS is invaluable in boats and for all off-road work.

Antennas give optimal gain on aircraft if mounted externally; high-wing Cessnas have very convenient struts. Obtaining aviation authority approval can be problematic, but mounts must in any case always be firm, fail-safe, and approved by the pilot. Selecting an understanding and cautious pilot is important too.

6.3.3 Software

Conservation questions may appear to be very simple, such as "do they survive if we release them here"? However, if data are collected in the right way, they can be used to answer further questions like "do they survive here significantly better than there" or "why didn't they stay here" or "what do these animals really need"? Alas, all too many projects look only at the simple questions. They invest much effort in setting up an experiment, yet omit to record information that could later explain why one treatment fails while another succeeds. However, a decision to use radio-tags is a good start toward discovering what really happens.

The next step is to collect data for analyses that add value to the fieldwork. In order to plan how best to collect data, it is important to access the analysis software in advance.

For example, the most sensitive tests require data to be collected systematically, in ways that make it easy to identify how long individuals survived, when they dispersed and how they used different areas. For testing whether performance relates to areas used by the animals, it will also be important to analyze location data. This too requires systematic data collection, in ways that are indicated by the software. Indeed, location analysis software should be used when starting the pilot phase of a project, to develop efficient field methods. Map data may also be needed to estimate the habitats available to different individuals. Software for survival analyses helps to estimate how many tags will be needed to show significant differences, which means using a computer before buying radio equipment. This is further explained below and in manuals (White and Garrott 1990; Kenward 2001). However, it is always best to consult a statistician before starting the work.

6.4 Approaches

6.4.1 Pilot studies

Harris *et al.* (1990) noted the importance of a pilot study when collecting location data. Before embarking on extensive work, it is wise to check techniques for capture, tagging, and data collection. Can the animals be marked without bias? Tagging nestlings is likely to minimize bias compared with trapping techniques that may select poor quality individuals. Tests for tag impacts can start with simple behavioral comparison with untagged birds, ideally in captivity and remembering that animals may always require a day or two to adjust to handling and tagging. How large a sample can be monitored in the field? That will depend on how easy it is to check individuals and move between them, which will improve with practice. A pilot study helps the pessimist to be more ambitious in a main study, and the optimist to avoid over-ambitious planning.

A pilot study should also address the issue of how often to monitor animals, typically by deliberate over-sampling so that analyses can define a minimum-effort protocol. That may mean that initial tracking is continuous, recording the location and time each time a bird feeds or flies, developing field skills to avoid disturbance and gaining behavioral insights that aid later work. This showed, for instance, that released naïve hawks were likely to survive once they had made 2–3 kills (Kenward *et al.* 1981). Continuous monitoring also reveals when animals are likely to be active, so that foraging observations can be planned for those times. Activity data can also be recorded by automated logging, for instance while tracking every hour to collect location data from 5–6 animals (i.e. a minimal sample for statistical tests).

For a study over more than 1 year, pilot work should also involve recording locations at 3 to 4 day intervals. This is convenient for recording dispersal and survival data, and for indicating the seasons in which animals tend to settle.

6.4.2 Recording locations

Location data reveal how and when animals move, disperse, and interact with resources, such as food and cover. Locations close in time also show how animals interact with each other, and hence space themselves or transmit disease. The data may be recorded automatically, or by close approach to animals and by triangulation techniques (details in Kenward 2001).

Continuous recording of locations in a pilot study builds a trajectory, which gives movements of dispersers in detail and gradually defines a home range if the animals are settled. However, it takes time to record each location, and a whole trajectory is not needed to define a home range. Indeed, it is a mistake to focus on collecting large numbers of locations, unless you need travel distances, because locations are not statistically independent records. Not only are adjacent records constrained by how far an animal can travel in a given time, but animals also tend to use the same roosts, routes, and foraging sites repeatedly, often at similar times of day. Comparisons between categories of animal, or the same animal in different seasons, are therefore based on a single index that represents, for example, the proportion of foraging locations in a particular habitat or the area encompassing a selected proportion of the locations.

When analyses are based on indices that represent each set of locations, the efficiency issue becomes "how few locations adequately estimate each index"? This standard number of locations should be established during the pilot study, by estimating with increasing numbers until the index becomes acceptably stable. For example, the area of a *convex* polygon plotted round 100% of the locations (X_{100}-defined for brevity by *unique letters* with a percentage inclusion) initially increases rapidly as locations are added, but eventually tends to an asymptote if the animal is settled in a seasonal home range (Figure 6.3(a)). If animals were monitored continuously, there may be hundreds of locations to reach, say, 95% of the maximum area. However, if locations are sampled at intervals of 1, 2, 3 h etc. a protocol can be developed to estimate X_{100} efficiently. With 2–5 locations sampled at intervals throughout the day, 30–50 locations are usually enough. If fewer locations are available for some animals, but the rate at which the asymptotic areas is approached is found to vary little among animals, then it may be possible to correct the incomplete samples by fitting an asymptotic regression.

Other approaches to this incremental analysis involve use of random instead of consecutive locations, and estimation of stability by overlap of outlines with

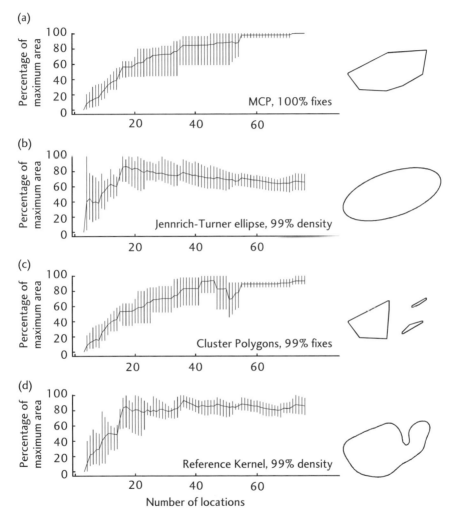

Fig. 6.3 The increase in range area as consecutive locations are added, estimated by (a) minimum convex polygons on peripheral locations, (b) ellipses encompassing 99% of the utilization distribution, (c) polygons round locations grouped by cluster analysis and (d) kernel contours for 99% of the utilization distribution. Lines show the mean and bars give the range of estimates for four Northern Goshawks *Accipiter gentilis* that were located 5–6 times daily. All methods indicate that range areas changed little after the first 10 days of tracking.

increasing numbers of random locations (Robertson *et al.* 1998). Whichever way one assesses the stability of outlines estimated round home ranges, the numbers of locations needed increases with the precision of fit to the locations. If an *ellipse* is plotted round, say, 95% of a bivariate distribution centered on the arithmetic mean of all the locations (E_{95}), stability can occur after 12–15 locations. In this case the area estimate can expand or contract, depending on whether the locations add peripherally or close to the center (Figure 6.3(b)). E_{95} gives a stable area estimate with a minimal number of locations, but has low precision in analyses of interactions with resources or other animals. At the other extreme, high precision is obtained by adding records in a grid of cells at the size of the tracking resolution, but hundreds of locations are required for stability.

Two methods that give reasonable precision with moderate numbers of locations are contours based on kernel functions (Worton 1989), for instance "*fixed*" *kernel Kf$_{95}$* (Figure 6.3(c)), and *cluster analysis* that defines *convex* polygons round groups of locations, for example, *Cx$_{95}$* (Figure 6.3(d)). In tests of ability to answer biological questions, the best results were usually with one of these methods (Kenward *et al.* 2001). It may therefore be wise to collect enough locations for both, because that is also enough for ellipses and X_{100}. Indeed, stability of X_{100} is quite a good criterion, because the other methods (except grid cells) have then gone through their initial increase phase (Figure 6.3). Provided the same standard number of locations is used throughout, there may be some added variation in the least stable methods, but there should not be bias.

Having decided how many locations are needed to estimate a seasonal home range, the frequency of interval-sampling will depend on how many different animals are to be tracked at once. If it takes all day to visit 10–20 animals, then intervals may be daily. It is then very important to vary the sampling order so that individuals that may be timetabling are not always recorded at the same time each day. Alternatively, locations of 10–20 animals that live densely may be recorded in 1–2 h, enabling 3–5 sample sessions each day. Autocorrelation analysis of spatio-temporal correlation between consecutive locations (Swihart and Slade 1985) may also have value for identifying optimal sampling intervals, although the original "time to independence" tends to be overestimated.

6.4.3 Using location data

Location data can be used for purposes other than home range definition. For example, resources can be estimated in circles or ellipses of a size that either reflect uncertainty or define availability (Arthur *et al.* 1996). Interactions can be determined solely from the distances between animals. However, home range outlines also provide overlap indices, identify neighbors for sociality analyses and estimate resource availability in a way needed for resource-area dependence analysis (Kenward 2001). Dispersal detection can be defined statistically as departure from a home range, and settling by the reverse process, though that will require records over longer periods than home ranges. Indeed, home ranges are likely to change from season to season. Techniques used in the short term to define a standard seasonal home range may also be used to define an annual home range.

With the development of Geographic Information Systems, detailed maps have become more available for analyzing use of resources. This is an important consideration when obtaining software for location analyses, because two options are available. One option is to attempt all analyses within industry-standard software, such as ESRI ArcView (www.esri.com). Animal Movement tools (www.absc.usgs.gov/glba/gistools) calculate some types of range outline, so that areas and resource use can be estimated. This software is free, but requires access to ESRI software, including the Spatial Analyst extension (which alone costs US$2495).

Alternatively, there is specialized software for radio locations, of which the most comprehensive is Ranges (www.anatrack.com, see the biotelemetry clearing house www.biotelem.org/software.htm for other software). Ranges (costing GB£300–590) is designed for use in pilot studies and provides automated autocorrelation, incremental, and dispersal analyses, with other home range and sociality analyses that are not present in the Animal Movement tools. Ranges 6 has comprehensive on-line help for rapid learning and, unlike ESRI software, also gives spreadsheet-ready results from automated repeating analyses on multiple sets of location data. Maps can be prepared in Ranges if they are simple or imported from ESRI or cheaper GIS (e.g. www.mapmaker.com, www.sbg.ac.at/geo/idrisi).

Maps can be based on categories (e.g. habitats) in line-bounded polygons or in an array of cells. If practical, they are best prepared initially as polygons (vector format), because these require less space for storage and convert readily to cells (rasters) at any scale. However, remote sensed maps (e.g. from Landsat images) come as rasters of a particular resolution, for example, 25 m for the 25 categories in the Landcover Map of Great Britain (Fuller *et al.* 1994).

6.4.4 Demography

The advent of reliable long-life tags enables rapid modeling of population structure, from age-specific survival and breeding data. Such models can be used for various purposes, including exploitation analyses (Chapter 13). Long-life tags also enable work on how dispersal, survival, and breeding relate to resource use and sociality. Survival rates can be estimated and compared with MARK (contact gwhite@cnr.colostate.edu) and other software (Chapter 5), and similar techniques are suitable for recording survival of nests of radio-tagged birds, and hence productivity (Chapter 3). For more details, and estimation of animal densities with radio-tags, see Kenward (2001) and Millspaugh and Marzluff (2001).

Pilot work is advisable for studies of density, dispersal, survival, and breeding, not least to check that tagging does not affect performance (Murray and Fuller 2000). There is something of a "Catch 22" in such tests, because power to detect differences is low in tests with small samples. This has two implications. One is that with a reporting rate of 5–10% (typical for ringing), it needs 10–20 times as many marked birds as radio-tags for estimates with comparable confidence limits. The second implication is that, even with the high reporting rate of radio-tags, well over a hundred tags may be required to detect small differences in demographic rates. Therefore, on the one hand it can be difficult to detect not only small impacts of radio tags but also the small differences in demographic rates that often occur between stable and declining populations. On the other hand, if there is a high re-sighting rate for colored markers, the cost of using radio tags merely to estimate annual survival rates would not be justified.

In order to exploit the main advantages of radio-tags over other demographic techniques, other preliminary tests will be needed. How frequently should animals be checked to reveal the seasonal details of survival and dispersal? When may home range characteristics correlate most strongly with demographic factors? What is the best time of day to check for presence or survival of tagged animals? It can be at night, when living birds perch high enough to be detected at optimal distances. How frequently should survival be checked if corpses are to be fit for cause-of-death analysis (Chapter 8) at different times of year? If cause-of-death is unimportant, thrice-yearly checks may suffice. One check can identify a wintering area, as a home-range-sized circle round roosts. A second check during incubation makes it easy to find nests, at least for females of single-brood species, and a third check during brood-rearing identifies the successful breeders (and prepares for tagging the next generation).

6.5 The future

The development of radio-tagging has depended on other technologies. Improvements have depended mostly on consumer or military requirements. The smallest raptor tags were provided initially for falconry, and it was these raptor enthusiasts who obtained US military support to reduce Doppler tags to a size for tracking migratory peregrine falcons. GPS units at wrist watch size have been developed for consumer-electronics. Miniaturized sensors (including vision) and reliable sources of high power-density will be required in future for military drones and could have applications in biology.

There will also be improved automation. Time-difference of arrival (TDOA) systems may become available for automated tracking of the smallest VHF tags. Analyses will become automated to minimize the need for decisions that now complicate them, leading ultimately to automation of complex modeling systems. Improvements in reliability and automated decision support will tend toward radio-tagging techniques that are developed initially by professionals, then adopted by trained volunteers. Spatially specific population modeling, based on thrice-annual checks by volunteers, is practical already.

References

Arthur, S. M., Manly, B. F. J., McDonald, L. L., and Garner, G. W. (1996). Assessing habitat selection when availability changes. Ecology, 77, 215–227.

Brittas, R. and Karlbom, M. (1990). A field evaluation of the Finnish 3-man chain: a method for estimating forest grouse numbers and habitat use. Ornis Fennica, 67, 18–23.

Evans, I.M., Summers, R.W., O'Toole, L., Orr-Ewing, D.C., Evans, R., Snell, N., and Smith, J. (1999). Evaluating the success of translocating Red Kites *Milvus milvus* in the UK. Bird Stud., 46, 129–144.

Fuller, M.R, Millspaugh, J.J., Church, C.E., and Kenward, R.E. (In press). Wildlife telemetry. In *Research and Management Techniques for Wildlife and Habitats*, 6th edn, ed. C. Braun. The Wildlife Society, Maryland.

Fuller, M.R., Secgar, W.S., and Howey, P.W. (1995). The use of satellite systems for the study of bird migration. *Israel J. Zool.,* 41, 243–252.

Fuller, R.M., Groom, G.B., and Jones, A.R. (1994). The Land Cover Map of Great Britain: an automated classification of Landsat Thematic Mapper data. *Photogrammet Eng Remote Sensing*, 60, 553–562.

Graber, R.R. and Wunderle, S.L. (1966). Telemetric observations of a Robin (*Turdus migratorius*). *Auk*, 83, 674–677.

Green, R. E., Hirons, G. J. M., and Cresswell, B. H. (1990). Foraging habitats of female Common Snipe *Gallinago gallinago* during the incubation period. *J. Appl. Ecol.,* 27, 325–335.

Goss-Custard, J.D. (1996). *The Oystercatcher: From Individuals to Populations*. Oxford University Press, Oxford.

Harris, S., Cresswell, W.J., Forde, P.G., Trewella, W.J., Woollard T., and Wray, S. (1990). Home-range analysis using radio-tracking data—a review of problems and techniques particularly as applied to the study of mammals. *Mam. Rev.*, 20, 97–123.

Higuchi, H., Ozaki, K., Fujita, G., Minton, J., Veta, M., Soma, M., and Mita, N. (1996). Satellite tracking of white-naped cranes migration and the importance of the Korean demilitarised zone. *Conserv. Biol.,* 10, 806–812.

Hill, I.F. (1998). Post-nestling mortality and dispersal in Blackbirds and Song Thrushes. D.Phil. thesis, University of Oxford.

Hill, I. F., Cresswell, B. H., and Kenward, R.E. (1999). Field-testing the suitability of a new back-pack harness for radio-tagging passerines. *J. Avian Biol.*, 30, 135–143.

Jouventin, P. and Weimerskirch, H. (1990). Satellite tracking of Wandering Albatrosses. *Nature,* 343, 746–748.

Kenward, R.E. (2001). *A Manual for Wildlife Radio-tagging.* Academic Press, London.

Kenward, R.E. and Marcström, V. (1988). How differential competence could sustain suppressive predation on birds. *Proc. Int. Ornitholo. Cong.,* 19, 733–742.

Kenward, R.E., Marquiss, M., and Newton, I. (1981). What happens to Goshawks trained for falconry. *J. Wildlife Manage.,* 45, 802–806.

Kenward, R.E., Hirons, G.J.M., and Ziesemer, F. (1982). Devices for telemetering the behavior of free-living birds. *Symp. Zool. Soc. Lond.,* 49, 129–137.

Kenward, R.E., Walls, S.S., Hodder, K.H., Pahkala, M., Freeman, S.N., and Simpson, V. R. (2000). The prevalence of non-breeders in raptor populations: evidence from rings, radio-tags and transect surveys. *Oikos,* 91, 271–279.

Kenward, R.E., Clarke, R.T., Hodder, K.H., and Walls, S.S. (2001). Density and linkage estimators of home range: nearest-neighbor clustering defines multi-nuclear cores. *Ecology,* 82, 1905–1920.

Meyburg, B.-U. and Meyburg, C. (1998). The study of raptor migration using satellite telemetry: some goals, achievements and limitations. *Biotelemetry,* XIV, 415–420.

Millspaugh, J.J. and Marzluff, J.M. (eds). (2001). *Radio Tracking and Animal Populations.* Academic Press, San Diego, U.S.A.

Murray, D.L. and Fuller, M.R. (2000). Effects of marking on the life history patterns of vertebrates. Pages 15–64. In Research Techniques in Ethology and Animal Ecology, eds. L. Boitani, T. Fuller, Columbia University Press, New York, U.S.A.

Newton, I. (1979). *Population Ecology of Raptors.* Poyser, Berkhamsted.

Patton, W.C., Zabel, C.J., Neal, D.L., Steger, G.N., Tilghman, N.G., and Noon, B.R. (1991). Effects of radio tags on spotted owls. *J. Wildlife Manage.,* 55, 617–622.

Raim, A. (1978). A radio transmitter attachment for small passerine birds. *Bird Band.,* 49, 326–332.

Rappole, J.H. and Tipton, A.R. (1991). New harness design for attachment of radio-transmitters to small passerines. *J. Field Ornithol.,* 62, 335–337.

Riley, J.R., Smith, A.D., Reynolds, D.R., Edwards, A.S., Osborne, J.L. Williams, I.H., Carreck, N.L., and Poppy, G.M. (1996). Tracking bees with harmonic radar. *Nature,* 379, 29–30.

Robertson, P.A., Aebischer, N.J., Kenward, R.E., Hanski I.K., and Williams, N.P. (1998). Simulation and jack-knifing assessment of home-range indices based on underlying trajectories. *J. Appl. Ecol.,* 35, 928–940.

Sutherland, W.J. (1996). *From Individual Behaviour to Population Ecology.* Oxford University Press, Oxford.

Swanson, G.A. and Keuchle, V.B. (1976). A telemetry technique for monitoring waterfowl activity. *J Wildlife Manage.,* 40, 187–189.

Swihart, R.K. and Slade, N.A. (1985). Testing for independence of observations in animal movements. *Ecology,* 66, 1176–1184.

Wallace, M.P., Fuller, M., and Wiley, J. (1994). Patagial transmitters for large vultures and condors. In *Raptor Conservation Today*, eds. B.-U. Meyburg, and R.D. Chancellor, pp. 381–387. World Working Group on Birds of Prey, Berlin.

White, G. C. and Garrott, R. A. (1990). *Analysis of Wildlife Radio-tracking Data*. Academic Press, New York.

Willebrand, T. and Marcström, M. (1988). On the danger of using dummy nests to study predation. *Auk,* 105, 378–379.

Wilson, R.P., Ducamp, J.-J., Rees, W.G., Culik, B.M., and Niekamp, K. (1992). Estimation of location: global coverage using light intensity. In *Wildlife Telemetry—Remote Monitoring and Tracking of Animals*, eds. I.G. Priede and S.M. Swift, pp. 131–143. Ellis Horwood, Chichester.

Wink, M., Staudter, H., Bragin, Y., Pfeffer, R., and Kenward, R. (1999). The use of DNA fingerprinting to estimate survival rates in the Saker Falcon (*Falco cherrug*). *J. für Ornithol.,* 140, 481–489.

Woodbridge, B., Finley, K.K., and Seager, S.T. (1995). An investigation of the Swainson's Hawk in Argentina. *J. Raptor Res.,* 29, 202–204.

Worton, B. J. (1989). Kernel methods for estimating the utilization distribution in home-range studies. *Ecology,* 70, 164–168.

7

Migration

Susanne Åkesson and Anders Hedenström

7.1 Introduction

The flights of some migratory birds are among the most impressive phenomena in nature. The migration of the Arctic Tern *Sterna paradisaea* between arctic breeding sites and Antarctic wintering areas (19,000 km one way) is a classic example, while the 12,000 km non-stop flight between Alaska and New Zealand by the Bar-tailed Godwit *Limosa lapponica* is perhaps even more astonishing. At the other end of the spectrum we find the trickle migration strategy adopted by many passerines, in which short flights are alternated by refueling episodes. Adaptations for migration are equally important in all migratory species, as an integral part of their life histories and annual cycles. They include the morphology of body and wings, flexibility of metabolic organs, accumulation of fuel (fat and protein), sensory capacities for direction finding (orientation and navigation), as well as an ability to make correct decisions about when to depart and when to stop. In flight, the bird has to know at what speed and altitude to fly for best economy, how to maintain its intended flight direction and how to deal with varying winds.

The questions asked by students of bird migration are very diverse and require an exclusive "toolbox" of techniques and approaches. Often the research is interdisciplinary, using techniques borrowed from fields such as mathematics, physics, physiology, sensory biology, and morphology. In this chapter we give some examples of questions asked and research techniques used in a modern bird migration laboratory.

7.2 Migration systems

Knowledge about population-specific breeding and wintering areas, and the migration routes between them, is fundamental to migration studies. The routes used are not necessarily the same in autumn and spring and may also differ between experienced (adult) and juvenile birds on their first migration. For conservation actions it may also be of interest to know the degree of 'migration connectivity' among different populations, that is, the movement of individuals between different summer and different winter populations (Webster *et al.* 2002). A number of approaches are available.

7.2.1 Mark-recapture

Most current knowledge about migration routes and wintering areas has been accumulated over the last century by national ringing (banding) programs, and recently presented as migration atlases by a few countries (e.g. Fransson and Pettersson 2001; Werham *et al.* 2002). Recent interesting results on ringing recoveries are usually published as annual reports by the different ringing schemes, such as the BTO recoveries published in *Ringing & Migration*. However, even though millions of birds have been ringed, our knowledge of certain species is still limited or lacking. By way of example, to get one recovery in Africa south of the Sahara of Willow Warblers *Phylloscopus trochilus* caught in Finland, no less than 16,000 had to be ringed (Hedenström and Pettersson 1987). Ringing recoveries not only reveal routes and wintering areas, but they are also useful when analyzing migration speed, strategies, and orientation mechanisms.

Because the slow rate of generating migration maps based on ringing recoveries, migrationists have invented supplementary techniques. In addition to a numbered metal ring, combinations of color bands are used for identification of birds using a telescope. This method is useful in birds like shorebirds using specific habitats along the migration routes. Larger birds, such as swans and geese, can be fitted with numbered neck-collars, likewise checked by telescope. Compared with ringing, this method allows repeated registration of individual birds along the migration route and hence improved temporal resolution in data on migration rate and stopover duration. However, neck-collars are limited to large birds and can accumulate ice during cold weather. Also, it is only possible to get records from locations visited by observers.

7.2.2 Morphology

Individuals from different populations and geographic origin typically vary in some respects, for example, color, size, wing length, and shape. Hence, morphological data can be used to distinguish populations at a migration site such as a bird observatory. However, since this approach is statistical it is a rather blunt research tool, but combined with genetic or stable isotope markers the number of individuals that can accurately be assigned to the correct population increases.

7.2.3 Genetic markers

From harmless blood samples (Chapter 9), molecular genetic techniques provide useful methods of assigning individuals to particular populations. A number of techniques are available and should be chosen according to the question asked. Mitochondrial DNA (mtDNA) haplotypes may vary among populations on regional scales, such as in Dunlins *Calidris alpina* across its arctic breeding range (Wennerberg 2001). Microsatellites—non-coding and highly repeated nuclear DNA sequences—show considerable variation among individuals and populations, but they appear to be relatively rare in the bird genome and have therefore not yet been extensively used (Webster *et al.* 2002). Nuclear markers, such as randomly amplified polymorphic DNA (RAPD) and amplified fragment length polymorphic DNA (AFLP) provide useful population markers. AFLP seems especially promising and has recently been used to distinguish different subspecies of the willow warbler across a hybrid zone in Sweden, where mtDNA and microsatellite markers failed to differentiate the populations (Bensch *et al.* 2002). See chapter 9 for further information on molecular genetic methods in avian research.

7.2.4 Stable isotopes

Naturally occurring elements often show clinal variation across a continental land surface with respect to ratios of stable isotopes. Some useful elements are carbon (C), hydrogen (H), nitrogen (N), and strontium (Sr) (Lajtha and Michener 1994). The processes behind systematic changes vary among the elements. In carbon the proportion $^{13}C : ^{14}C$ (conventionally known as $\delta^{13}C$) is determined by the relative abundance of C3 and C4 plants and hence largely determined by the composition of plant communities. The $\delta^{13}C$ is translated through the food chain from plants, or through phytophagous insects to birds. Therefore, the location of a bird during molt is reflected by the stable isotope signature laid down in its feathers, which is preserved after the feather has finished growing. By analyzing bird feathers using mass spectrometry the stable isotope ratios can be determined and compared with known regional variation. Using the variation of $\delta^{13}C$ and $\delta^{15}N$, the different wintering areas (west and east Africa) could be confirmed in two populations of Willow Warblers (Chamberlain *et al.* 2000) (see Figure 7.1). In another study, Marra *et al.* (1998) used $\delta^{13}C$ to show that arrival time from spring migration in American Redstarts *Setophaga ruticilla* correlated with the quality of their wintering habitat. Depending on the season of molt, different species are suited for summer or winter population differentiation, while the unique biannual molt in the Willow Warbler makes this species particularly useful for studying both summer and winter areas. In some species molt is

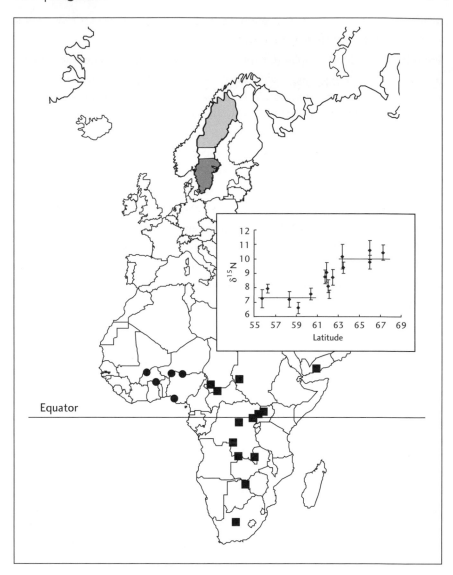

Fig. 7.1 Map showing recoveries of willow warblers ringed in Sweden. Recoveries from birds ringed in southern Sweden are shown as filled circles and those from northern Sweden by squares. The breeding distribution shows the approximate location of the hybrid zone where there is a migration divide. The diagram shows the $\delta^{13}N$ signature from southern Sweden, through the hybrid zone and northern Sweden. These data represent the location of the winter molt taking place in Africa and hence show the differing wintering areas for the two populations (based on Chamberlain *et al.* 2000).

divided between seasons and locations, providing the possibility of using stable isotopes to make a crude map of movements.

7.3 Migration behavior and strategies

The process of migration is typically divided into periods of refueling and flights between consecutive stopovers. Even though the flights are sometimes spectacular because of their length and altitude, they only take about one-seventh of the total migration time in small passerines (Hedenström and Alerstam 1997). Flight is energetically very expensive but due to the relatively fast transport, and hence short time required, only about one-third of the total energy consumption is flight cost, while the remaining two-thirds is spent while on the ground (Hedenström and Alerstam 1997). Studies on the timing of migratory flights, flocking, flight directions, speed, and wind drift, are often carried out at migration hotspots where large numbers of migrants can be observed. As always, the question asked dictates the method used. Because the overall speed of migration is to a large extent determined by the rate of energy accumulation (Alerstam and Hedenström 1998), we often want to monitor the rate of mass (fuel) gain and stopover duration of individual birds. Information used for orientation and navigation is probably gathered before flight departure, and hence behavior pertaining to orientation can be obtained using caged or radio-tagged birds. In this section we present some widely used methods for study of the behavior and physiology of migrating birds.

7.3.1 Counting and observing migrants

Terrestrial birds tend to migrate over land as far as possible and therefore concentrate at certain migration hotspots before inevitable sea-crossings or narrow land bridges. Examples of such locations where masses of migrants concentrate are Panama, Falsterbo in Sweden, Gibraltar, Bosphorous in Turkey, and Eilat in Israel. Daily counts of migrating birds passing such hotspots reveal the seasonal timing of migration among species. Concentrations of migrants also occur at inland sites where birds follow leading lines in the landscape (mountains, lakeshores, rivers, etc). Ageing and sexing of birds on the wing (possible in many raptors; Kjellén 1992) provides information on differential migration and timing among sex and age classes. Annual migration counts over many years are also used to monitor population numbers.

Moon watching is a low-tech method to record direction and intensity of migration, where an observer uses a telescope to register nocturnal migrants passing the face of the moon. By recording entry and exit of bird silhouettes as if the moon was a watch, the flight direction can be derived (Nisbet 1959). The apparent speed is mainly determined by the distance from the observer. Moon watching has been used on a continent wide scale (North America) to give a snapshot of the overall migration intensity and flight directions (Lowery and Newman 1966). The drawbacks are that it can only be used near full moon in clear weather and the observation cone has a relatively small angle (on average 0.52°). Using a telescope with 40 × magnification about 50% of the birds are detected at 1.5 km distance and zero at 3.5 km (Liechti *et al.* 1995).

By pointing a ceilometer (a strong directed light) toward the sky and observing birds blipping past the light using binoculars information can also be obtained on flight direction (Gauthreaux 1969). Ceilometer observations are mainly useful for studying low flying birds (up to 500 m), but can be used in overcast conditions. The technique is also useful to study orientation in relation to local topography (Åkesson 1993).

7.3.2 Tracking migrants

With an optical range finder (e.g. Leica Vector), the distance to a bird can be measured, and furnishing the instrument with azimuth and elevation scales provide polar coordinates to the bird, which can easily be converted to space coordinates (x, y, z) (Hedenström and Alerstam 1996). Multiple registrations of positions allow reconstruction of flight tracks and analysing the data in relation to wind speed and direction at each altitude, the flight speed and direction in relation to the air are obtained (Figure 7.2). The wind profile can be obtained by tracking ascending helium-filled weather balloons using the range finder. If the data are fed directly to a computer the instrument should be referred to as an "ornithodolite" (Pennycuick 1982).

Using radio-transmitters with ground-based receivers (see Chapter 6), migratory birds can be tracked during stopover and at departure, when timing of flights and vanishing directions are obtained (Åkesson *et al.* 1996). The radio

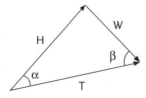

Fig. 7.2 Triangle of velocities. The track vector (T) is the sum of the heading (H) and wind (W) vectors. The angle between track and heading α shows the amount of drift with constant heading, while the angle β shows the wind direction in relation to the track direction.

signal from a small transmitter (e.g. 0.7 g) can be picked up at 20 km distance from a departing and climbing bird. This method has shown that even though most nocturnally flying birds initiate their flights shortly after sunset, birds may depart at any time during the night (Åkesson *et al.* 1996, 2001*b*).

The use of satellite telemetry has revolutionized the tracking of individual birds. Positions of birds are obtained from the Argos satellite system, which is a polar-orbiting-based system (Argos 1996). Argos provides position data including accuracy classifications for each position obtained. One should be aware that the precision of locations may be considerably lower than what manufacturer's data suggest. The life span of transmitters is mainly determined by the size of the battery and hence the weight (minimum about 18 g), but the duration can be extended using small solar panels. Thereby, the entire round-trip migration including the wintering period of an individual can be tracked. The data obtained are ideal for analyzing overall migration speed, stopover locations and duration, ground speed of flight, orientation, and navigation abilities. The only drawback is that the transmitters are still too heavy for use on small birds and the technique is therefore restricted to medium sized and large birds, such as geese and eagles.

7.3.3 Remote sensing: infrared device

Because birds are relatively warm objects in relation to the ambient air, a sensitive infrared sensor can detect the radiation from birds flying overhead across the sky. By pointing a thermal imaging device of 1.45° opening angle to the sky, migrating birds can be detected from 300 m up to 3000 m (Zehnder *et al.* 2001). Flight tracks are recorded on video and targets are classified into size classes to estimate flight altitudes. Infrared sensors work best at night under clear skies. The method detects birds at higher altitudes than ceilometer observations.

7.3.4 Remote sensing: radar

Radars are the most powerful tools available for tracking migrating birds. A radar emits short pulses of radio waves and records echoes of these from targets, whether birds and aeroplanes. Since the radio waves travel by the (constant) speed of light the distance between the radar and target is determined from the time delay between emitting a pulse and receiving the echo. A great advantage is that radars can be used in overcast conditions, and at any time of day and night. Accounts of the principles and technical basis of radars are given by Alerstam (1990) and Bruderer (1997).

Surveillance radars are mainly used for air traffic control at airports. They have a fan-beam of wide vertical angle (10–30°) and a narrow angle in the horizontal plane (≤2°). By rotating the radar antenna the sky is scanned for echoes

Fig. 7.3 Radar-based map of nocturnal bird migration over the United States within the altitudinal band of 430–1075 m for the period 26–30 April 2000. Arrows show mean direction of migration and different pattern of arrows indicates the mean number of birds (km^{-3}). From Gauthreaux *et al.* (2003).

with a high horizontal resolution but no altitudinal resolution. Surveillance radars are therefore used when studying migration intensity and general migration direction. By using a continent-wide network of weather surveillance radars, bird migration traffic can be quantified continent-wide and related to weather conditions (Figure 7.3). This approach replaces the moon-watching method, is less weather dependent, and requires no field observers (Gauthreaux *et al.* 2003).

A tracking radar emits a narrow "pencil beam" (1.3–2°) by which individual birds or flocks can be tracked. When operated in automatic tracking mode the radar records repeat measurements of distance, elevation, and azimuth angles to the target, from which speed and direction can be calculated. Wind profiles are obtained by using the radar to track ascending weather balloons carrying an aluminium foil for maximum reflectance. Heading and airspeed are then calculated from the tracking data against the wind (Figure 7.2). The radar echo often shows rhythmic fluctuations that can be recorded and used to estimate the wing-beat frequency of the bird—a useful measure of size when the target could not be identified. Tracking radar studies should be combined with visual observations for specific identification of birds being tracked, although this is sometimes impossible under cloud cover and at long distances.

7.3.5 Stopover

An important part of the migration process and strategy is the stopover periods during which energy is accumulated. Methods of estimating the duration of stopover phases and to test if animals of different groups (age and sex classes) differ in their stopover behavior require appropriate statistical modeling (Schaub *et al.* 2001). Simultaneous estimates of fuel accumulation can be made using individually color-marked birds and a remote operated electronic balance baited with food or from recaptures of ringed birds. Fuel deposition rate is given by the fuel deposition divided by stopover duration.

7.4 Physiology of migration

The migration process involves alternate flight and stopover periods during which energy is consumed and accumulated, respectively. During flight the metabolic rate is among the highest encountered in animals, and can be sustained for tens of hours by birds crossing ecological barriers. Also, the fueling interludes require efficient foraging and metabolic machineries to cope with high rates of food intake, assimilation, and conversion to adipose tissue in birds flying long nonstop flights these alternate performance requirements—flight versus fueling—are associated with rapid physiological changes in the alimentary tract and associated organs (Piersma and Lindström 1997). During refueling the stomach is enlarged, the intestine is elongated and surface area increased, and the liver is enlarged. In flight, however, all unnecessary payload is costly to carry, and therefore organs used during the fueling period can be reduced to a minimum for best flight economy.

7.4.1 Body composition

The most notable change in migratory birds is the accumulation of fat; sometimes a bird about to undertake a long flight where it cannot feed can double its body mass due to fat accumulation. Subcutaneous fat (adipose tissue) can be scored visually (see Chapter 4), with knowledge of body mass and help from regression analysis, the lean mass and fat mass can be estimated (although the fat class scoring does not reflect the amount of fat linearly). On dead birds (see Chapter 8), the fat mass can be determined by chemical extraction using a Sohxlet apparatus and petroleum ether in a mixture of ethyl alcohol (typically 3 : 1 mixture of 95% ethyl alcohol and petroleum ether is used for fat extraction). However, it is often of interest to study the rate of fat (and protein) accumulation in individuals, in which case working with dead birds is not feasible. A technique that looked promising was the measuring of total body electrical conductivity

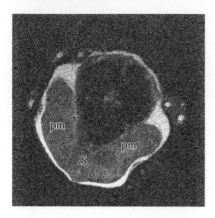

Fig. 7.4 MRI of a sanderling *Calidris alba*, showing a 2 mm cross section. The white layer is fat, the flight muscles are shown as gray (pm) and the sternum (s) is indicated.

(TOBEC®; Scott *et al.* 2001) or bioelectrical impedance (BIA; Lichtenbelt 2001), which are related to the amount of body fat. However, TOBEC requires calibration against dead birds of known fat content, and the result also depends on the distribution of the fat (Unangst and Merkley 2002). Dual-energy X-ray absorptiometry is a rather new, but expensive, technique that could potentially be used to estimate fat and lean mass, as well as total bone mineral density (Nagy 2001).

Magnetic resonance imaging (MRI) has recently been used for measuring fat deposits in White Storks *Ciconia ciconia* (Berthold *et al.* 2001). This technique can also be used on small birds and provides data on the amount of fat and its spatial distribution (Figure 7.4). MRI is of limited availability and expensive, however, because most machines are used for human diagnosis.

Ultrasound scanning is also a promising method for estimating the size of digestive organs and muscles (Starck *et al.* 2001). It has been used to show how flight muscle diameter varies in parallel with fuel reserves in Red Knots *Calidris canutus* (Lindström *et al.* 2000).

7.4.2 Energetics

The basal metabolic rate (BMR) is the metabolic rate at rest, during night, in darkness and in post-absorptive condition. BMR is typically measured in a respirometry chamber, where the rate of oxygen consumption or carbon dioxide production is measured. The metabolic scope is often related to BMR (e.g. as $5 \times BMR$), and determines the potential for fuel deposition. The proportion of fat and protein used can be determined from the respiratory quotient. Another method to measure energy consumption is the doubly labeled water technique. The theory and assumptions for this method is explained in Speakman *et al.* (2001) and Chapter 9. It requires the bird to be handled at the beginning and end of the measurement period.

Which of the alternative fuel substrates are catabolized can be determined from concentrations of blood plasma metabolites. Free fatty acids and glycerol are products of fat metabolism from adipose tissue, while uric acid is the end product from protein catabolism and hence indicates the use of protein, such as the consumption of flight muscle mass. Triglycerids and VLDLs indicate the formation of adipose tissue (fat accumulation). For an example see Jenni-Eiermann *et al.* (2002).

7.4.3 Endocrinology

Many behavioral and physiological processes during migration are proximally under hormonal control (see Chapter 9 for suitable techniques).

7.5 Flight in wind tunnels

Birds can be tracked in the field using the different methods described above. However, for experiments under controlled conditions, a wind tunnel is required. A wind tunnel creates a smooth (laminar) airflow in a test section where birds are trained to fly (Figure 7.5). There are a few wind tunnels dedicated for bird flight research in Europe and North America, but especially two new wind tunnels of recirculating design at Lund University, Sweden and the Max Planck Institute, Seewiesen, Germany, give very low turbulence in air flow (Figure 7.5; Pennycuick *et al.* 1997). This is important in order to generate a natural situation reflecting flight through nonturbulent air. If one is interested of studying the

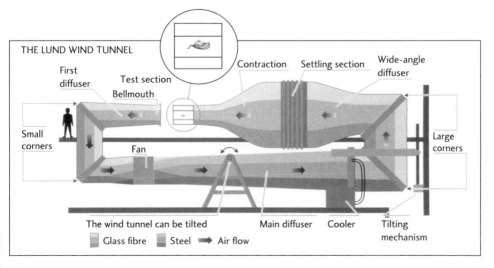

Fig. 7.5 The Lund University wind tunnel for bird flight research.

effect of turbulence on flight, turbulence can easily be created by inserting nets or objects upstream from the test section.

One important role for wind tunnels is the validation or testing of flight mechanical theory, which among other things, predicts a U-shaped relationship between power output and flight speed. Measurements of metabolic rate, using a respirometry mask or doubly labeled water, have often been used for evaluating flight mechanical theory, but this measures power input. These measurements are only valid if the conversion efficiency of fuel energy to mechanical work is constant at all speeds, which might not be true. Direct measurement of mechanical power output is difficult, but can be made by inserting a strain gauge and measuring the force applied to the humerus by the flight muscle (Dial *et al.* 1997; Tobalske *et al.* 2003), variations in the vertical acceleration of the body combined with wing-beat kinematics (Pennycuick *et al.* 2000), or from estimating the impulse associated with vortex wake structures.

The wind tunnel can also be used for studying wing-beat kinematics and flight style by using high-speed video cameras.

7.6 Orientation and navigation

Orientation refers to compass orientation or directed movement, while the term navigation is usually restricted to the theory and practice of charting a course to a distant goal to which the animal has no direct sensory contact.

7.6.1 Emlen funnels

Compass orientation has traditionally been studied in passerine migrants by recording their directed migratory activity (*Zugunruhe*) in circular cages, so called Emlen funnels (e.g. Emlen and Emlen 1966; Figure 7.6). The technique is suitable to record orientation in small passerines, such as the European Robin *Erithacus rubecula*. Modified and enlarged cages can be used for other species as well, such as waders (Sandberg and Gudmundsson 1996). During the experimental period, lasting between 1 h and a complete night period, the bird's activity is recorded by its claw scratches in the pigment of Tipp-Ex paper covering the sloping walls of the cage, or by having an ink pad in the bottom of the cage and white paper on the sloping walls to record the bird's movements. Circular cages with automatic computer registration are now used in several laboratories.

7.6.2 Manipulating sensory input

The main benefit of studying migratory orientation in cages is that the external information perceived by the bird can be manipulated. By producing an artificial

Fig. 7.6 Adult white-crowned sparrow in orientation cage (Emlen type) at the magnetic North Pole. The sloping walls are covered by type-writer correction paper. Photo: Susanne Åkesson.

magnetic field using large magnetic coils, such as modified Helmholz coils (Wiltschko and Wiltschko 1995), the perception of the geomagnetic field can be manipulated. In a similar way sun compass orientation has been investigated by shifting the position of the sun using mirrors, and by using filters to shift the alignment of polarized light and to depolarize the incoming light. Opaque Plexiglas sheets placed on top of the cages are used to screen off visual cues. By shifting the birds' internal time sense relative to the natural dark–light cycle using an artificial dark–light cycle, the function of the birds time-compensated sun compass can be investigated.

In studies of the functional characteristics of the birds' magnetic sense various techniques have been applied, such as exposure to strong (0.5–1 T) magnetic pulses and thereafter observation of the birds' orientation in cages, as well as neurophysiological recordings during magnetic field manipulations (for review see Wiltschko and Wiltschko 1995). In the search for a magnetic sensor containing magnetite, histology techniques, magnetic force microscopy, and a Superconducting Quantum Interference Device (SQUID) are used.

In conditioning experiments a bird is trained to detect a feeder associated with a particular stimulus (magnetic, visual), and then the bird's ability to use this cue is challenged with only the stimuli present at randomized locations.

7.6.3 Displacement experiments

Birds and other animals have been suggested to navigate by using either a combination of two geomagnetic parameters, field intensity and the angle of inclination (bi-coordinate magnetic navigation) varying across the Earth's surface or by celestial information. Both the geomagnetic and celestial parameters (elevation angle to certain star configurations and sky rotation) can be manipulated in the laboratory by using large magnetic coils (see above) and a planetarium sky. Studies have been performed with passerines where course shifts have been recorded as a response to simulated geomagnetic and geographical displacements.

Large-scale displacement experiments with ringed birds, performed mainly during 1930–70, have been used to study navigation abilities. The main aim is to find whether displaced birds maintain the same heading and end up in the "wrong" place (expected in clock-and-compass orientation), or whether they change their heading in accordance with the displacement and end up in the correct place (i.e. true navigation to a specific goal). Some of the most spectacular experiments involved over 15,000 starlings transported from the Netherlands in autumn to release sites in Switzerland and Spain. In more recent years the orientation of caged passerine migrants has been studied during lateral displacements by ship relative to their intended migration route (e.g. Åkesson *et al.* 2001*a*), asking the same type of questions.

7.6.4 Selection experiments

Both *Zugunruhe* (restless behavior shown at migration times) and direction are encoded in the birds' genetic migration program (Berthold 1996), and are probably exposed to strong selection. The length and intensity of nocturnal migratory activity can be studied in cages recording the bird's jumping activity. The inheritance of migratory activity has been studied both by selection- and cross-breeding experiments demonstrating that the phenotypic character can be changed in only a few generations. Furthermore, cross-breeding experiments with migratory Blackcaps *Sylvia atricapilla* from European populations with different migratory directions (SW- and SE- in autumn) show that the migratory direction is inherited in an intermediate fashion in the offspring (Berthold 1996).

7.6.5 Circular statistics

A special type of statistics called "circular statistics" is required to analyze circular data (e.g. Batschelet 1981), such as departure directions of migrating birds recorded by radio-telemetry or migratory activity recorded in Emlen funnels. Since we deal with directions that can be represented on a circle we cannot use linear statistics to treat these data. This can be illustrated by the angular difference between 10° and 350° being only 20°, with a mean direction of 0° (north), while the arithmetic mean of 180° would erroneously indicate a mean direction toward south. Vector addition is used to calculate the mean vector for a group of

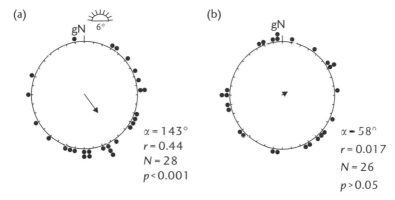

Fig. 7.7 Circular diagram showing orientation of white-crowned sparrows *Zonotrichia leucophrys gambelii* at the magnetic North Pole. α is the mean direction, *r* is the length of the mean vector, *N* is sample size and *p* is level significance of a Rayleigh test (Batschelet 1981). (*a*) Visual cues available to the birds, (b) an opaque sheet covers the orientation cage and restricts the birds to the use of magnetic cues only, and they become disoriented because the magnetic field lines are vertical and provide no directional information. Based on Åkesson *et al.* (2001 *a*).

directions, and the length of this vector indicates the concentration. From the mean vector and sample size, suitable test statistics can be calculated (Batschelet 1981). Data recorded in Emlen-funnels are usually divided into sectors and the mean orientation for each bird in a test is calculated. Results from different individuals of an experimental condition can later be pooled for which the mean orientation is calculated. Examples of funnel experiments performed at the Geomagnetic North Pole are given in Figure 7.7.

7.7 Modeling migration

Migration behavior, including stopover duration, optimal fuel loads, the use of winds and physiological flexibility can be predicted using simple optimality models (Alerstam and Hedenström 1998 for review). Predictions derived from different currencies, such as energy or time minimization and maximizing survival, often differ and which strategy is adopted can be tested. Predictions regarding the optimal flight behavior have also been derived. At present, there are more theories than there are critical tests of their assumptions and predictions, and so there are plenty of research opportunities.

Dynamic or state-dependent optimization is used to derive optimal migration polices and the annual routine programs of migratory birds (Clark and Mangel 2000; Houston and McNamara 2000), including the scheduling of migration, breeding, and molt. Evolutionary population models have been used for understanding the evolution of migration and to explore the effects of habitat loss in migratory birds (Sutherland 1996).

7.8 Concluding remarks

The study of bird migration is a truly interdisciplinary research effort including ideas and techniques from many different disciplines. It requires both theoretical and experimental development and a great challenge for new students. We now understand a great deal about the evolution and ecology of migration, but there are still many mysteries awaiting solution. These problems can be tackled either in the laboratory or in the field.

Acknowledgements

We are very grateful to R. Green, I. Newton, and W. Sutherland for comments on the manuscript.

References

Åkesson, S. (1993). Coastal migration and wind compensation in nocturnal passerine migrants. *Ornis Scand.*, 24, 87–94.

Åkesson, S., Hedenström, A., and Alerstam, T. (1996). Flight initiation of nocturnal passerine migrants in relation to celestial orientation conditions at twilight. *J. Avian Biol.*, 27, 95–102.

Åkesson, S., Morin, J., Muheim, R., and Ottosson, U. (2001*a*). Avian orientation at steep angles of inclination: experiments with migratory white-crowned sparrows at the magnetic North Pole. *Proc. R. Soc. Lond. B*, 268, 1907–1913.

Åkesson, S., Walinder, G., Karlson, L., and Ehnbom, S. (2001*b*). Reed warbler orientation: initiation of nocturnal migratory flights in relation to visibility of celestial cues at dusk. *Anim. Behav.*, 61, 181–189.

Alerstam, T. (1990). *Bird Migration*. Cambridge University Press, Cambridge.

Alerstam, T. and Hedenström, A. (1998). The development of bird migration theory. *J. Avian Biol.*, 29, 343–369.

Argos (1996). User's manual 1.0. CLS/Service Argos. Landover, Maryland, USA.

Batschelet, E. (1981). *Circular Statistics in Biology*. Academic Press, New York.

Bensch, S., Åkesson, S., and Irwin, D.E. (2002). The use of AFLP to find an informative SNP: genetic differences across a migratory divide in willow warblers. *Mol. Ecol.*, 11, 2359–2366.

Berthold, P. (1996). Control of Bird Migration. Chapman & Hall, London.

Berthold, P., Elverfeldt, D., Fiedler, W., Hennig, J., Kaatz, M., and Querner, U. (2001). Magnetic resonance imaging and spectroscopy (MRI, MRS) of seasonal patterns of body composition: a methodological pilot study in white storks (*Ciconia ciconia*). *J. Ornithol.*, 142, 63–72.

Bruderer, B. (1997). The study of bird migration by radar. Part 1: The technical basis. *Naturwissenschaften*, 84, 1–8.

Chamberlain, C. P., Bensch, S., Feng, X., Åkesson, S., and Andersson, T. (2000). Stable isotopes examined across a migratory divide in Scandinavian willow warblers (*Phylloscopus trochilus trochilus* and *Phylloscopus trochilus acredula*) reflect their African winter quarters. *Proc. R. Soc. Lond. B*, 267, 43–48.

Clark, C.W. and Mangel, M. (2000). *Dynamic State Variable Models in Ecology: Methods and Applications*. Oxford University Press, New York.

Dial, K., Biewener, A.A., Tobalske, B. W., and Warrick, D.R. (1997). Mechanical power output of bird flight. *Nature*, 390, 67–70.

Emlen, S.T. and Emlen, J.T. (1966). A technique for recording migratory orientation of captive birds. *Auk*, 83, 361–367.

Fransson, T. and Pettersson, J. (2001). *Swedish Bird Ringing Atlas*. Vol. 1. Örebro: Naturhistoriska Riksmuséet and Sveriges Ornitologiska Förening.

Gauthreaux, S.A. (1969). A portable ceilometer technique for studying low-level nocturnal migration. *Bird Banding*, 40, 309–320.

Gauthreaux, S.A., Jr., Bleser, C.G., and Van Blaricom, D. (2003). Using a network of WSR-88D weather surveillance radars to define patterns of bird migration at large spatial scales. In *Avian Migration*, eds. P. Berthold, E. Gwinner, and E. Sonneschein, pp. 335–346. Berlin: Springer.

Hedenström, A. and Pettersson, J. (1987). Migration routes and wintering areas of willow warblers. *Phylloscopus trochilus* (L.) ringed n Fennoscandia. *Ornis Fennica*, 64, 137–143.

Hedenström, A. and Alerstam, T. (1996). Optimal flight speeds for flying nowhere and somewhere in the skykark. *Alauda arvensis*. *Behav. Ecol.*, 7, 121–126.

Hedenström, A. and Alerstam, T. (1997). Optimum fuel loads in migratory birds: distinguishing between time and energy minimization. *J. Theor. Biol.*, 189, 227–234.

Houston, A. I. and McNamara, J. M. (2000). *Models of Adaptive Behaviour: An Approach Based on State*. Cambridge University Press, Cambridge.

Jenni-Eiermann, S., Jenni, L., Kvist, A., Lindström, Å., Piersma, T., and Visser, H.G. (2002). Fuel use and metabolic response to endurance exercise: a wind tunnel study of a long-distance migrant shorebird. *J. Exp. Biol.*, 205, 2453–2460.

Kjellén, N. (1992). Differential timing of autumn migration between sex and age groups in raptors at Falsterbo, Sweden. *Ornis Scand.*, 23: 420–434.

Lajtha, K. and Michener, R. H. (eds) (1994). *Stable Isotopes in Ecology and Environmental Science*. Blackwell, Oxford.

Lichtenbelt, W. D. van Marken (2001). The use of bioelectrical impedance analysis (BIA) for estimation of body composition. In *Body Composition Analysis of Animals: A Handbook of Non-destructive Methods* ed. J.R. Speakman, pp. 161–187. Cambridge University Press, Cambridge.

Liechti, F., Bruderer, B., and Paproth, H. (1995). Quantification of nocturnal bird migration by moonwatching: comparison with radar and infrared observations. *J. Field Ornithol.*, 66, 457–468.

Lindström, Å., Kvist, A., Piersma, T., Dekinga, A., and Dietz, M.W. (2000). Avian pectoral muscle size rapidly tracks body mass changes during flight, fasting and fueling. *J. Exp. Biol.*, 203, 913–919.

Lowery, G.H. and Newman, R.J. (1966). A continentwide view of bird migration on four nights in October. *Auk*, 83, 547–586.

Marra, P.P., Hobson, K.A., and Holmes, R. T. (1998). Linking winter and summer events in a migratory bird by using stable-carbon isotopes. *Science*, 282, 1884–1886.

Nagy, T.R. (2001). The use of dual-energy X-ray absorptiometry for the measurement of body composition. In *Body Composition Analysis of Animals: A Handbook of Non-Destructive Methods*, ed. J. R. Speakman, pp. 211–229. Cambridge University Press, Cambridge.

Nisbet, I.C.T. (1959). Calculation of flight directions of birds observed crossing the face of the moon. *Wilson Bull.*, 71, 237–243.

Pennycuick, C.J. (1982). The ornithodolite an instrument for collecting large samples of bird speed measurements. *Phil. Transact. R. Soc. Lond.* B, 300, 61–73.

Pennycuick, C.J., Alerstam, T., and Hedenström, A. (1997). A new low-turbulence wind tunnel for bird flight experiments at Lund University, Sweden. *J. Exp. Biol.*, 200, 1441–1449.

Pennycuick, C.J., Hedenström, A., and Rosén, M. (2000). Horizontal flight of a swallow (*Hirundo rustica*) observed in a windtunnel, with a new method for directly measuring mechanical power. *J. Exp. Biol.*, 203, 1755–1765.

Piersma, T. and Lindström, Å. (1997). Rapid reversible changes in organ size as a component of adaptive behaviour. *Trends Ecol. and Evolut.*, 12, 134–138.

Sandberg, R. and Gudmundsson, G.A. (1996). Orientation cage experiments with Dunlins during autumn migration in Iceland. *J. Avian Biol.*, 27, 183–188.

Schaub, M., Pradel, R., Jenni, L., and Lebreton, J.-D. (2001). Migrating birds stop over longer than usually thought: an improved capture-recapture analysis. *Ecology*, 82, 852–859.

Scott, I., Selman, C., Mitchell, P. I., and Evans, P. R. (2001). The use of total body electrical conductivity (TOBEC) to determine body composition in vertebrates. In *Body Composition Analysis of Animals: A Handbook of Non-destructive Methods,* ed. J.R. Speakman, pp. 127–160. Cambridge University Press, Cambridge.

Speakman, J.R., Visser, G.H., Ward, S., and Król, E. (2001). The isotope dilution method for the evaluation of body composition. In *Body Composition Analysis of Animals: A Handbook of Non-destructive Methods,* ed. J.R. Speakman, pp. 56–98. Cambridge University Press, Cambridge.

Starck, J.M., Dietz, M.W., and Piersma, T. (2001). The assessment of body composition and other parameters by ultrasound scanning. In *Body Composition Analysis of Animals: A Handbook of Non-destructive Methods,* ed. J.R. Speakman, pp. 188–210. Cambridge University Press, Cambridge.

Sutherland, W.J. (1996). *From Individual Behaviour to Population Ecology*. Oxford University Press, Oxford.

Tobalske, B.W., Hedrick, T.L., Dial, K.P., and Biewener, A.A. (2003). Comparative power curves in bird flight. *Nature*, 421, 363–366.

Unangst, E.T. Jr. and Merkley, L.A. (2002). The effects of lipid location on non-invasive estimates of body composition using EM-SCAN technology. *J. Exp. Biol.*, 205, 3101–3105.

Webster, M.S., Marra, P.P., Haig, S.M., Bensch, S., and Holmes, R.T. (2002). Links between worlds: unraveling migratory connectivity. *Trends Ecol. Evol.*, 17, 76–83.

Wennerberg, L. (2001). Breeding origin and migratory pattern of dunlin (*Chalidris alpina*) revealed by mitochondrial DNA analysis. *Mol. Ecol.*, 10, 1111–1120.

Wernham, C., Toms, M., Marchant, J., Clark, J., Siriwardena, G., and Baillie, S. (2002). *The Migration Atlas: Movements of the Birds of Britain and Ireland*. T. & A. D. Poyser, London.

Wiltschko, R. and Wiltschko, W. (1995). *Magnetic Orientation in Animals*. Springer, Berlin.

Zehnder, S., Åkesson, S., Liechti, F., and Bruderer, B. (2001). Nocturnal autumn bird migration at Falsterbo, South Sweden. *J. Avian Biol.*, 32, 239–248.

8

Information from dead and dying birds

John E. Cooper

8.1 Introduction

From time to time the avian biologist has access to dead or dying birds. If properly investigated, these birds can provide information that is either of biological importance (e.g. morphometrics, molt patterns, samples for DNA) or of relevance to studies on health and disease (e.g. presence of pathological lesions, toxic residues).

Insofar as the latter category is concerned, dead or dying birds can be used: (a) to ascertain the cause of disease or death in that individual or population—a "diagnostic" investigation, or (b) to provide data on background health status, for example, presence/absence or numbers of parasites, underlying pathology, body condition—so called "health monitoring."

These two activities may sound similar but they differ in orientation and value. Diagnosis is essentially a *veterinary* task, aimed at trying to detect and determine a disease, often with a view to treatment or control. Health monitoring, on the other hand, is a more broadly based concept that is concerned with developing a database of factors that might be influencing the survival of individuals or the status of a population while not necessarily causing clinical disease or death. Such monitoring is usually an *interdisciplinary* task, with an input from biologists and others as well as from veterinarians (Cooper 1989, 2002) and is increasingly relevant to conservation programs for threatened species (Woodford 2001).

The investigation of birds following an unexplained "die-off" (mortality on a large scale) is often a combination of a "diagnostic" investigation and a study aimed at "health monitoring." The former, diagnostic work, may provide a cause of death, and often this is not the most important finding from an ecological point of view. Thus, many wild birds die of starvation which is fairly readily diagnosed by an experienced veterinary or wildlife pathologist, but the important question in such cases is, usually, what led to the starvation? Often the answer to the latter is complex. Thus, initial investigation of a group of waders found dead on the sea coast may reveal that the proximate cause of death was emaciation and hypothermia, but more detailed investigation may indicate that the underlying factors leading to that state of affairs include large numbers of parasites in the intestine, renal (kidney) disease, and exposure to unusually severe weather. An alternative scenario is that a bird is found to be heavily infested with internal parasites but this is because it has had to change its diet—perhaps because of climatic change or paucity of the usual prey species—and as a result has acquired the parasites from a novel food item. Here the parasites have caused death but they were secondary to other factors. In such cases, all available information needs to be obtained from the carcasses and must then be put together with relevant background history in order to build up a picture. Understanding the death of wild birds often resembles a jig-saw puzzle; no one piece provides the full answer, and the puzzle is only complete when all the portions have been put together.

Whether one is involved in diagnostic pathology or *postmortem* health monitoring, the ability to examine a dead bird correctly, in a systematic and reproducible way, to describe what is found and then to take and submit specimens, is of the utmost importance. This point is reiterated later. Dead birds can be of any age. Nestlings and fledglings present various challenges. The investigation of eggs and embryos needs particular skill and experience. In this chapter all these procedures, together with the collection and storage of samples, will be discussed.

There are many sources of dead and dying birds. Sometimes individuals are submitted by concerned members of the public. On other occasions, the morbidity or mortality rate is so high that a concerted effort is made to collect specimens for examination. From time to time researches request carcasses for specific studies—for instance, to search for viruses such as that which causes West Nile disease, or to investigate causes of death and pathology of a particular species (Cooper 1993a). Live birds are sometimes taken into wildlife rehabilitation centers and these too can provide useful information if subjected to systematic examination and appropriate sample-taking (Cooper 2003a).

Clearly, in such cases, there is often bias in sampling. As a general rule, members of the public are more inclined to report dying or dead birds of a popular or rare species than pests, but it is always important to obtain as much background information as possible. There is merit in visiting the site oneself and collecting bodies or samples rather than relying on material that is sent in. The presence of live birds, even if they are moribund, may permit the taking of blood and other *antemortem* samples and this, coupled with full necropsy and other laboratory investigations, can provide much valuable information (see Tables 8.4 and 8.5).

The killing of birds for examination and sampling is generally discouraged, especially in countries such as Britain where, in addition to legal protection for most avian species, there is a strong body of public opinion about animal issues and also considerable confusion about what does or does not compromise welfare. However, there are always exceptions. If birds are dying in large numbers, particularly if they appear to be in pain or distress, there may be public support for killing them on humanitarian grounds. In some countries, pest species are culled in large numbers—well-known examples are the control programs for Galah Cockatoos *Cacatua rosiecapilla* in Australia and Quelea *Quelea quelea* in Africa—and often relatively little public concern is expressed. In these and similar instances, obtaining freshly dead specimens is usually not difficult. More often, however, the investigator has to make use of opportunistic sampling, collecting birds as and when they become available or relying on chance submissions.

8.1.1 Terminology—definitions and explanation

Before discussing *postmortem* examination and associated techniques, it is important to define terms. Ornithologists do not always use words and phrases relating to health, disease, and pathology in the same way as do veterinarians and other medically trained personnel. Such differences can lead to confusion, especially in interpreting findings that are reported in literature, and some such misunderstandings may take years to resolve (Cooper 1993*b*).

First and foremost, a "*postmortem* examination" is synonymous with a "necropsy" and the two terms will be used interchangeably in this chapter. The word "autopsy" is, in English, usually reserved for the *postmortem* examination of a human being: in French and certain other languages it can be used for any species. Other important terms that have specific meanings and are relevant to this chapter are listed in Table 8.1. Some others appear later in the text.

8.1.2 Methodology—an overview of techniques

Three main aspects of methodology relate to the examination of dead and dying birds:

(i) the killing (euthanasia) of sick, sometimes moribund, birds;
(ii) the gross *postmortem* examination of dead adult or immature birds and of eggs;
(iii) the taking of samples from birds or eggs for laboratory examination.

As was pointed out above, the killing of birds is sometimes necessary, either on humanitarian grounds or because material is needed for investigation. Whether

Table 8.1 *Some medical and pathological terms*

Medical term	Meaning
Clinical signs	The features of a disease that can be observed—for instance, lameness, diarrhea, etc. (see other examples below)
Symptoms	The features of a disease that are experienced and can be recounted by a human patient—for instance, giddiness, abdominal pain. DO NOT USE FOR ANIMALS
Anorexia	Absence of appetite
Dyspnea	Difficult breathing
Tachypnea	Rapid breathing
Dysphagia	Difficulty in swallowing
Diarrhea	Loose feces
Dysentery	Blood in feces
Edema	Abnormal accumulation of fluid under the skin or elsewhere
Hyperemia	Increase in blood supply
Atrophy	Decrease in size of a tissue or organ
Hypertrophy	Increase in size of a tissue or organ
Prognosis	Forecast of the probable course of a disease
Pathology	The study of disease
Etiology	The study of the cause of the disease
Disease	Disordered state of an organism or organ
Infection	The entry of an organism into a susceptible host in which it may persist, but detectable clinical or pathological effects may or may not be apparent
Infectious disease	A disease caused by the actions of a living organism (virus, bacterium, etc.), as opposed to physical injuries or endocrinological disorders or genetic abnormalities
Latent infection	An inapparent infection in which the pathogen persists within a host, but may be activated to produce clinical disease by such factors as stress or impaired host resistance
Pathogen	An organism capable of producing disease
Lesion	An abnormality caused by disease of a tissue; usually it is characterized by changes in appearance of that tissue, for example, a raised nodule on the skin
Focus (plural "foci")	A small, usually distinct, lesion such as a micro-abscess in the liver
Toxemia	A toxin (poison) is present in the blood
Bacteremia	Bacteria are present in the blood
Viremia	A virus is present in the blood
Parasitemia	A parasite is present in the blood
Septicemia	Multiplication of organisms in the blood, usually with pathological effects on organs
Incubation period	The time between acquisition of infection and onset of clinical signs
Mortality rate	The proportion of deaths during a given time
Morbidity rate	The proportion of clinical cases during a given time
Incidence	The number of new cases of a particular disease during a stated period of time
Prevalence	The total number of cases of a particular disease at a given moment of time

in the case of free-living (wild) birds this can be done without a licence or express permission from the relevant authorities depends upon the country or region, the species and the circumstances Whatever the background, the aim must be to kill the bird in a way that is: (a) humane, that is, causes minimal pain, stress or fear; (b) legal; (c) aesthetically acceptable; and (d) least likely to have an adverse effect on subsequent *postmortem* or laboratory investigations (Cooper 1987).

Before embarking on a program of euthanasia (culling), a protocol should be prepared following consultation with others. Relevant published information should be consulted and effort made to ensure that the method followed—and the circumstances under which it is used—conforms to the above four requirements. Whether a method is legally acceptable will relate not only to where the killing is to be carried out but also to the purpose. Thus, for example, in the United Kingdom birds that are part of a research project under The Animals (Scientific Procedures) Act 1986 may only be euthanased using a method listed in Schedule I of the Act. This does not, however, preclude the use of other methods if the birds are not part of such a licensed project—as would apply, for example, to sick or injured specimens that are being culled on welfare grounds.

Table 8.2 lists some methods of killing birds, with comments on each technique. Most types of euthanasia require experience if they are to be carried out humanely, with minimal danger to the handler. It is wise to perfect techniques first on a dead bird. Training in methods of euthanasia can usually be obtained from a veterinary surgeon, a veterinary nurse, an animal technician, or sometimes an aviculturist. In the UK most veterinary practices will euthanase a wild bird free-of-charge, so long as they are not also expected to dispose of the carcass.

Further information about euthanasia is to be found in the References and Further Reading.

8.2 The *postmortem* examination

It is always important to plan well. A hastily performed necropsy, with inadequate notes or unrepresentative samples, may yield data that are confusing, useless, or even erroneous.

The steps that should be taken in preparing for a *postmortem* examination are as follows:

1. Decide why the necropsy is to be carried out. The various types of *postmortem* examination, which have different objectives, are summarized in Table 8.3.

Table 8.2 *Methods of euthanasia of birds*

	Method	Comment
Physical[a]	Dislocation of neck	Generally for larger birds (waterfowl, herons, etc.) of up to 3 kg only Needs training or experience
	Pressure on sternum	For small birds (less than 100 g in weight) only
	Striking the cranium on a hard surface	Must be immediate. For small birds (up to 23 g)
	Other	Large birds can be killed by striking on the head with a suitable heavy instrument but skill is needed
Chemical[b]	Overdose of injectable anesthetic agent such as sodium pentobarbitone	For best results a barbiturate should be given intravenously but this requires skill. Intraperitoneal (intracoelomic) administration is easier but death is not instantaneous and internal organs may be damaged
	Overdose of an inhalation anesthetic agent, for example, halothane, isoflurane	Requires the use of an anesthetic chamber for small birds or a mask for larger birds
	Exposure to 100% carbon dioxide	As above (chamber). Suitable for large numbers of small birds

[a] All physical methods cause damage, which may hamper *postmortem* examination, and may prove aesthetically unacceptable.
[b] The majority of chemical methods require the possession of potent anesthetic agents, most of which can only be obtained on veterinary prescription. Argon gas is now being used instead of carbon dioxide in UK poultry hatcheries for reasons of human health.

2. Check that appropriate facilities and equipment are available, including protective clothing and other means of reducing the risk of spread of infectious disease (see later).
3. Be sure that the person carrying out the *postmortem* examination is sufficiently knowledgeable about the techniques.
4. Do whatever "homework" is possible beforehand—for example, by obtaining relevant information about the normal anatomy of the species (Harcourt-Brown 2000: King and McLelland 1984) or its biology and natural history (Cooper 2003*b*). Seek advice if necessary.

Table 8.3 *Types of postmortem examination*

Purpose	Category	Comment
To determine the cause of death	Diagnostic	Routine diagnostic techniques are followed
To ascertain the cause of ill-health (not necessarily the cause of death)	Diagnostic/ health monitoring	Usually routine—but detailed examinations and laboratory tests may be needed to detect nonlethal changes
To provide background information on supposedly normal birds on the presence or absence of lesions, parasites, or of other factors, such as fat reserves or carcass composition	Health monitoring	As above. Must be methodical if information is not to be missed
To provide information for a legal case or similar investigation—for instance, on the circumstances of death or the possibility that the bird suffered pain or distress while it was alive	Forensic/legal	Can be very different from the categories above. The approach depends upon the questions asked. There must be a proper "chain of custody" and all material must be retained
For research purposes, such as removal of tissue samples or examination of organs	Investigative	Depends upon the requirements of the research worker

8.3 Health and safety

Dead and dying birds can present hazards to those who are involved in the investigations. Sometimes the dangers are physical—for example, the risk of injury while capturing live birds or retrieving dead ones from marshes—or chemical, because of contact with formaldehyde—but the most important category are the "zoonoses." These are often defined—for example, by the World Health Organization (WHO)—as "diseases and infections that are naturally transmissible between vertebrate animals and humans" but increasingly there is a tendency to consider zoonoses as any disease or infection that can be acquired by humans from animals (for a review of zoonotic infections, see Cooper 1990 and more recent specific publications, relating to "new" hazards such as West Nile virus). A useful general reference text on zoonoses, which includes data on both animals and humans, is the book edited by Palmer *et al.* (1998).

With zoonoses the picture is constantly changing. Infectious agents that were once not considered to be important in humans are now recognized as being potentially pathogenic. Many of these "opportunists" take advantage of a debilitated host—in particular, a person who is immunosuppressed. Immunosuppression in humans can result from an infectious disease (HIV/AIDS is the best known example but there are many others, such as malaria), malnutrition or some form of medication which is reducing the immune response. It is therefore wise to assume that sick or dead birds represent a source of pathogens for humans. If this precautionary approach is followed, and appropriate safeguards are taken, the risks involved in carrying out an examination of the bird will usually be low.

The specific precautions that should be implemented to minimize the spread of zoonotic infections depend on the circumstances. In some countries health and safety legislation may require the employer of those embarking upon *postmortem* examinations or sample-taking to compile a risk assessment before the work is done. The avian biologist or veterinarian who is likely to be involved in such work will need to follow the rules and to take appropriate precautions. In other countries, adherence to the same level of risk assessment and protection of staff will probably not be possible. Nevertheless, the scientist who is involved in such work, regardless of the country involved, has a moral responsibility for assistants and other staff, and it is therefore wise to draw up and adhere to a code of practice aimed at minimizing the risk of infection. A similar approach is usually needed in the field, where good facilities for the examination of birds are seldom available and improvisation is necessary. Again, in some countries there are health and safety obligations on employers and employees relating to field work but elsewhere these may not exist and instead a voluntary code of practice may need to be instigated (Cooper 1996). This is further discussed later, under 8.7 (Legal Aspects).

Basic rules that will help to reduce the spread of zoonoses are as follows:

- Be aware of the risks from birds, by reading the literature and by consulting veterinarians, physicians, and others who have the appropriate knowledge
- Familiarize colleagues and others involved in the study with the possible hazards
- Draw up a risk assessment, even if this is not a statutory requirement where you are working; follow this with a Standard Operating Procedure (SOP) in order to reduce the risks to a minimum
- Ensure that you and your colleagues understand how microorganisms can be spread and how best such spread may be prevented
- Operate the "clean/dirty" principle; whereby during a *postmortem* examination one person keeps "clean" and handles only the note-books, tape recorder, and the outside of specimen bottles, while the other is "dirty" and comes into direct contact with the dead bird and its tissues.

- Seek professional advice where the risk appears to be substantial—if for example, chlamydiosis (chlamydophilosis) is suspected or if the picture suggests a new or unusual infectious agent.
- Always follow basic hygienic precautions, such as the wearing of gloves and the use of disinfectants, and be prepared to invest in specialized equipment and facilities (which can range from facemasks to safety cabinets) if this is deemed necessary.
- If, during the course of a *postmortem* examination, an accident occurs (e.g. splashing of possibly infected material on to the face), seek medical advice. Likewise, if clinical signs that may be suggestive of a zoonotic infection occur following a necropsy, report this and detail the work that you have been doing.
- Consider producing a card that can be shown to and read by a member of the medical profession, explaining that the holder comes into contact with sick or dead birds and, therefore, in the event of an unexplained fever, or other clinical signs, might be infected with an avian pathogen.
- Remember if working overseas or as part of an international team that there may be barriers on account of language, especially amongst support staff. Consider using diagrams and other visual aids in protocols and literature as well as the written word.

8.4 *Postmortem* examinations (necropsies)

Many methods have been advocated for the *postmortem* examination of birds. Some have been devised by veterinarians, usually specifically for the diagnosis of disease (Wobeser 1981; Hunter 1989; Cooper 2002), while others have originated from ornithologists who have been either interested in mortality amongst wild birds or have developed necropsy methods in order to obtain samples for research (van Riper and van Riper 1980). A basic technique developed for those working in the field, especially in areas where access to professional advice is limited, was published by ICBP (now Bird Life International) 20 years ago (Cooper 1983).

The actual method of necropsying birds is not important so long as it is efficient and reproducible. *A postmortem* examination is not just a matter of "opening up the body"; it is a structured process, which involves both external and internal observation and, if the circumstances warrant, detailed investigation of some organs and tissues. Young birds and eggs/embryos require special techniques (see later).

A comprehensive necropsy examination will encompass features of both a "diagnostic" and a "health monitoring" investigation. As a result, it will encompass a whole range of tests and analyses, in addition to weighing and detailed measuring (see later). It can be time consuming. There is a place for such detailed work—for instance, when birds of a rare or threatened species are involved—but often it is not feasible for the avian biologist or veterinarian to carry out a lengthy examination on each bird that is presented.

A *postmortem* examination form, designed to be comprehensive, is given in Appendix I.

Such a comprehensive examination involves a whole range of investigations. It may be best carried out by a team of people but, however, this is often not feasible, especially in the field.

An abbreviated *postmortem* examination protocol, which usually proves successful for preliminary investigation, is given below:

1. On reception of the specimen, record the history and give the bird a reference number—always good practice and an essential precaution (to facilitate "chain of custody") if legal action is possible.
2. Examine the bird externally (including beak, buccal cavity, auditory canal, preen gland, and cloaca)—record (and quantify) any parasites, lesions, or abnormalities. Comment on plumage and molt, using a standard system—for example, that advocated by the British Trust for Ornithology (Ginn and Melville 1983).
3. Weigh and record standard measurements such as wing chord (carpus), tarsus, culmen, combined head and bill length and sternum (see also Chapter 4). A bird's bodyweight (mass) is of limited value without a measurement of linear dimensions.
4. Open the bird from the ventral surface by lifting or removing the sternum—examine superficial internal organs (see Figure 8.1). Record any lesions or abnormalities.
5. Remove and set aside heart, liver and gastrointestinal tract (ligate oesophagus and rectum to prevent spillage of contents) and examine deeper internal organs (see Figure 8.2). Record any lesions or abnormalities.
6. Fix in buffered formalin (10%) small portions of lung, liver, kidney plus any organ or tissue showing an abnormality (see Table 8.5).
7. Open portions of intestine and look with naked eye or hand lens for food, other material (e.g. pellets), lesions and parasites. Examination is facilitated if the material is placed in a Petri dish with a little saline. Save any parasites and make an effort to quantify them, for example, by estimating the proportion of the intestine examined and counting the number of parasites seen.

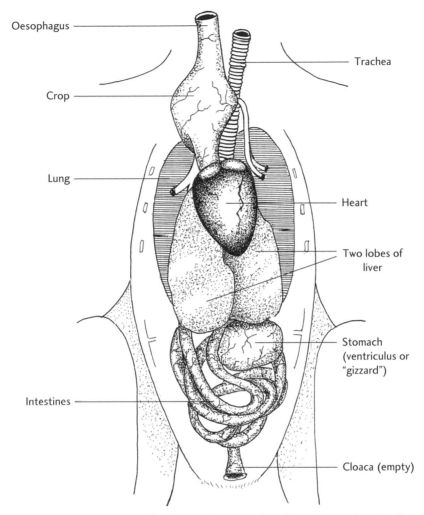

Oesophagus

Trachea

Crop

Lung

Heart

Two lobes of liver

Stomach (ventriculus or "gizzard")

Intestines

Cloaca (empty)

Fig. 8.1 Appearance of superficial internal organs when the sternum is first lifted (drawing by Pamela Athene Smith).

8. After examination save the bird's carcase, frozen or fixed in formalin (see later).
9. Record how and where the body and samples have been saved and include a reminder that they may need to be processed or discarded at a later date.

Again, health and safety considerations apply—see earlier. Some hazards are physical (e.g. a cut from a scalpel), others biological (e.g. inhalation of bacteria

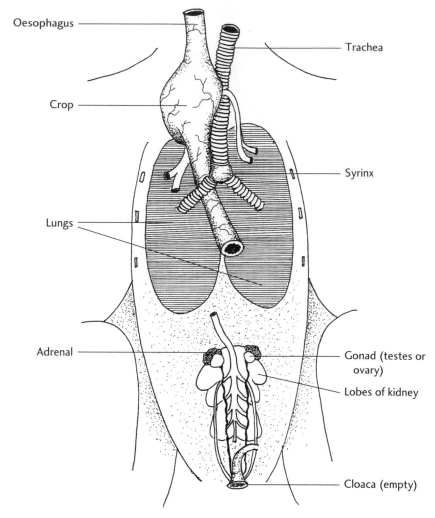

Oesophagus

Trachea

Crop

Syrinx

Lungs

Adrenal

Gonad (testes or ovary)

Lobes of kidney

Cloaca (empty)

Fig. 8.2 Appearance of deeper internal organs, following removal of heart, liver, and gastrointestinal tract (drawing by Pamela Athene Smith).

from an infected carcase), a few chemical (formalin is an irritant to eyes, mucous membranes, and skin and may be carcinogenic). Appropriate precautions must be followed.

Appropriate equipment should be used when carrying out a *postmortem* examination. The basics are a scalpel with blade, scissors, and forceps but for optimal results these should be of an appropriate size and type. Ophthalmological instruments may be needed when necropsying a warbler whereas heavy duty equipment is more likely to prove serviceable for a large seabird. Rat-toothed forceps are ideal for grasping tissues during dissection but can damage samples destined for the histology laboratory. A handlens or dissecting loupe is invaluable for the investigation of small birds or tiny lesions.

The examination of young, "neonatal," birds is not always straightforward. They are not just small or immature versions of the adult. The immune system is only just starting to develop and to respond to antigens in the environment. Powers of thermoregulation are usually poorly developed, especially in nidicolous species. Therefore, susceptibility to certain infectious agents, as well as to physical factors such as cold, may be enhanced. The person carrying out the examination must be aware of these differences from the outset, and investigation of the young bird should follow standard techniques for "neonates," similar to those originally developed for domestic poultry and now widely extended to psittacine and raptorial birds (Cooper 2002). An important feature of the necropsy of young birds is the examination, measuring, and sampling of the bursa of Fabricius: this organ, which lies adjacent to the cloaca, is a key part of the immune system and its investigation is vital if mortality and morbidity in young birds is to be fully investigated. The bursa and also the thymus should be examined, measured, and fixed in formalin for subsequent examination. If in doubt over the examination of young birds, the advice of an experienced avian pathologist should be sought—and this approach applies also to the investigation of eggs and embryos (see below).

The comprehensive examination of eggs is a highly specialized field. Much information has been gained over the past few decades from studies on the domestic fowl and other galliform birds, but also from other captive birds, including passerines, parrots, raptors, and owls (Cooper 2002, 2003*b*). The examination of eggs of wild birds often does not follow a standard protocol, and toxicologists tend to take samples in a different way from those who are interested in infectious disease, developmental abnormalities, or incubation failures. Space does not permit a detailed description of the specific techniques that should be followed in order to obtain maximum information from eggs and embryos, but a protocol is given (Appendix II), together with a specimen report form (Appendix III).

Measuring eggshell thickness is an important part of assessing eggs, whether or not they are fertile. Various methods can be employed; a useful index was described by Ratcliffe (1967) and this and other methods were explored and compared by Green (2000).

Key features of any *postmortem* examination are (a) the recording of all that is seen or done, (b) the taking of samples, and (c) the retention of material for subsequent study.

The objective of the person who is carrying out a *postmortem* examination should be to observe and to record. There are inherent dangers in attempting to interpret findings at an early stage or as the *postmortem* progresses. Something that may appear to be significant initially, such as damage to a pectoral muscle or a pale liver, may subsequently prove to be of little consequence because, for instance, bacteriological examination (the results of which may not be received for 3–4 days) reveals that the bird died of an overwhelming infection. It is therefore best to reserve judgment until all tests are complete. Too many investigations of avian mortality in the past have been hamstrung by premature judgements, based on inadequate information, regarding the likely cause of morbidity or mortality.

The assessment of "condition" has for long been controversial and yet is considered an important index in studies of survival and reproductive success (Bowler 1994; Fox *et al.* 1992; Moser and Rusch 1988). Those examining birds *post mortem* are usually anxious to give a "condition score" because this is part of evaluation and provide some indication as to the likely cause of death. Methods of assessing condition include:

1. Relating bodyweight (mass) to standard morphometrics—an important reason for measuring, as well as weighing, birds *post mortem* (but it must be remembered that carcasses suffer from gradual evaporation, and hence weight loss).
2. Assessing and scoring the amount of fat, both subcutaneous and internal.
3. Measuring muscle (especially pectoral) size, both macroscopically and histologically.
4. Whole body measurements, for example, the TOBEC system (Samour 2000).

Each of these methods—and, indeed, others—has its own devotees and which, if any, is used will depend upon the protocol being followed. However, it is most important that some assessment is made that might aid others in relating findings from one bird to another. Thus, measurements of carpus, tarsus, etc. should be routine, as well as bodyweight calculations, and there should be some scoring system for as many as possible of the other parameters, such as the quantity of fat that is visible or the size of pectoral muscles (for further discussion of body condition assessment, see Chapter 4).

Space does not permit detailed discussion of all body systems, but mention should be made of the reproductive tract because of the importance in ornithological studies of assessing and measuring breeding success (Newton 1998). Close examination of the reproductive system is therefore often useful. Sexing a bird should not be difficult but sometimes, especially if the bird is immature, if there is *postmortem* change, or if it is a non-breeder, the gonads may be difficult to see. The use of a handlens and strong reflected light can help but if this fails, a portion of the relevant tissue can be taken for sex determination by histological examination (the section often also contains adrenal gland, which can sometimes provide information about stress and endocrine function).

During necropsy, always note the appearance of the ovary or testes. In species such as falconiforms, the presence or absence of a vestigial right ovary should also be recorded. Whenever possible—and always when a series of birds is being examined—the size of the gonad(s) should be noted, by measuring, weighing, or scoring. The state of activity of the ovary (follicle development) is also important. The color of the testes should also be recorded: sometimes they are pigmented.

Other observations on the reproductive tract can provide useful information. Readily visible, well developed, oviducts usually indicate that a bird has laid eggs. For many species reliable data are lacking so again the size of the organ should be recorded by measuring, weighing or giving it a score.

Gross examination of the reproductive system can always be supplemented by histological examination. The gonad and tract, or parts of them, should be fixed in buffered formalin 10% and hematoxylin and eosin—stained sections prepared. Where there is particular interest in breeding history, reproductive organs can, after measuring and weighing, be fixed in formalin for study at a later date.

Weighing of organs—especially liver, heart, spleen, kidney, and brain is to be encouraged whenever possible. Organ : bodyweight ratio change can be a feature of some infectious and noninfectious (especially toxic) diseases.

A gross *postmortem* examination does not *per se* provide the answer to the bird's death or its failure to thrive, and the answer (the final pieces of the jig-saw, as described earlier) is the carrying out of laboratory tests, such as toxicology and bacteriology. The taking of samples is discussed in the next section.

The retention of material following *postmortem* examination is an important part of any investigation for several reasons:

1. It may be necessary to go back to the carcass in order to carry out additional investigations. This can happen, for example, if laboratory tests indicate that a bacteriological infection is involved, in which case samples taken can be cultured in order to identify those bacteria.
2. Carcasses or other material may be required for legal purposes—if, for example, a court action is to be bought in connection with the bird's death.
3. Material may be needed for research purposes. The requirements can range from whole bodies, study skins, or skeletons for museums, to the retention

of relevant samples for morphometrics or study of gross or microscopical morphology. In some cases, the bird's carcass and or tissues may be required for a "Reference Collection"—see later. The likely fate of carcasses, tissues, and specimens should, therefore, be assessed *before* the examination is carried out. Appropriate containers will be needed and a decision made as to how best to preserve the material. The latter is an important consideration; thus, for example, tissues for histology can be safely stored in 10% buffered formalin but this method of fixation may prove deleterious for studies on DNA. Freezing will preserve most microorganisms and poisons but will hamper subsequent histology or electronmicroscopy. Plastic bottles may produce erroneous results if used to store samples for certain toxicological analyses.

Facilities for storage of bodies and tissues may be limited, in which case a decision has to be made as to what is retained and for how long. As a general rule, following a *postmortem* examination, the bird's carcass and tissues can be kept in a refrigerator ($+4°C$) for up to 5 days, after which, if still needed, they should either be frozen ($-20°C$) or fixed in formalin or ethanol. Often it is wise to save aliquots of tissue using a combination of fixation methods, for the reasons cited above.

Material should be retained for future reference or retrospective studies whenever feasible (Cooper and Jones 1986; Cooper, Dutton, and Allchurch 1998). If a specific Reference Collection exists, the carcass should be fixed *in toto* in formalin other than small portions of tissue, for example, liver, which should be frozen or (less satisfactory) fixed in ethanol for DNA work. Complete carcasses are sometimes used for analysis, to measure body composition, as part of assessing "condition."

8.5 Laboratory investigations

Laboratory investigation of samples is a key part of *postmortem* examination of dead birds and investigation of sick or dying ones. A whole range of tests can be used and the choice depends upon:

(a) The indications in the field: birds found dead after a spillage of chemicals are likely to warrant toxicological examination rather than culture for bacteria, fungi, or viruses.

(b) The resources available. Many laboratory techniques are expensive and the cost of some may be prohibitive. Sometimes funding permits a selection of tests to be done on a proportion of the birds. Often samples have

Table 8.4 *Testing procedures for use on live and dead birds*

	Live birds	Dead birds
Clinical examination	+	−
Postmortem examination	−	+
Radiology	+	+
Hematology	+	+/−
Clinical chemistry	+	+/−
Microbiology	+	+
Toxicology	+/−	+
Histology	+/−	+
Electron microscopy	+/−	+
Chemical analysis of carcasses	−	+

Note: +/− = of limited value only.

to be stored in the hopes that they can be analyzed at a later date (see earlier). Those who work with birds should ensure that they have the correct containers and chemicals available and know how to store and use them properly. Some of this is relevant to human health and safety, as well as to practical considerations: thus, for example, glutaraldehyde (see Table 8.5), which can be hazardous, has to be stored at 4°C and will soon deteriorate. Alcohols and formaldehyde, which present different hazards, are more stable.

Lists (not comprehensive) of investigations that may be carried out on live and dead birds are given in Table 8.4.

Some of the main laboratory tests that can be used in avian work are listed in Table 8.5. For detailed descriptions the various references should be consulted. Many of the techniques that are listed need experience. Training in the preparation of, say, cytological preparations can be sought from an experienced veterinary or medical pathologist.

A difficulty is often how to decide which specimens to keep and how they should be preserved. The diagram that follows illustrates the range of possibilities and the varied methods used. When samples are small or limited, a "triage" system may need to be followed. Thus, if the history of dying waterfowl suggests a poisoning such as botulism, it may be wise to use scarce intestinal samples for toxin studies and to forego parasitological examination of them.

Table 8.5 *Laboratory tests on birds*

Samples	Available from	Comments
Blood in appropriate anticoagulant for hematological and clinical chemical analysis and detection of hemoparasites	Usually only from live birds, occasionally small samples can be retrieved from birds that are very recently dead	Various blood tests can be carried out on birds, and increasingly databases of reference values are being established. The subject is a specialized one and reference should be made to a standard text (Campbell 1995; Hawkey and Dennett 1989). Blood smears can be of value but, again, experience is needed to produce good preparations and the possibilities of error, especially when looking for and quantifying hemoparasites, are high (Cooper and Anwar 2001; Feyndich *et al.* 1995; Godfrey *et al.* 1987)
Blood without anti-coagulant (serum) for serological investigation	Usually only from live birds, occasionally small samples can be retrieved from birds that are recently dead	Serology, usually to detect antibodies to viruses and other organisms, has an important part to play both in disease diagnosis and health monitoring. It has recently, for example, been used to demonstrate neutralizing antibodies to West Nile virus in Britain (Buckley *et al.* 2003). Various serological tests are available and each demands skill in performance and interpretation. A rise in antibody titer is usually considered indicative of exposure to a specific organism. Such a rise usually takes time and may not be apparent in birds that have only recently contracted an infection
Tissues fixed in 10% formalin (preferably buffered) for histology	Dead birds, occasionally live (biopsies)—the latter usually only where the lesion is on the skin or is readily accessible surgically	Fixed tissues can be stored indefinitely and examined at a later stage. The general rule should be to take lung, liver, and kidney (LLK), plus any organs that show abnormalities or which are considered important because they may provide useful information (e.g. bursa of Fabricius and thymus of young birds, which can yield data on immune status—see text)

Sample type	Birds/source	Comments
Tissues fixed in glutaraldehyde for transmission electronmicroscopy (TEM)	As above	Samples should usually not exceed 20 mm × 20 mm and there must be at least 10 times the volume of fixative as there is tissue. Small carcasses can be fixed whole, following opening for processing Generally as above but only tiny samples are taken. Scanning electron microscopy (SEM) employs different techniques and is not considered here
Cytological preparations	As above (histology)	Easy to take, cheap to process (readily done in any veterinary practice or in the field), and produce rapid results. Usually consist of touch preparations/impression smears which can give valuable information about tissues within a few minutes. The samples must first be blotted on filter paper in order to remove excess blood
Swabs, organ/tissue samples, and other specimens for microbiological and other investigations	Live birds (superficial lesions, mouth or cloacal swabs, or dead birds but also samples from internal organs)	Usually comprise swabs (in transport medium if they are to be sent elsewhere), portions of tissue, or exudates/transudates (Hunter 1989; Scullion 1989). If culture proves impossible for financial or other reasons, an impression smear stained with Gram or other stains will often provide some useful information
Tissues for toxicological examination	Mainly dead birds but some small samples can be taken from live birds, for example, blood or muscle biopsies for certain pesticide analyses, feathers (for heavy metal and other analyses)	Toxic chemicals may have been the cause of death or could have contributed to the bird's ill-health, either directly or by increasing its susceptibility to infectious disease. Samples from wild bird casualties can be taken and stored routinely for toxicological analysis Samples for toxicology are usually kept frozen and can be analyzed at a later date. As with formalin-fixed samples, such specimens should be taken and stored even when there is no immediate prospect of their being analyzed

Table 8.5 *(continued)*

Samples	Available from	Comments
Droppings (mixture of feces and urates, as voided) for parasitology and other tests	Both live birds (recently voided droppings) and dead birds (removed from the cloaca *postmortem*)	Droppings provide a means of diagnosing some diseases and obtaining health monitoring data with minimal disturbance to the live bird. Droppings will often be passed when a bird is handled or restrained in a bag or net. The fecal component can be used to detect internal parasites, to provide information on other changes in the intestine (e.g. presence of blood, undigested food, etc.) or to investigate the origin of recently ingested food items. Feces can also be used to detect bacteria, fungi, and viruses. Molecular techniques, for example, PCR, are increasingly being used to detect the antigens of pathogenic organisms and to provide other information based on DNA technology. The urate component of feces can be used to investigate kidney function and may also yield parasites associated with the renal system. In all cases fresh samples provide the most reliable results
Stomach (or crop) contents	Usually from dead birds but stomach/crop washings can be obtained from live birds—or regurgitation can be stimulated by physical or chemical means. The cast pellets of birds of prey and certain other species can provide valuable information	As above (feces): also provide information on diet. Stimulation of regurgitation must be carried out with care and in occordance with any legal or ethical requirements

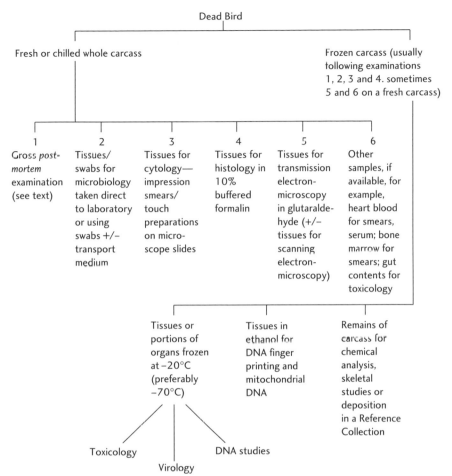

Sample taking during *postmortem* examination of birds.

8.6 Interpretation of findings

When investigating dying or dead birds, remember that a "diagnosis" is not necessarily the objective. As was explained earlier, the proximate cause of death may not be the most important finding, but the endogenous or exogenous factors, that *contributed* to the bird's demise. Apparently background findings of, for example, parasitism or inactive gonads may also be of relevance, especially when monitoring the health of an avian population.

Care must, therefore, be taken over terminology. A "diagnosis" is one thing, the "cause of death" another. All findings need to be interpreted in the context of the background, history, the circumstances under which the birds were found, the species (and sex/age ratio) and a multiplicity of other aspects (e.g. weather, reproductive activity) that may have played a part.

Sometimes the interpretation of findings presents few difficulties. For example, a swan that has flown into a power-line and broken its neck will show characteristic gross *postmortem* lesions—mainly hemorrhage. At other times, however, interpretation can be problematic or require lateral thinking. The swan that has struck a power-line may have done so because of an underlying infectious disease, such as avian tuberculosis, which has made it more susceptible to accident. For this reason, biologists often separate the proximate from the ultimate causes of death.

The finding of micro- or macroparasites on or in a bird can be misleading. Sometimes they have been acquired from elsewhere—for instance, prey species (e.g. lice from corvids on falcons) or contamination from other carcasses in the *postmortem* room. Even when such organisms are *bona fide* isolates, their relevance may not be clear. Intestinal worms associated with ulceration of a bird's intestine, or bacteria isolated from a swollen eye, are clearly of some significance, but what about the findings of these organisms *without* associated lesions? Are they of significance or not? Much remains to be learned about the biology of pathogens (Reece 1989) and host–parasite relations in birds (Cooper 2001) and, until reliable data are available, the best that can usually be done is to record the finding, both qualitatively and quantitatively, and to attempt to relate it to the bird's body condition and systemic health. Data on captive birds have proved of some value, as have findings in wild bird causalities (Cooper 2003*a*).

Useful publications regarding interpretation of laboratory findings in birds include those on histopathology by Randall and Reece (1996), on hematology (Campbell 1995; Hawkey and Dennett 1989) and on microbiology by Scullion (1989) and Cooper (2000).

Situations where birds are found dead or dying and where gross *postmortem* findings and laboratory investigations are unrevealing or confusing, often present a dilemma. For instance, gross necropsy, together with standard laboratory tests, may offer no specific diagnosis (cause of death) for a group of parrots found dead in a South American suburb. Additional tests may also provide no specific cause of death; heavy metal values, for example, might be elevated, but not to the extent that they can be considered lethal. In such cases, careful analysis of all the findings, together with extraneous factors such as climate, are essential but might still be inconclusive. The available data should be stored—as should a selection of samples—because later studies, perhaps using more sophisticated investigations, may provide an answer.

Interpretation of findings is also often hampered by the lack of reliable reference values. For example, although in recent years there have been great advances in our knowledge of the hematology and blood biochemistry of birds, the data available largely relate to species that are kept in captivity or have been subjected to detailed study in the wild. For the vast majority of the world's nearly 10,000 species of birds, there are no reliable reference values. Likewise, toxicological investigations can be thwarted because of a paucity of information on what are "normal" background values, what are sublethal and what are lethal for a given species. Extrapolation is sometimes possible and the best line of approach, but it is far from ideal.

The absence of some very basic data is a cause for concern. For instance, the normal ranges of organ weight and organ/body weight ratios of most species of birds are not known and yet such information could so easily be gathered if proper records were kept and findings freely disseminated. There is a great need to involve scientists of all disciplines, undergraduate and postgraduate students, and "amateur" naturalists in filling such gaps in our knowledge.

Comprehensive databases on different taxa of birds are much needed. These should encompass basic biological parameters as well as information about organisms (both macroparasites and microparasites) that have been associated with the species, diseases that have been diagnosed, and published and unpublished observations on pathology. A valuable initial step is the compilation of checklists of parasites of different species, genera, families, or orders, especially if linked with studies on the birds' biology (Storer 2000). Few publications on general ornithology include reference to the increasingly important role that infectious agents appear to play in free-living avian populations. Some otherwise authoritative works include a minimum amount of information about the health and disease of birds and often even fail to include references to standard texts on this subject (Marzluff and Sallabanks 1998).

8.7 Legal aspects

There is nothing to stop anyone from carrying out a *postmortem* examination. In some countries, however, including the United Kingdom and many other European and Commonwealth countries, the making of a formal diagnosis, even as a result of examining a dead bird, is restricted by law to the veterinary profession. This is another reason for advocating that those doing *postmortem* examinations of birds concentrate on recording what they see and not rushing into an interpretation or "diagnosis."

There are other aspects of examination that have a legal dimension and those involved in such work should be aware of this. Health and safety legislation may dictate how and where a *postmortem* examination is performed. Where a zoonotic disease is suspected, the legislation may demand that a risk assessment is carried out (see earlier) and, perhaps, that the necropsy is only performed if either: (a) appropriate protection—clothing, equipment, and facilities—is available for all those involved, and (b) the personnel are appropriately experienced or trained in the handling of dangerous microorganisms and disposal of clinical waste.

The number of zoonoses known to be contractable from birds is increasing, partly as a result of the rise in numbers of immunosuppressed human beings (see earlier). For example, the disease cryptococcosis, due to *Cryptococcus neoformans*, an organism harbored and disseminated by some wild birds, especially pigeons, is now much more frequently reported in humans, primarily those that are immunocompromised. Therefore, the assumption should be that an organism *is* a risk to humans rather than the converse.

Other legislation is also relevant to the examination of dying and dead birds. Many of these concern the movement of carcasses or specimens (Cooper 1987, 2000). In-country regulations usually relate primarily to postal requirements for adequate packing and transportation of what might be pathogenic material. When moving samples from one country to another, the situation becomes more complex because conservation legislation, especially CITES, may apply. The appropriate Ministry (Department) of Agriculture of the receiving country is likely to require documentation describing the type of material that is being transported, particularly its likely pathogenicity, and if the birds or samples in question are covered by CITES, there will be an additional need for permits.

The situation regarding the movement of small specimens, such as blood smears, or tissues for DNA study, is a cause of frustration for many who are involved in avian research who wish to send samples to colleagues or laboratories in other countries. Even the smallest sample can fall into the category of a "recognized derivative" under the CITES Regulations and thus require appropriate documentation and authorization. There have been strong moves in recent years to obtain an exemption for such material, especially if it is required for important diagnostic, forensic, or similar purposes. The Conference of the Parties of CITES continues to debate the issue and, at the time of writing, following the CITES Conference of the Parties in Santiago, Chile, 2002, the likely outcome seems to be the introduction of a "fast-track" system for small, but urgent, samples.

It is important that those involved with the investigation of dying and dead birds are familiar with the relevant legislation and work within its limitations. The reputation of ornithology is not served by ignoring or breaking the law, however tedious and inconvenient the rules may seem.

In many countries of the world, legislation relating to birds, their protection, health and safety, and movement of samples, is nonexistent or is poorly enforced. In such circumstances it is good practice to work to "in-house" protocols and to develop and use guidelines that, although not legally binding, help to ensure high standards. Such an approach, using tested codes of practice, does much to enhance the reputation of those involved in avian research (Cooper 1996).

8.8 Conclusions

Perhaps the most important point being made in this chapter is that the investigation of dying or dead birds requires careful planning, a systematic approach, collection and collation of *all* data, and close collaboration between researchers in different disciplines. Investigation of morbidity and mortality in birds has, in the past, frequently been hampered by a lack of liaison between avian pathologists and veterinarians. As a result, there has often been a divergence of techniques, of terminology and of research methods. Fortunately that situation is now changing (Cooper 1993*b*; Greenwood 1996).

Much remains to be learned about causes of morbidity and mortality in wild birds. Dying and dead specimens provide invaluable information. Such data must be used wisely and relevant material should be retained for subsequent examination if opportunities are not to be missed. It is hoped that this chapter will play a part in encouraging a more concerted approach in the future.

Acknowledgements
I am grateful to the Editors of this volume for inviting me to contribute. I should like to express my particular appreciation to Ian Newton, for his help and support over the years and for encouraging me to work with him and his colleagues at Monks Wood on diverse problems related to wild birds.

I am indebted to Pamela Smith for typing and preparing the manuscript and for the line drawings (Figs. 1 and 2) depicting internal anatomy. My wife, Margaret Cooper, contributed information on the law.

An early draft of this paper was read and commented upon by Dick Best, Nigel Harcourt-Brown, and Tony Turk. I am most grateful to them for their advice and suggestions.

References and further reading

Bowler, J.M. (1994). The condition of Bewick's Swans *Cygnus columbianus bewickii* in winter as assessed by their abdominal profiles. *Ardea*, 82(2), 241–248.

Buckley, A., Dawson, A., Moss, S.R., Hinsley, S.A., Bellamy, P.E., and Gould, E.A. (2003). Serological evidence of West Nile Virus, Usutu virus and Sindbis virus infection of birds in the UK. *J. General Virol.*, 84, 2807–2817.

Campbell, T.W. (1995). *Avian Haematology and Cytology.* Iowa State University Press, Ames.

Cooper, J.E. (1983). *Guideline Procedures for Investigating Mortality in Endangered Birds.* International Council for Bird Preservation, Cambridge.

Cooper, J.E. (1987). Pathology. In *Raptor Management Techniques Manual*, eds. B.A. Giron Pendleton, B.A. Millsap, K.W. Cline, and D.M. Bird. National Wildlife Federation, Washington DC.

Cooper, J.E. (1989). *Editor Disease and Threatened Birds.* Technical Publication No. 10. International Council for Bird Preservation (now Bird Life International), Cambridge.

Cooper, J.E. (1990). Birds and zoonoses. *Ibis*, 132, 181–191.

Cooper, J.E. (1993a). Pathological studies on the barn owl. In *Raptor Biomedicine*, eds. P.T. Redig, J.E. Cooper, J.D. Remple, and D.B. Hunter. University of Minnesota Press, Minneapolis.

Cooper, J.E. (1993b). The need for closer collaboration between biologists and veterinarians in research on raptors. In *Raptor Biomedicine*, eds. P.T. Redig, J.E. Cooper, J.D. Remple, and D.B. Hunter. University of Minnesota Press, Minneapolis.

Cooper, J.E. (2001). Parasites and birds: the need for fresh thinking, new protocols and co-ordinated research in Africa. *Ostrich Suppl.*, 15, 229–232.

Cooper, J.E. (2002). *Birds of Prey: Health & Disease.* Blackwell Science, Oxford.

Cooper, J.E. (2003a). Principles of clinical pathology and post-mortem examinations. In *Manual of Wildlife Causalities*. British Small Animal Veterinary Association, Gloucester.

Cooper, J.E. (2003b). *Captive Birds*. World Pheasant Association, Fordingbridge Hampshire.

Cooper, J.E. and Anwar, M.A. (2001). Blood parasites of birds: a plea for more cautious terminology. *Ibis*, 143, 149–150.

Cooper, J.E. and Jones, C.G. (1986). A Reference Collection of Endangered Mascarene Specimens. *Linnean*, 2, No. 3, 32–37.

Cooper, J.E., Dutton, C.J., and Allchurch, A.F. (1998). Reference collections in zoo management and conservation. *Dodo*, 34, 159–166.

Cooper, M.E. (1987). *An Introduction to Animal Law*. Academic Press, London and New York.

Cooper, M.E. (1996). Community responsibility and legal issues. *Semin. Avian Exot. Pet. Medi.*, 5(1), 37–45.

Cooper, M.E. (2000). Legal considerations in the international movement of diagnostic and research samples from raptors—conference resolution. In *Raptor Biomedicine III* eds. J.T. Lumeij, J.D. Rample, P.T. Redig, M. Lierz, and J.E. Cooper. Zoological Education Network, Lake Worth, Florida.

Feyndich A.M., Perce, D.B., and Godfrey, R.D. (1995). Hematozoa in thin blood smears. *J. Wildlife Dis.*, 31(3), 436–438.

Fox, A.D., King, R., and Watkin, J. (1992). Seasonal variation in weight, body measurements and condition of free-living teal. *Bird Stud.*, 39, 53–62.

Fudge, A.M. ed. (2000). *Laboratory Medicine. Avian and Exotic Pets*. WB Saunders, Philadelphia.

Ginn, H.B. and Melville, P.S. (1983). *Moult in Birds*. British Trust for Ornithology, Tring.

Godfrey, R.D., Fedynich, A.M., and Perce, D.B. (1987). Quantification of hematozoa in blood smears. *J. Wildlife Dis.*, 23(4), 558–565.

Greenwood, A.G. (1996). Veterinary support for *in situ* avian conservation programmes. Bird Conserv. Int. 6, 285–292.

Green, R.E. (2000). An evaluation of three indices of eggshell thickness. *Ibis*, 142, 676–679.

Harcourt-Brown, N. (2000). Birds of Prey. Anatomy, Radiology and Clinical Conditions of the Pelvic Limb. CD Rom. Zoological Education Network, Lake Worth, Florida, USA.

Hawkey, C.M. and Dennett, T.B. (1989). *A Colour Atlas of Comparative Veterinary Haematology*. Wolfe, London.

Hunter, D.B. (1989). Detection of pathogens: monitoring and screening programmes. In *Disease and Threatened Birds* ed. J.E. Cooper. Technical Publication No. 10. International Council for Bird Preservation (now Bird Life International), Cambridge.

King, A.S. and McLelland, J. (1984). *Birds, their Structure and Function*. Ballière Tindall, London.

Marzluff, J.M. and Sallabanks, R. eds. (1998). *Avian Conservation*. Research and Management. Island Press, Washington DC.

Moser, T.J. and Rusch, D.H. (1988). Indices of structural size and condition of Canada geese. *J. Wildlife Manage.*, 52(2), 202–208.

Newton, I. (1998). *Population Limitation in Birds*. Academic Press, London.

Palmer, S.R., Soulsby, Lord, and Simpson, D.I.H. eds. (1998). *Zoonoses*. Oxford University Press, Oxford.

Randall, C.J. and Reece, R.L. (1996). *Color Atlas of Avian Histopathology*. Mosby-Wolfe, London and Baltimore.

Ratcliffe, D.A. (1967). Decrease in eggshell weight in certain birds of prey. *Nature*, 215, 208–210.

Reece, R.L. (1989). Avian pathogens: their biology and methods of spread. In *Disease and Threatened Birds*, ed. J.E. Cooper. Technical Publication No. 10. International Council for Bird Preservation (now Bird Life International), Cambridge.

Samour, J.H. (2000). *Avian Medicine*. Mosby, London.

Scullion, F.T. (1989). Microbiological investigation of wild birds. In *Disease and Threatened Birds* ed. J.E. Cooper. Technical Publication No. 10. International Council for Bird Preservation (now Bird Life International), Cambridge.

Storer, R.W. (2000). The Metazoan Parasite fauna of Grebes (Aves: Podicipediformes) and its Relationship to the Birds' Biology. 90 pages. Miscellaneous Publications Museum of Zoology, University of Michigan no. 188. Ann Arbor: University of Michigan, USA.

van Riper, C. and van Riper, S.G. (1980). A necropsy procedure for sampling disease in wild birds. *Condor*, 82, 85–98.

Wobeser, G.A. (1981). Necropsy and sample preservation techniques. In *Diseases of Wild Waterfowl* ed. G.A. Wobeser, Plenum Press , New York.

Woodford, M.H. ed. (2001). Quarantine and Health Screening Protocols for Wildlife prior to Translocation and Release into the Wild. OIE, VSG/IUCN. Care for the Wild International and EAZWV. Paris, France.

Appendix I

PostMortem Examination (Necropsy) Form

Species _____   Reference No _____

Date of Submission_____   Origin _____

Ring (band) number _____   Any other identification _____

Relevant history / circumstances of death:

Request - diagnosis (cause of death/ill-health), health monitoring, forensic
investigation/research/other

Any special requirements re. techniques to be followed, instructions regarding fate of
body/samples

Submitted by _____   Date _____

Received by_____   Date _____

Measurements: Carpus   Tarsus   Other   Bodyweight (Mass)

Condition score: Obese or fat / good / fair or thin / poor

State of preservation: Good / fair / poor / marked autolysis

Storage since death: Refrigerator / ambient temperature / frozen / fixed

} A number ("score") can be used for these

EXTERNAL OBSERVATIONS, including preen gland, state of molt, ectoparasites, skin condition, lesions, etc.

MACROSCOPIC EVALUATION ON OPENING THE BODY, including
position and appearance of organs, lesions, etc.

ALIMENTARY SYSTEM

MUSCULOSKELETAL

CARDIOVASCULAR

RESPIRATORY

URINARY

REPRODUCTIVE

LYMPHOID (including bursa and thymus)

NERVOUS

} This section can be expanded, subheadings can be inserted, including checklist of organs and tissues

OTHER SAMPLES TAKEN

_____	Bact	Paras	Hist	DNA	Cytology	Other	(e.g. serology)
_____	Bact	Paras	Hist	DNA	Cytology	Other	(e.g. serology)
_____	Bact	Paras	Hist	DNA	Cytology	Other	(e.g. serology)
_____	Bact	Paras	Hist	DNA	Cytology	Other	(e.g. serology)
_____	Bact	Paras	Hist	DNA	Cytology	Other	(e.g. serology)
_____	Bact	Paras	Hist	DNA	Cytology	Other	(e.g. serology)
_____	Bact	Paras	Hist	DNA	Cytology	Other	(e.g. serology)

LABORATORY FINDINGS

Date: _____ Initials: _____ Reported to whom: _____

PRELIMINARY REPORT (based on gross findings and immediate laboratory results, for example, cytology)

Reported to _____ Date _____ Time _____

FINAL REPORT (based on all available information)

FATE OF BODY / TISSUES
destroyed / frozen / fixed in formalin(other) / retained for Reference Collection / sent elsewhere

FATE OF RING/BAND (if appropriate)

PM examination performed by: _____ Date _____ Time _____
Assisted by _____

Appendix II

Protocol for Examination of Unhatched Eggs

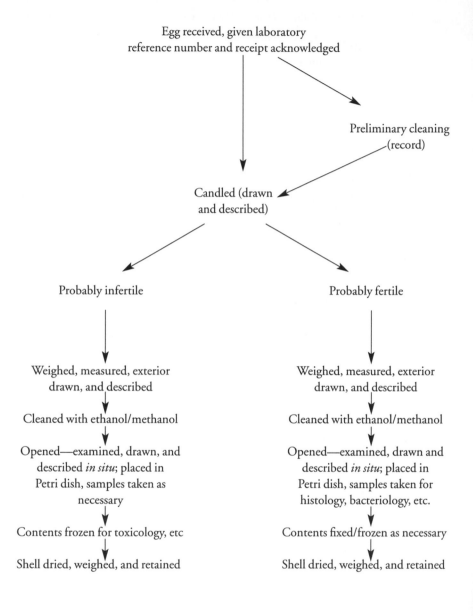

Egg received, given laboratory
reference number and receipt acknowledged

Preliminary cleaning
(record)

Candled (drawn
and described)

Probably infertile

Probably fertile

Weighed, measured, exterior
drawn, and described

Cleaned with ethanol/methanol

Opened—examined, drawn, and
described *in situ*; placed in
Petri dish, samples taken as
necessary

Contents frozen for toxicology, etc

Shell dried, weighed, and retained

Weighed, measured, exterior
drawn, and described

Cleaned with ethanol/methanol

Opened—examined, drawn and
described *in situ*; placed in
Petri dish, samples taken for
histology, bacteriology, etc.

Contents fixed/frozen as necessary

Shell dried, weighed, and retained

Appendix III

Examination of Eggs / Embryos

Reference number:
Received: (date) _____ (by) _____
Receipt acknowledged by: _____   Date _____
Method of packing/wrappings:
History:

EGG / EMBRYO EXAMINATION
(to be completed for each specimen)

Species: _____
Owner / Origin: _____
Weight of whole unopened egg: _____   __   Length: _____ Width: _____
External appearance (see Figure 8.1)
Appearance on candling (see Figure 8.2)
    Embryo
    Air cell
    Blood vessels
    Fluids
Appearance when opened (see Fig. 3)
Contents:
Embryo: Length (crown-rump)
    Amniotic cavity
    Allantoic cavity
    Yolk sac
Other comments:
Microbiology:
Histopathology:
Other tests:
Samples sent elsewhere:
Weight of dried eggshell: _____ Thickness (measurement or index): _____
Samples stored
COMMENTS

Date: **Signature:**

9

Techniques in physiology and genetics

Alistair Dawson

9.1 Introduction

This chapter is concerned with fields of enquiry that may necessitate invasive techniques to provide samples or to manipulate physiology. The aim is to provide some guidance in sampling techniques and a consideration of the associated ethical and legal procedures. I have made no attempt to describe the technologies (e.g. radioimmunoassy, DNA fingerprinting) in detail, merely to provide a guide to more detailed sources of information.

9.2 Sampling techniques

9.2.1 Ethical considerations

Much ornithological research will inevitably compromise the well-being of individual birds. This may range from disturbance as a result of observational studies in the field through to physiological and/or psychological pain resulting from invasive techniques in laboratory-based physiological studies. Consequently, ornithologists must make ethical decisions when designing studies (e.g. Bekoff 1993; Emlen 1993). They must balance the likely scientific or conservation gain (new or useful information) against the cost to the bird (suffering of the individual). This is particularly true for physiological studies. Everyone would agree that unnecessary pain is unacceptable. But there is no simple way to define what potential scientific gain justifies a particular degree of suffering. In some countries it is a legal requirement to address this ethical issue and to justify proposed procedures. Many ornithological journals have an ethical policy (e.g. Ibis, Volume 137 pp 457–458) that must be complied with before a paper can be considered for publication.

9.2.2 Legal considerations—catching wild birds for research

Legal restrictions on catching wild birds vary widely between countries. In the United Kingdom, wild birds are protected by the Wildlife and Countryside Act (1981). The British Trust for Ornithology (BTO) licenses the catching of wild birds for the purpose of marking with conventional metal or plastic leg rings, through authority delegated to it by the statutory agencies. However, deliberately removing any samples of blood, feathers, or other tissues from live wild birds is not permitted under this licence. Any sampling procedure requires a separate licence issued by the relevant country agency (English Nature, Scottish Natural Heritage, Countryside Council for Wales). In addition, such sampling may require a separate licensing procedure from the Home Office. The purpose of the sampling determines whether a Home Office licence is required. If the sampling is for scientific or other experimental purposes, for example, DNA analysis for a study on population dynamics, then Home Office licensing is required. If the sole purpose is to identify the bird, for example, establishing the provenance or sex of the bird through DNA analysis then a licence is not required from the Home Office provided that the procedure causes no more than momentary pain or distress and no lasting harm. If in doubt, the Home Office Inspectorate should be consulted. Keeping wild birds in captivity, for whatever purpose, requires a licence from the appropriate country agency. If birds are kept in captivity for scientific research, this will also require Home Office licensing.

In other European countries, the degree of protection afforded to wild birds varies, from rigorous protection as in Sweden, where the capture of any wild bird requires permission from the Swedish Environmental Protection Agency, to comparatively low levels of protection in some other countries.

Within the USA and Canada, wild birds are also given rigid legal protection through The Migratory Bird Treaty Act. There are additional numerous and complex laws, regulations, and policies among administrative authorities at various levels (national, state, county). Any research that involves disturbing, handling, collecting, holding captive, or in any way manipulating wild birds requires written approval from the appropriate regulatory authorities. Details regarding permit applications and wildlife protection are given by Little (1993) and can be obtained directly from the US Fish and Wildlife Service (USFWS) regional offices or the Canadian Wildlife Service. Permits may also be required from landowners, for example US National Park Service, USFWS (National Wildlife Refuges) and US Forest Service.

In New Zealand birds are protected under the New Zealand Wildlife Act 1953. Permits to take wild birds for scientific research are obtained from the

Department of Conservation. In Australia, permits must be obtained from the State and Territory conservation authorities.

9.2.3 Legal considerations—scientific experiments on birds

In the United Kingdom, experiments on animals are regulated by The Animals (Scientific Procedures) Act 1986, which is administered by the Home Office. The Act itself is quite short but there is extensive Guidance on the Operation of the Act (Figure 9.1). This, and the Application Forms for the different licences required, can be obtained from the Home Office website—www.homeoffice. gov.uk/animalsinsp/reference/index.htm

The Act regulates any experimental or other scientific procedure applied to a "protected animal" that may have the effect of causing that animal pain, suffering, distress, or lasting harm. The Act requires that, before any regulated procedure is

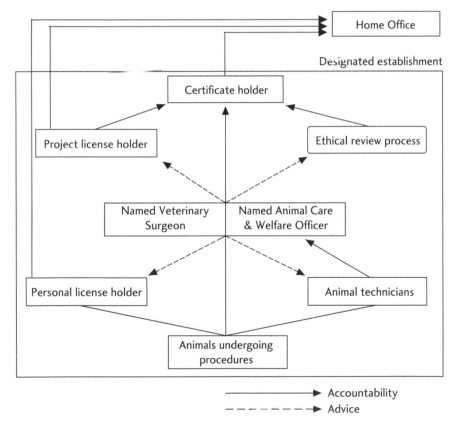

Fig. 9.1 Operation of The Animals (Scientific Procedures) Act in the United Kingdom.

carried out, it must be part of a program specified in a project license and carried out by a person holding appropriate personal license authorities. In addition, work must normally be carried out at a designated scientific procedure establishment. Regulated procedures can only be authorized and performed if there are no scientifically suitable alternatives that *replace* animal use, *reduce* the number of animals needed, or *refine* the procedures used to cause less suffering—these are known as the 3Rs. In addition, the likely benefits (to humans, other animals, or the environment) must be weighed against the likely welfare costs to the animals involved. The Act defines a "protected animal" as any living vertebrate, other than man and so, obviously, it includes birds. In the case of bird embryos, protection extends from halfway through the incubation period. A "regulated procedure" is defined as "any experimental or other scientific procedure applied to a protected animal which may have the effect of causing that animal pain, suffering, distress, or lasting harm." This encompasses any material disturbance to normal health (defined as the physical, mental, and social well-being of the animal). This includes physiological or psychological discomfort, whether immediately or in the longer term.

Control of regulated procedures is exercised by licensing at three levels.

Personal licenses. This is an endorsement that the holder is a suitable and sufficiently competent person to carry out specified regulated procedures on specified classes of animal as part of a program of work authorized by a project license. Satisfactory completion of an appropriate training course is required before a personal license is issued.

Project licenses. These must weigh the likely adverse effects on the animals involved against the benefit (to humans, other animals, or the environment) likely to accrue from the program of work, establish that there are no alternatives to the use of animals, and ensure that full use will be made of reduction and refinement strategies to minimize the number of animals used and the likely pain, suffering, distress, or lasting harm to be caused. The program of work has to be set out and each experimental procedure described in detail. Records of the numbers of animals used must be kept. These are collated by the Home Office annually. [The United Kingdom is the only country that records the numbers of birds used in scientific experiments.] Satisfactory completion of an appropriate training course is required before a personal license is issued.

Certificates of designation. These are issued to individuals responsible for compliance with the provisions of the Act at establishments where laboratory animals are used and for the provision of appropriate standards of accommodation and care. The certificate holder is required to nominate one or more Named Animal

Care & Welfare Officers (NACWOs) to be responsible for the day-to-day care of the animals and one or more Named Veterinary Surgeons. The certificate holder must establish a local ethical review processes.

At the end of an experiment, birds normally have to be killed by a humane method. However, the Act does allow for release to the wild in certain circumstances, for example, following minor procedures on free-living birds (plucking a feather for DNA analysis or taking a blood sample for a doubly labeled water study). In this case, this has to be stipulated in the Project License and written permission must be obtained from a veterinary surgeon stating that, following such a procedure, the animal is fit to be released.

Normally, scientific studies will be done at the establishment holding the Certificate of Designation. Exceptionally, a place specified in the personal and the project license may be a place other than a designated establishment (PODE). This will often be the case with ornithological research. In such cases, the project license holder is required to notify the Home Office prior to any procedure being performed. This allows a Home Office inspector to be present when the work is carried out, should he or she so wish.

The Animals (Scientific Procedures) Act in the United Kingdom implements the requirements of the European Directive 86/609/EEC on the approximation of laws, regulations, and administrative provisions of the Member States regarding the protection of animals used for experimental or other scientific purposes. In theory, legislation in other European countries should be similar to that in the United Kingdom. This is true for several countries. For example, in Sweden, there is a national committee for experiments on animals (Centrala Försöksdjursnämnden), which coordinates the activity of seven "local ethical committees for experiments on animals." In some other European countries, however, legislation is considerably less stringent.

In the United States, birds (and rodents) are (at the time of writing) excluded from animal welfare legislation. Nevertheless, there are welfare procedures that must be followed when birds are used in scientific experiments. The institution where the work is to be done has to be accredited by the Association for the Assessment and Accreditation of Laboratory Animal Care and then it is self-policed by the Institutional Animal Care and Use Committee (IACUC). In general inspections occur every few months and each procedure on every grant has to be approved by the committee. Each person performing a procedure has to be trained and has to pass a test, although the test varies from institution to institution. A protocol has to be written for each procedure and this has to be approved by the IACUC before work can start. Alternatives have to be sought, especially for surgical procedures. Often, sterile surgical techniques are required. Appropriate analgesics and post-operative care must be provided. Any deaths have to be reported and explained.

In Canada there is no legislation (yet) covering the use of animals in research. Rather the Canadian Council on Animal Care (CCAC) has responsibility "to ensure, through education, assessment and persuasion, that the use of animals for research . . . employs care to acceptable scientific standards." CCAC is funded by public funds from major research councils. There is a "voluntary control program" administered at the institution level, but committed to implementing the guiding principles of the CCAC as an independent advisory body. All research Institutes, Universities, and other bodies involved in research have an Animal Care Committee (ACC). Studies have to conform to CCAC guidelines for the work to be approved by the ACC. The system is enforced because research councils require confirmation of animal care permits before research grants are funded.

9.2.4 Housing and husbandry

There are two detailed and useful publications which provide a wealth of informa-tion regarding the welfare of birds used in research—"Guidelines to the use of wild birds in research," edited by Abbot S Gaunt and Lewis W Oring (1999) available at—www.nmnh.si.edu/BIRDNET/GuideToUse/index.html, and Laboratory Animals Volume 35, Suppl. 1 (2001) "Laboratory birds: refinements in husbandry and procedures," which is also available at www.rsm.ac.uk/pub/la.htm. Among the very many recommendations of the latter are:-

- identify behavioral requirements—husbandry protocols should encourage a range of behaviors similar to those seen in the wild
- simulate appropriate wild conditions whenever possible
- include compatible conspecifics (for social species)
- allow sufficient space for exercise—flapping flight should be made possible
- provide good quality space including perches and refuges
- encourage foraging behavior
- Promote good health. High health standards do not necessitate sterile housing.

As well as dealing with welfare in general, this publication has detailed recommen-dations for a range of species: seabirds, ducks, and geese, domestic fowl, turkey, quail, pigeons, parrots, crows, starling, and finches.

A good standard of well-being and welfare requires appropriate housing, husbandry, and care. Wherever possible and appropriate, birds should be kept in outdoor aviaries. Where that is incompatible with the aims of the research, they should be housed in indoor aviaries. Cages should be avoided whenever possible.

A good standard of welfare not only benefits the birds—results obtained from well-cared for birds will be more biologically meaningful.

9.2.5 Blood sampling

Blood samples can be collected from one of three veins—the right jugular vein (the left jugular vein is small), the ulnar (wing) vein, and the medial metatarsal vein (leg). The right jugular is a large vessel alongside the trachea. It is clearly visible beneath the skin after the overlying feathers have been parted (plucking is not normally necessary). The feathers can be dampened with surgical spirit. A thumb can be used to apply light pressure to the vein—this prevents it from moving and causes the vein to swell slightly. Blood is withdrawn into a syringe. Use the smallest gauge needle compatible with syringe size and blood volume (i.e. 25G for small birds). Apply light pressure on cotton wool to the venepuncture site as the needle is withdrawn. This prevents further blood loss and helps to prevent the formation of hematoma. Carefully remove the cotton wool to avoid disrupting the newly formed clot and ensure that bleeding has ceased. The ulnar vein can be seen as it passes over the ventral side of the elbow. In this case, plucking some feathers may improve visibility. Again, light pressure applied to the vein proximal to the elbow causes the vein to swell slightly. With larger birds, a syringe can be used to collect blood. With smaller birds (<100 g) an alternative approach is to prick the vein with a needle and to collect blood into micro-hematocrit capillary tubes as it emerges onto the skin surface. Pressure applied to cotton wool again prevents excessive bleeding and hematoma formation. The medial metatarsal vein is only appropriate for large birds (>1 kg). For further details see Morton *et al.* (1993).

Blood can be kept in one of three ways. If whole blood is required, for example, for DNA extraction, clotting can be prevented by prior treatment of the syringe or capillary tube, and the storage vessel, with an anti-coagulant such as heparin. Similarly, if plasma is required, again an anti-coagulant should be used, and the sample then needs to be centrifuged to precipitate the erythrocytes. Plasma can be aspirated with a pipette and stored. Alternatively, the sample can be stored as serum. In this case, no anti-coagulant is used, the sample is allowed to form a clot, which will then contract. After a few hours, the sample is centrifuged and serum can be aspirated and stored. The difference between plasma and serum is that the latter does not contain fibrinogen, which can be an advantage in some procedures.

The volume of blood taken should be the minimum required. In the United Kingdom the Home Office guidelines suggest that the maximum volume removed in any 28-day period is 15% of total blood volume. The blood volume of birds is approximately 7 ml per 100 g body weight. This equates to 1 ml of blood per 28-day period for a 100 g bird. Ideally, a single sample should not exceed 0.5 ml per 100 g bird. If repeated sampling is necessary, sampling sites can be changed, for example, alternating between the ulnar vein of the left and right wings. Repeated sampling from the same site may damage the vein or even cause it to become occluded, leading to necrosis. Lubjuhn *et al.* (1998) found that blood sampling free-living Great Tits *Parus major* had no effect on survival or breeding success.

Venepuncture can be used to provide a blood sample for DNA extraction for molecular studies. However, current technology permits the use of very small samples, and sufficient can often be obtained from the base of a feather plucked during molt, when the feather is still growing. This is clearly quicker and less traumatic, but, in the United Kingdom, this still requires licensing if it is for a scientific purpose.

Techniques in modern physiology and genetics can require biopsy of any of several tissues. Those most commonly sampled (in addition to blood) are adipose tissue, muscle, liver, and gonad. Such biopsies require surgery and hence anesthesia (see below).

9.2.6 Administration of substances

There are four commonly used routes for substance administration (Morton *et al.* 2001). Substances can be administered orally if this is appropriate. Ideally this can be done by dissolving the substance in drinking water or mixing with food. In cases where a precise dose is required, for example, toxicological studies, oral gavage (inserting directly to stomach through a tube) may be necessary. In this case, care must be taken to avoid obstructing the trachea. The recommended maximum dose is 10 ml kg^{-1} body weight.

Intravenous injection. The veins that can be used are the same as for blood sampling, that is, the right jugular vein, the ulnar vein, and the medial metatarsal vein. Choice will depend largely on the size of the bird. The maximum injected volume is 5 ml kg^{-1}.

Subcutaneous injection. This is particularly suitable for birds because the skin is only loosely attached to underlying tissue, and the loose skin at the nape of the neck is an ideal site. Maximum volume is 2–5 ml kg^{-1}.

Intramuscular injection. Ideally, this should only be used if other routes are inappropriate. Intramuscular injections can be painful and they can lead to bleeding and may cause necrosis. Care must be taken to avoid nerves and blood vessels. Maximum volume is 0.05 ml kg^{-1}.

Intraperitoneal injections. This should be avoided in birds because substances are likely to enter the air sacs and so may affect respiration.

Substances may also be administered in slow release implants—this requires surgery and hence anesthesia (see below).

9.2.7 Anesthesia

General anesthesia can be induced using a gaseous anesthetic or by injection. Inhalatory anesthetics have the advantage that dose can be changed during the procedure and, because of the "one-way" flow through system of the avian respiratory system, clearance of the gas, and hence recovery, can be rapid. The disadvantage of gaseous anesthetics is that they require specialist delivery systems. Isoflurane is the recommended agent. It induces anesthesia rapidly, and recovery is also rapid. Ether should not be used and halothane is not as safe for birds as it is for mammals. Perhaps the best agent for injection is propofol. This produces a rapid induction of anesthesia with good muscle relaxation. Recovery is rapid and non-traumatic. Propofol must be given be intravenous injection at a dose of $10–14$ mg kg^{-1}. However, propofol-induced anesthesia is of short duration and this may not be appropriate for longer procedures. Ketamine has been widely used for birds ($20–50$ mg kg^{-1}, given subcutaneously or intramuscularly). It is a good sedative but a poor anesthetic. Muscle relaxation is poor, recovery is slow and it is also violent—birds can injure themselves if not cared for appropriately. Ketamine is better used in combination with other agents such as diazepam, midazolam, or medetomidine. Post-operative analgesia is recommended. For further details on anesthesia, analgesia, and post-operative care, see Laboratory Animals Volume 35, Suppl. 1 (2001) "Laboratory birds: refinements in husbandry and procedures," which is also available at www.rsm.ac.uk/pub/la.htm, Ritchie *et al.* (1994) and Altman *et al.* (1997).

9.2.8 Implants

Many experiments involve injecting substances into birds, for example, to induce anesthesia, to investigate the acute effects of exposure to a pollutant or pesticide, or for studies using doubly labeled water. In other cases, long-term treatment is necessary, for example, to investigate the role of a particular hormone or the effects of chronic exposure to a pollutant. Repeated injections should be avoided wherever possible—the treatment is less scientifically valid and can cause undue stress. In theses cases, implants can be used. Two types of implants are commonly used.

Silastic implants. Silastic tubing (manufactured by DowCorning) is permeable to small non-polar molecules. The tubing can be cut to the desired length, one open end is sealed with silicone adhesive, the substance of concern is packed into the tubing and the other end is then sealed. One advantage of this technique is that dose rate remains constant and is directly proportional to the length of tubing used. For example, doubling the dose rate can be achieved by doubling the length of the implant or by using two implants of the same length.

Miniosmotic pumps. ALZET® osmotic pumps (Alza corporation, Palo Alto, California) continuously deliver test substances at controlled rates. They are available in three sizes with 100 μl, 200 μl, and 2 ml reservoirs and operate for various periods, from 1 day to 4 weeks. Delivery rate, in terms of volume, is determined by the model. The required dose rate can be achieved by calculating the appropriate concentration of substance in the reservoir. These pumps are best suited to substances that can be delivered as aqueous solutions.

Both types of implant should be positioned subcutaneously—intraperitoneal implants often become encapsulated in connective tissue and they may cause hemorrhage. Subcutaneous implants should not be placed on the back because they can rupture the skin. They can be implanted on the flank or side of the thorax. Gaunt and Oring (1999) recommend that the back of the neck should not be used because the implants can penetrate the thoracic cavity. Personally, I have not found this to be a problem. Subcutaneous implants in birds are relatively easy because the skin is only loosely attached to the underlying tissue. A small incision (slightly longer than the diameter of the implant) is made in the skin, the implant is passed through and the incision is sutured. In free-living birds, care should be taken that long-term treatment does not interfere with vital functions. For example, short-term use of testosterone implants can be useful to examine the behavioral effects of testosterone, but long-term use may prevent molt or migration. Ideally, the implants should be removed at the end of the experiment. If this is not possible, the implant should be designed so that all of the test substance has diffused before it can have negative effects.

9.2.9 Laparotomy

Laparotomy is a major surgical procedure and should be carried out under general anesthesia. In the United States laparotomy of free-living birds is sometimes done without anesthesia, or with local anesthesia. The justification is that, for birds that are to be released quickly back to the wild, general anesthesia causes more trauma than the surgery. In the United Kingdom, such a procedure would not normally be permitted. Laparotomy was frequently used to determine sex in monomorphic species, but this has largely been superseded by the use of molecular markers in blood or other tissue samples. The procedure is still used to assess sexual maturity by measuring testicular size or the size of ovarian follicles. The procedure has been described by Ingram (1978). A more recent refinement is that tissue adhesives such as Vetbond™, can be used to close the wound. A noninvasive technique—magnetic resonance imaging—has been used to visualize internal organs of birds (Romagnano et al. 1996) and to assess testicular maturity (Czisch et al. 2001). However, the current availability of suitable MRI machines and their costs mean that this technique is unlikely to become widely used in the near future.

9.3 Ecotoxicology

Toxicology is concerned with the harmful effects of chemicals in man and other species. Two species of birds are used in standard OECD testing protocols—Japanese quail and bobwhite quail. Historically, the toxicity of chemicals has been measured as the median lethal dose (LD_{50}) following acute exposure to the chemical. More recently, acute exposure has been used to assess the no observed effect dose (NOED)—the highest dose that produces no lethal effects. Alternatives to acute toxicity tests, which merely classify chemicals as harmful toxic or very toxic, are now preferred because these require the use of far fewer animals (Timbrell 1995).

Ecotoxicology is concerned with how the harmful effects of chemicals on individuals impact on populations and ultimately upon ecosystems (Walker et al. 1996). These chemicals may be anthropogenic (pollutants, pesticides) or natural chemicals occurring at toxic concentrations as a result of human activity (contaminants). In its widest sense, ecotoxicology ranges from molecular effects to effects on ecosystems. Consequently it encompasses a wide range of technologies. Most studies in ecotoxicology involve measurements of pollutant residues in tissues or assessments of the physiological changes caused by pollutants. Pollutant residues can be measured in dead birds, but only if the pollutant is fairly stable and not rapidly metabolized. For example, residue levels of organochlorine pesticides, polychlorinated biphenyls, and mercury have been monitored in birds of prey in the United Kingdom since the early 1960s (Newton et al. 1993). Marked declines in the populations of birds of prey coincided with the introduction of organochlorine pesticides, an effect later attributed eggshell thinning (Ratcliffe 1970; Peakall 1993) caused by DDE, a metabolite of DDT, and to direct lethal effects of the more toxic organochlorines. Tissue residues of labile or rapidly metabolized pollutants cannot easily be measured. However, it may be possible to measure the physiological effects of such pollutants to assess exposure. Such responses are called biomarkers (Peakall 1992; Walker et al. 1996).

One of the most commonly used biomarkers in avian ecotoxicology is the inhibition of cholinesterase (Thompson 1991). Cholinesterase metabolizes the neurotransmitter acetylcholine in nerve synapses. Organophosphorus compounds and carbamates are the most widely used insecticides and birds frequently suffer nontarget exposure. Both types of insecticide act by inhibiting acetylcholinesterase (AChe) resulting in an accumulation of acetylcholine in the synapses (Mineau 1991). AChe inhibition is usually measured in the brain, since this is the principle site of action and brain AChe inhibition can be related directly to behavioral effects (Hart 1993). Measuring brain AChe inhibition is obviously destructive. Nondestructive assessment, using inhibition of blood AChe (or butyrylcholinesterase) may be more acceptable, and can be used on free-living birds (e.g. Parsons *et al.* 2000). However, the relationship to inhibition of brain AChe is complex. A further complication is that exposure to other pesticides can act synergistically to enhance the toxic effects of some insecticides (Johnston 1995; Johnston *et al.* 1996).

Another commonly used biomarker is the induction of the heme containing enzymes known as cytochromes P_{450} (so-called because their spectral peak is at 450 nm). These form a large family of monooxygenases with a wide range of functions including biosynthesis of endogenous compounds such as steroid hormones, and metabolism of a range of endogenous and exogenous compounds. The latter function is of relevance to ecotoxicology—they detoxify many anthropogenic compounds. A wide range of chemicals induces P_{450} activity in birds, including polycyclic aromatic hydrocarbons (PAH), polychlorinated biphenyls (PCB), organochlorine pesticides and ergosterol biosynthesis inhibiting fungicides (e.g. Ronis *et al.* 1998; Schlezinger *et al.* 2000). This makes P_{450} a useful biomarker to detect pollutant exposure. Conversely, it does not reveal the specific causative pollutant. The most commonly used tissue for estimating P_{450} activity is the liver, which means that sampling is destructive.

An area of current interest in ecotoxicology is endocrine disruption—exogenous chemicals that interfere with the normal functioning of the endocrine system. There is clear evidence that fish exposed to phytooestrogens in pulp mill effluent and to human-derived oestrogens in sewage outflows suffer endocrine disruption. There is also evidence that synthetic chemicals can also have endocrine disrupting effects, but it is uncertain whether this can be caused by environmentally realistic levels of exposure. Whether there are any examples of endocrine disruption in free-living birds (and other terrestrial vertebrates) remains controversial (Dawson 2000).

9.4 Endocrinology

Most organisms live in environments that fluctuate on a predictable schedule (seasonal cycles). Individuals must therefore adjust to maximize their survival over a wide range of environmental conditions. The annual cycle comprises a series of life-history stages, with each stage largely devoted to a particular activity at the optimal time of year, for example, breeding, molt, migration, and overwintering. In birds, as in other vertebrates, hormones regulate morphology, physiology, and behavior appropriately (Jacobs 2000; Wingfield *et al.* 2002). Because hormones, by definition, are transported in blood, assessing their concentration in blood samples offers a convenient nondestructive approach to investigate endocrine control mechanisms. Investigating the roles of particular hormones, for example, gonadal steroid hormones, such as testosterone, can be achieved by monitoring seasonal changes (e.g. Wingfield and Farner 1978*a,b*; Dawson 1983), implanting the hormone (e.g. Ketterson *et al.* 1996, 2002) or removing the testes (e.g. Dawson and Goldsmith 1984).

In recent years, the response of adrenocortical hormones to a standardized stressor has been used to study adaptation to environment and to monitor species in potentially disturbed habitats (e.g. Wingfield *et al.* 1994). To do this requires holding the individual for a period of 30–60 min and collecting a small blood sample at intervals for measurement of hormones. The standard stress is simply capture, handling, and restraint—it is assumed that all individuals of all species will regard capture and handling as stressful. Between samples, birds can be held in cloth bags, which allow adequate ventilation but prevent injury if the bird struggles. These should be placed in a secure place in the shade and sheltered from direct effects of weather. This stress series protocol provides useful information on hormonal changes in response to stress and birds are released unharmed. Care should be taken to ensure that breeding birds are not withheld from their nests for too long. At other times the 30–60-min holding period is not a problem, unless the individual becomes separated from a flock, or could potentially lose a territory. Investigator discretion is required. Caution should be exercised in interpreting the results. The protocol shows the magnitude of the stress response. Often, the first sample is assumed to represent a "base-line" value. This will not normally be the case because the sampling procedure takes longer than the stress response (Dawson and Howe 1983).

For basic information on vertebrate endocrinology, see Norris (1997) and for further information on current topics in avian endocrinology see Dawson and Chaturvedi (2000), Harvey and Etches (1997), and Sharp (1993).

The most important tool in endocrinology is the measurement of hormone concentrations in serum or plasma by radioimmunoassy (RIA). See Chard (1995) for details of the methodology. Briefly, this involves competition between radioactively labeled hormone and unlabeled hormone for a limited number of binding sites on an antibody specific to the hormone being assayed. After reaching equilibrium, the bound and unbound fractions of hormone are separated and the radioactivity remaining in one of the fractions is counted. The amount of antibody and labeled hormone is constant and the unknown amount of unlabeled hormone in the sample is calibrated against a standard curve of known amounts of unlabeled hormone. Steroid hormones are identical in all vertebrates and so radioactively labeled hormones and antibodies can be readily purchased. In the case of birds, the volume of blood samples will often be small. Chromatography can be used to separate different steroid hormones allowing multiple measurements to be made on each sample (Wingfield 1975). Peptide hormones, for example, luteinizing hormone (LH) or prolactin differ slightly between species and so purified preparations for the standard curve and for radiolabeling used to have to be extracted (e.g. Follett *et al.* 1972) but are now more easily prepared by recombinant techniques (e.g. Talbot and Sharp 1994). Changes in steroid hormones can be monitored noninvasively, although less accurately, by measuring the concentration in feces (Cockrem and Rounce 1994; Goymann *et al.* 2002). This is particularly useful for endangered species (Cockrem and Rounce 1995). An alternative to RIA is ELISA, which uses a colorimetric end-point rather than radioactive counting.

9.5 Energetics

The determination of metabolic rates and energy expenditure is a key aspect of many studies on birds. Basal metabolic rate (BMR) is defined as the metabolic rate of an animal at rest, but not asleep, in a post-absorptive state within the thermoneutral zone. This is comparatively easy to assess in humans. With birds it is obviously impossible to ensure that they remain resting but awake. Other definitions therefore need to be used that define the parameters during which metabolic rate is assessed: fasting metabolic rate, least observed metabolic rate, and resting metabolic rate (Blaxter 1989; Speakman 2000). These "basal" metabolic rates are normally measured by respirometry. The bird is kept in a respirometry chamber in which temperature is accurately controlled. Air is passed through the chamber at a known rate. The difference in the concentrations of oxygen and carbon dioxide in air entering the chamber and air leaving the chamber is used to calculate the metabolic rate. Nudds and Bryant (2001) provide a recent detailed description of the methodology.

Animals living in their natural environment will expend energy at a greater than basal rate almost all of the time. Free-living energy use is defined as daily energy expenditure (DEE) or field metabolic rate (FMR). This can sometimes be estimated indirectly and noninvasively by calculating a time and energy budget, but there are many assumptions and inaccuracies associated with this method.

The most commonly used direct assessment is using doubly labeled water (Speakman 1997, 2000). The method depends on the fact that isotopes of oxygen in body water are in complete equilibrium with oxygen in respiratory carbon dioxide. An isotopic label of oxygen introduced into body water will be eliminated as water and as carbon dioxide. In contrast, an isotopic label of hydrogen will be eliminated only as water. If both isotopes are introduced as water ($^2H_2^{18}O$— hence doubly labeled water) at the same time the relative difference in their elimination will reflect production of carbon dioxide. In practice, a bird is injected with doubly labeled water, a blood sample is taken at about the time that the doubly labeled water has equilibrated with unlabeled body water (30–60 min) and a further sample(s) taken later. The isotopic ratios in the samples are measured by mass spectrometry.

In addition to metabolism, the nutritional status of a bird is important. This can be assessed from body mass and size, and from fat score (see Chapter 4). Change in mass or condition can be assessed only by re-capturing the birds and repeating the measurements. Measuring metabolites of lipids in a single blood sample may prove to be a useful indicator of change in body mass (Williams *et al.* 1999; Guglielmo *et al.* 2002). Similarly, measurement of yolk precursors in the blood may be useful to assess the reproductive status of female birds (Vanderkist *et al.* 2000).

9.6 Molecular genetics

Recent advances in molecular biology have resulted in methodologies that can be used to determine the sex of individual birds, paternity and kinship, geographical structuring within species, phylogenetic relationships among species and the timing of speciation events. See Brown (2001) for further information on DNA technologies.

DNA can be obtained from a wide range of sources. From dead birds, any tissue can be used. From living birds, blood is the most practical, and this can be obtained as a blood sample following venepuncture, or from the base of a plucked feather. An advantage of birds over mammals is that their red blood cells are nucleated and so only a small sample is required. DNA can also be obtained directly from shed feathers (Eguchi and Eguchi 2000; Bello *et al.* 2001). DNA is extracted from the tissue sample using protein-denaturing agents, salt and solvents.

Identifying individuals and relatives. In a major breakthrough that led to the process of DNA fingerprinting or profiling, Jefferies *et al.* (1985*a*) discovered that specific nucleotide sequences occurred in a repeated order, called tandem repeats, with high levels of variation meaning that individuals differ in their numbers of repeats. Thus by determining the lengths of these repeated sequences it is possible to discriminate between different individuals. To achieve this, restriction enzymes are added that cut the DNA whenever a given sequence of bases (e.g. GACCAT) occurs, so providing a large numbers of fragments of DNA. When these fragments are added to a gel with an electric current the shorter fragments move more rapidly. Probes are added that attach (hybridize) to the specific repeated sequence and make them visible. The different length fragments thus produce different bands. As chromosomes occur in pairs there will be two bands for each repeat sequence region and as the tandem repeats occur in a number of places in the genome the result is a series of bands (the DNA fingerprint) that is unique to an individual (Jeffreys et al. 1985b; Burke and Bruford 1987). See Burke (1989a,b), Burke et al. (1991), and Krawczak and Schmidtke (1998) for methodological details.

Repetitive sequence elements that occur in tandem are known as minisatellite sequences (typically 10 to 100 bases long) or microsatellite sequences (less than 10 bases). In practice minisatellites are used for DNA fingerprinting as described above. However, this method requires good quality DNA and can be difficult to analyze. As a result there is an increasing use of microsatellites and "single locus probes." These use exactly the same concept, but analyze just a single region of tandem repeats so an individual will have just two bands, one from each chromosome (or a single band if homozygous). This method typically incorporates polymerase chain reaction (PCR), which makes multiple copies of a few stands of DNA, so allowing the analysis of extremely small samples.

Each individual will have two bands at each microsatellite locus, one inherited from each parent. By combining many loci it is possible to estimate relatedness among individuals, for example, Höglund and Shorey (2003) used microsatellites to determine the frequency of full sib and half sib relationships on a White-bearded Manakin Manacus manacus lek. This method is routinely used to determine the parentage of offspring and the frequency and source of dumped eggs.

Relatedness among species and populations. Because mitochondrial DNA (mt DNA) is passed down the maternal line, it is of no value in establishing paternity. However, sequencing regions of the mitochondrial genome can be used to investigate phylogenetic relationships among populations of the same species and among species. Mitochondrial DNA is thought to have a mutation fixation rate several times greater than nuclear DNA, making it extremely variable and has the further advantage of not being recombined during meiosis so giving a clear line of descent. It is easy to work with because it is single copy gene (one allele per individual) yet has multiple copies in terms of number of molecules per cell. For example, by comparing sequences of yellow wagtails across the Palearctic it has been possible to determine the phylogeny, assess differentiation within and between regions and show evidence for bottlenecks and rapid expansion (Ödeen and Björklund 2003).

Human genetic studies have increasingly used single nucleotide polymorphisms (SNPs), which determine single base differences at a range of locations across the genome. This technique has recently been applied to birds (Primmer *et al.* 2002). These have a number of advantages such that they occur at a high frequency across the genome and this multilocus approach probably gives more reliable results than just comparing one sequence. Another advantage is that SNPs can be analyzed using automated processes.

Sex determination. A high proportion of bird species are sexually monomorphic and are therefore difficult or impossible to sex, except by laparotomy to examine the gonads. Nestlings or embryos are obviously difficult to sex. A DNA test that can be used to establish the sex of most species of birds (Griffiths *et al.* 1998) is based on two conserved chromo-helicase-DNA-binding (CHD) genes that are located on the sex chromosomes. Unlike mammals, in birds the females are heterozygous (ZW) and males are homozygous (ZZ). The *CHD-W* gene is located on the W chromosome and is therefore unique to females. *CHD-Z* is on the Z chromosome and therefore occurs in both sexes. The test involves PCR with a single set of primers. It amplifies homologous sections of both genes which incorporate introns whose lengths usually differ. When examined on a gel, there is a single band in males (CHD-Z) but in females there is a distinct second band (CHD-W). Sexing can be done for nestlings and even embryos, but may be unreliable in eggs that have not yet developed a visible embryo (Kalmbach *et al.* 2001).

Prey species. Another potential use of molecular techniques is to identify prey species in the gut contents, feces or regurgitated pellets of predator species (Symondson 2002). This is carried out by amplification of the prey DNA using PCR and then comparing sequences with online DNA databases of previously studied genes.

Acknowledgements
George Bentley, Tony Williams, Shelley Hinsley, Bengt Silverin, Sara Goodacre, and Brent Emerson provided useful information for this chapter.

References
Altman, R.B., Clubb, S.L., Dorrestein, G.M., and Queensberry, K. (1997). *Avian Medicine and Surgery*. WB Saunders, Philadelphia.

Bekoff, M. (1993). Experimentally induced infanticide: the removal of birds and its ramifications. *Auk*, 110, 404–406.

Bello., N, Francino, O., Sanchez, A. (2001). Isolation of genomic DNA from feathers. *J. Vet. Diag. Invest.*, 13, 162–164.

Blaxter, K. (1989). *Energy Metabolism in Animals and Man*. Cambridge University Press, Cambridge.

Brown, T. (2001). *Gene Cloning and DNA Analysis*. Blackwell, Oxford.

Burke T. (1989*a*). DNA fingerprinting and other methods for the study of mating success. *Trends Ecol. Evol.*, 4, 139–144.

Burke T. (1989*b*). DNA fingerprinting and RFLP analysis. *Trends Ecol. Evol.*, 4, 140–141.

Burke T. and Bruford, M.W. (1987). DNA fingerprinting in birds. *Nature*, 327, 149–152.

Burke T., Dolf, G., Jeffreys, A.J., and Wolff, R. (1991). *DNA Fingerprinting: Approaches and Applications*. Birkhäuser Velag, Basel.

Chard, T. (1995). *An Introduction to Radioimmunoassy and Related Techniques*. Elsevier, New York.

Cockrem, J.F. and Rounce J.R. (1994). Faecal measurements of oestradiol and testosterone allow the non-invasive estimation of plasma steroid concentrations in the domestic fowl. *Br. Poult. Sci.*, 35, 433–443.

Cockrem, J.F. and Rounce J.R. (1995). Non-invasive assessment of the annual gonadal cycle in free-living Kakapo (*Strigops habroptilus*) using fecal steroid measurements. *Auk*, 112, 253–257.

Czisch, M., Coppack, T., Berthold, P., and Auer, D.P. (2001). In vivo magnetic resonance imaging of the reproductive organs in a passerine bird species. *J. Avian Biol.*, 32, 278–281.

Dawson, A. (1983). Plasma gonadal steroid levels in wild Starlings (*Sturnus vulgaris*) during the annual cycle and in relation to the stages of breeding. *Gen. Compar. Endocrinol.*, 49, 286–294.

Dawson, A. (2000). Mechanisms of endocrine disruption with particular reference to occurrence in avian wildlife: a review. *Ecotoxicology*, 9, 59–69.

Dawson, A. and Chaturvedi, C.M. (2002). *Avian Endocrinology*. Narosa Publishing House, New Delhi.

Dawson, A. and Goldsmith, A.R. (1984). Effects of gonadectomy on seasonal changes in plasma LH and prolactin concentrations in male and female Starlings (*Sturnus vulgaris*). *J. Endocrinol.*, 100, 213–218.

Dawson, A. and Howe, P.D. (1983). Plasma corticosterone in wild Starlings (*Sturnus vulgaris*) immediately following capture and in relation to body weight during the annual cycle. *Gen. Compar. Endocrinol.*, 51, 303–308.

Eguchi, T. and Eguchi, Y. (2000). High yield DNA extraction from the snake cast-off skin or bird feathers using collagenase. *Biotechnol. Lett.*, 22, 1097–1100.

Emlen, S.T. (1993). Ethics and experimentation: hard choices for the field ornithologist. *Auk*, 110, 406–409.

Follett, B.K., Scanes, C.G., and Cunningham, F.J. (1972). A radioimmunoassay for avian luteinizing hormone. *J. Endocrinol.*, 52, 359–378.

Gaunt, A.S. and Oring, L.W. (1999). Guidelines to the use of wild birds in research. www.nmnh.si.edu/BIRDNET/GuideToUse/index.html.

Griffiths, R., Double, M.C., Orr, K., and Dawson, R.J.G. (1998). A DNA test to sex most birds. *Mol. Ecol.*, 7, 1071–1075.

Goymann, W., Mostl, E., and Gwinner, E. (2002). Non-invasive methods to measure androgen metabolites in excrements of European Stonechats, *Saxicola torquata rubicola*. *Gen. Compar. Endocrinol.*, 129 80–87.

Guglielmo, C.G., O'Hara, P.D., and Williams, T.D. (2002). Extrinsic and intrinsic sources of variation in plasma lipid metabolites of free-living Western Sandpipers (*Calidris mauri*). *Auk*, 119, 437–445.

Hart, A.D.M. (1993). Relationship between behavior and the inhibition of acetylcholinesterase in birds exposed to organophosphorus pesticides. *Environ. Toxicol. Chem.*, 12, 321–336.

Harvey, S. and Etches, R.J. (1997). *Perspectives in Avian Endocrinology. Journals of Endocrinology Ltd.*, Bristol.

Höglund, J. and Shorey, L. (2003). Local genetic structure in a white-bearded manakin population. *Mol. Ecol.*, 12, 2457–2463.

Ingram, K.A. (1978). Laparotomy technique fore sex determination of psittacine birds. *J. Am. Vet. Med. Assoc.*, 173, 1244–1245.

Jacobs, J.D. and Wingfield, J.C. (2000). Endocrine control of life-cycle stages: a constraint on response to the environment. *Condor*, 102, 35–51.

Jefferies, A.J., Wilson, V., and Thein, S.L. (1985*a*). Hypervariable minisatellite regions in human DNA. *Nature*, 314, 67–73.

Jefferies, A.J., Wilson, V., and Thein, S.L. (1985*b*). Individual-specific fingerprints of human DNA. *Nature*, 316, 76–79.

Johnston, G.O. (1995) The study of interactive effects of pesticides in birds—a biomarker approach. *Aspects Appl. Biol.*, 41, 25–31.

Johnston, G.O., Dawson, A., and Walker, C.H. (1996). Effects of prochloraz and malathion on the Red-legged partridge—a semi-natural field study. *Environ. Pollut.*, 91, 217–225.

Kalmbach, E., Nager, R.G., Griffiths R., and Furness, R.W. (2001). Increased reproductive effort results in male-biased offspring sex ratio: an experimental study in a species with reversed sexual dimorphism. *Proc R Soc, Lond B Biol Sci.*, 268, 2175–2179.

Ketterson, E.D., Nolan, V., Cawthorn, M.J., Parker, P.G., and Ziegenfus, C. (1996). Phenotypic engineering: using hormones to explore the mechanistic and functional bases of phenotypic variation in nature. *Ibis*, 138, 70–86.

Ketterson, E.D., Nolan, V., Casto, J.M., Buerkle, C.A., Clotfelter, E., Grindstaff, J.L., Jones, K.J., Lipar, J.L., McNabb, F.M.A., Neudorf, D.L., Parker-Regna, I., Schoech, S.J., and Snajdr, E. (2002). Testosterone, phenotype and fitness: a research program in evolutionary behavioural endocrinology. In Avian Endocrinology, eds. A. Dawson and C.M. Chaturvedi, pp. 19–40. Narosa Publishing House, New Delhi.

Krawczak, M. and Schmidtke, J. (1998). *DNA Fingerprinting 2nd edn.* BIOS Scientific Publishers, Oxford.

Lubjuhn, T., Brun, J., Winkel, W., and Muth, S. (1998). Effects of blood sampling in Great tits. *J. Field Ornithol.*, 69, 595–602.

Little, R. (1993). Controlled Wildlife. Vol. I. Federal permit procedures; Vol. II. Federally protected species; Vol III. State permit procedures. Association of Systematic Collections, Washington D.C.

Mineau, P. (1991). Cholinesterase-inhibiting insecticides: their impact on wildlife and the environment. Elsevier, New York.

Morton, D.B., Abbot, D., Barclay, R., Close, B.S., Ewbank, R., Gask, D., Heath, M., Mattic, S., Poole, T., Seamer, J., Southee, J., Thompson, A., Trusswell, B., West, C., and Jennings, M. (1993). Removal of blood from laboratory animals and birds. *Lab. Ani.*, 27, 1–22.

Morton, D.B., Jennings, M., Buckwell, A., Ewbank, R., Godfrey, C., Holgate, B., Inglis, I., James, R., Page, C., Sharman, I., Verschoyle, R., Westall, L., and Wilson, A.B. (2001). Refining procedures for the administration of substances. *Lab. Anim.*, 35, 1–41.

Newton, I., Wyllie, I., and Asher, A. (1993). Long-term trends in organochlorine and mercury residues in some predatory birds in Britain. *Environ. Pollut.*, 79, 143–151.

Norris, D.O. (1997). *Vertebrate Endocrinology*. Academic Press, San Diego.

Nudds, R.L. and Bryant D.M. (2001). Exercise training lowers the resting metabolic rate of Zebra Finches, Taeniopygia guttata. *Funct. Ecol.* 15, 458–464.

Parson, K.C., Matz, A.C., Hooper, M.J., and Pokras, M.A. (2000). Monitoring wading bird exposure to agricultural chemicals using serum cholinesterase activity. *Environ. Toxicol., Chem.*, 19, 1317–1323.

Peakall, D.B. (1992). *Animal Biomarkers as Pollution Indicators*. Chapman & Hall, London.

Peakall, D.B. (1993). DDE-induced eggshell thinning: an environmental detective story. *Environ. Rev.* 1, 130–20.

Primmer, C.R., Borge, T., Lindell, J., and Sætre, G.-P. (2002). Single-nucleotide polymorphism characterisation in species with limited available sequence information: high nucleotide diversity revealed in the avian genome. *Mol. Ecol.* 11, 603–612.

Ratcliffe, D.A. (1970). Changes attributed to pesticides in egg breakage frequency and eggshell thickness in some British birds. *J. Appl. Ecol.*, 7, 67–115.

Ritchie, B.W., Harrison, G.J., and Harrison, L.R. (1994). *Avian Medicine: Principles and Application*. Wingers Publication, Lakeworth, Florida.

Ronis, M.J.J., Celander, M., and Badger, T.M. (1998). Cytochrome P450 enzymes in the kidney of the Bobwhite Quail (*Colinus virginianus*): induction and inhibition by ergosterol biosynthesis inhibiting fungicides. *Compar. Biochem. Physiol. C-Pharmacol. Toxicol. Endocrinol.*, 121, 221–229.

Schlezinger, J.J., Keller, J., Verbrugge, L.A., and Stegeman, J.J. (2000). 3,3',4,4'-tetrachlorobiphenyl oxidation in fish, bird and reptile species: relationship to cytochrome P450 1A inactivation and reactive oxygen production. *Compar. Biochem. Physiol. C Pharmacol. Toxicol. Endocrinol.*, 125, 273–286.

Sharp, P.J. (1993). *Avian Endocrinology. Journals of Endocrinology Ltd.*, Bristol.

Speakman, J.R. (1997). *Doubly Labelled Water: Theory and Practice*. Chapman & Hall, London.

Speakman, J.R. (2000). The cost of living: field metabolic rates of small mammals. *Adv. Ecol. Res.*, 30, 177–297.

Symondson, W.O.C. (2002). Molecular identification of prey in predator diets. *Mol. Ecol.*, 11, 627–641.

Talbot, R.T. and Sharp, P.J. (1994). A radioimmunoassay for recombinant-derived chicken prolactin suitable for the measurement of prolactin in other avian species. *Gen Compar. Endocrinol.*, 96, 361–369.

Thompson, H.M. (1991). Serum "B" esterases as indicators of exposure to pesticides, pp 109–126. In *Cholinesterase-Inhibiting Insecticides: Their Impact on Wildlife and the Environment*, ed. P. Mineau, Elsevier, New York.

Timbrell, J.A. (1995). *Introduction to Toxicology*. Taylor & Francis, London.

Vanderkist, B.A., Williams, T.D., Betram, D.F., Logheed, L.W., and Ryder, J.L. (2000). Indirect, physiological assessment of reproductive state and breeding chronology in free-living birds: an example in the Marbled Murrelet (*Brachyramphus marmoratus*). *Funct. Ecol.*, 14, 758–765.

Walker, C.H., Hopkin, S.P., Sibly, R.M., and Peakall, D.B. (1996). Principles of Ecotoxicology, pp. 321. Taylor & Francis, London.

Williams, T.D., Guglielmo, C.G., Egeler, O., and Martyniuk, C.J. (1999). Plasma lipid metabolites provide information on mass change over several days in captive sandpipers. *Auk*, 116, 994–1000.

Wingfield, J.C. and Farner, D.S. (1975). The determination of five steroids in avian plasma by radioimmunoassay and competitive protein-binding. *Steroids*, 26, 311–327.

Wingfield, J.C. and Farner, D.S. (1978*a*). The endocrinology of a natural breeding population of the White-crowned Sparrow (*Zonotrichia leucophrys pugetensis*). *Physiol. Zool.*, 51, 188–205.

Wingfield, J.C. and Farner, D.S. (1978*b*). The annual cycle of plasma irLH and steroid hormones in feral populations of the White-crowned Sparrow, *Zonotrichia leucophrys gambelii*. *Biol. Reprod.*, 1046–1056.

Wingfield, J.C., Suydam, R., and Hunt, K (1994). The adrenocortical responses to stress in Snow Buntings (*Plectrophenax nivalis*) and Lapland Longspurs (*Calcarius lapponicus*) at Barrow, Alaska. *Comparat. Biochem. Physiol. C-Pharmacol. Toxicol. Endocrinol.* 108, 299–306.

Wingfield, J.C., Soma, K.K., Wikelski, M., Meddle, S.L., and Hau, M. (2002). Life cycles, behavioural traits and endocrine mechanisms. In Avian Endocrinology, eds. A. Dawson and C.M. Chaturvedi, pp. 3–17. Narosa Publishing House, New Delhi.

References

10

Diet and foraging behavior

William J. Sutherland

10.1 Introduction

Studies of diet and foraging behavior can help answer a wide range of questions. What does a species eat? Why does it prefer certain areas for feeding? Is there competition for food among particular species or among age classes or sexes within a species? How many individuals can a site sustain? Is the food depleted during a season? What are the consequences of habitat change? Answering some of these questions also requires a parallel study of prey abundance (see Chapter 11).

Many of the methods described here involve storing food materials in alcohol. Seventy percent ethyl or propyl alcohol is usually used but anything in the range 60–80% is suitable. Alcohol extracts water from tissues so either the material should be a small fraction of the total volume or the alcohol should be replenished after a day and again after a few days. Alcohol is highly flammable and needs to be stored in fireproof containers. Thus, take only small quantities in the field in leakproof unbreakable containers (not glass) or take the fresh samples back to the lab. Alcohol also dissolves fats, so should not be used for preserving material to be used to obtain dry masses or fat contents.

The diet varies through the day, through the year, and between years, sites, sexes, or age groups and even between individuals. The sampling needs to be designed to reflect this variability.

10.2 Diet composition

10.2.1 Direct observation

Useful information can often be obtained by directly observing where birds feed, what techniques they use, and what their captured prey looks like. Such information is especially useful when combined with an examination of the range of prey available. For example, observations may show that the bird is feeding from leaf litter in damp areas in the forest and taking long thin items about the same length as the bird's bill. Sampling beneath the leaves in the same habitat shows the only two prey groups that both occupy this habitat and fit this description are leeches and worms. Further observations will often distinguish between them due to differences in color, shape, or predator behavior (e.g. leeches can be picked up but worms may have to be pulled from their burrows). Alternatively, techniques such as dropping analysis can then be used to make the final assessment. Because analysis of droppings relies on the identification of food fragments that are resistant to digestion, direct observation is particularly useful for seed-eating birds that remove the hard seed coat (testa) before eating the seed and birds that feed on invertebrates with few hard parts.

On muddy areas it is sometimes possible to examine footprints and peck marks to see where the bird pecks. This can sometimes be used to see if it pecks at features such as holes in the mud or to link the footprints to food remains.

Studying habitat choice can facilitate an assessment of diet. Discovering that the species prefers feeding on one patch or one tree species, or feeding in an area at a particular time of day, especially if accompanied by examination of potential food items.

The usefulness of direct observations varies. Utilization of seeds and fruits taken direct from the plant can often be determined unambiguously. The diet of birds on mudflats is often easy to determine from observation and mud sampling, because there are relatively few prey species and they differ markedly in appearance. Even for species where direct observation seems impossible (e.g. such as canopy dwellers), it is useful to know if they take food from tree trunks, underneath leaves, or if they catch prey in mid-air. In most cases, observation of feeding methods and feeding places can narrow the range of possible items but other techniques are necessary to fully determine the diet. Food hoarders, such as many tits (chickadees) and corvids, may appear to be foraging when actually searching to retrieve cached items. Finding the caches provides a way for the observer to record the food.

Taking captive-hatched tame chicks of precocial species to selected locations in the field and observing what they eat may seem a useful way to study the diet. However, this method should be used with caution. In many species, such as galliforms, the mother draws the attention to specific types of prey items by picking them up in the bill and dropping them and uttering special calls ("tit-bitting") and may also make prey available to chicks by scratching them up from the litter layer or knocking down insects from tall vegetation. This behavior by the mother probably has a large influence on what the chicks eat and may make the diet of captive-hatched chicks feeding on their own a poor guide to that of wild chicks.

10.2.2 Nest observations

The food carried in the bill to the nestlings or incubating partners can often be identified but may differ from that taken by the foraging adults themselves. Foraging adults often swallow small items they find, but take large prey back to the nest.

Watching from a hide overlooking the nest can be useful. Excellent data on the diet of chicks can be obtained using nest cameras with an infrared beam fixed so the bird triggers the camera as it returns to the nest. This equipment is available commercially. For birds using nest boxes, the camera can be placed inside so that entering birds are photographed. The camera obviously needs a motordrive and some can be adapted to take long series of exposures (e.g. 250). Slide film is probably easiest for subsequent identification; an alternative is to use a video. The camera can be hidden within a box and a car battery used as a power supply. If a clock is placed in view of the camera, then nest daily patterns and provisioning rates can be accessed. Many cameras can print the date and time of exposure onto every frame. A ruler can be placed at the same distance so that prey size can be estimated from the photographs. Alternatively the prey can be related to bill length. Combining data from nests with observations of foraging behavior means that the diet can be linked to the feeding habitat. Analyzing films and especially video can be tedious, so consider the time that this will take before starting the study and allow for the fact that the camera may not work well all the time. Observations are needed to see whether this apparatus affects the behavior of the birds.

10.2.3 Remains and signs

Birds often leave evidence of food items they have taken. For herbivorous species look for bite marks in vegetation, while for seed-eaters look for discarded husks. The remains of vertebrates that are not eaten whole can be examined to determine species, age, sex, condition, or parasite load (see Chapter 8). Vertebrates captured by mammals often show bites on the bones and can thus be distinguished from those captured by birds (which may have "v" shaped pieces missing from the edge of the sternum). Eggs preyed upon by birds often contain a pool of yolk as they cannot lick the remains clean. Some species remove heads or wings of insects or leave the shell of molluscs and crustaceans. Observations are necessary to see what proportion of items are swallowed whole. The number of leaf miners removed by insectivorous birds can be determined by the presence of tear marks on the mine. Crossbills, woodpeckers, and squirrels all tackle tree cones in different identifiable ways. Collecting and identifying pollen from the throats and foreheads of captured nectivorous birds can be used to determine the flowers visited.

10.2.4 Dropping analysis

Birds produce droppings (a mixture of feces and urine), which can be examined for hard remains. Although identification of remains is more difficult than stomach contents, stomach flushings, or nestling ligatures, the low level of intrusion makes dropping analysis useful. Droppings can often be collected from birds caught for ringing. This is easiest if the birds are held in clean bird bags (otherwise droppings can be mixed across individuals), and held over a polythene sheet during ringing. Chicks within the nest often produce dropping sacs for the parents to carry away, and these are often produced when the bird is handled. These sacs can be picked up by tweezers.

Droppings can often also be collected in the field. In some cases, such as under roosts, nests, or perches, we can be confident which species or even which individual produced which dropping. For some species, the dispersion of droppings can be used to help identify them. Lark droppings can sometimes be difficult to distinguish from those of large finches and sparrows, but groups of 5–10 closely spaced droppings of the right size on the ground in an open area are very likely to represent the roost site of a lark because, in most European habitats at least, species with similar droppings roost off the ground. Rain can rapidly destroy the droppings of small birds. If you want large samples it is best to search after a spell of dry weather, but if you particularly want fresh droppings search 1–2 days after the end of a wet spell.

In most cases it is necessary to watch the adults to be confident that the droppings are of the species of interest. However this is often surprisingly difficult. One technique is to have two people, with one watching the bird using a telescope. Once a dropping is deposited, one observer keeps staring through the telescope at the location of the dropping. The collector walks to behind the dropping and then toward the observer who directs the collector to the location using arm movements.

Each dropping is then usually scraped into a separate bottle, and frozen or preserved in 70% alcohol. If dried the contents are harder to tease apart during analysis. Freezing solid droppings (e.g. from geese or gamebirds) has the advantage that each dropping stays separate and they can be kept together while if placed in alcohol they disintegrate and must be kept separate. Placing droppings into 20% potassium hydroxide solution for 20 min will remove uric acid and particles (Green and Tyler 1989), they can then be washed through a sieve and stored in 70% alcohol.

For assessing the diet it is necessary to identify parts that are retained in the droppings but for which there are only a few (ideally one) per individual. Mouthparts are usually hard and are often used for this purpose with legs as an alternative. A particular pair of legs is most characteristic for some insect taxa and are therefore best to count. For example, the front legs of carabid beetles have characteristic notch-like indentations. In some cases it is not possible to identify such rare and distinctive features and instead it is necessary to count the number of broken pieces of a large body part (e.g. pieces of snail shell or millepede exoskeleton) or body parts of which each prey individual has many (e.g chaetae of earthworms, wing scales of moths).

For the microscopic examination of droppings samples it is best to spread the sample in a sufficient volume of ethanol that pieces do not obscure one another, but it is inconvenient to handle a large volume of liquid by making temporary mounts on microscope slides or attempting to scan the whole of a petri dish. Figure 10.1 shows one method of overcoming these problems. The dish can be inscribed with radial marks (say at 10 degree intervals) to provide sub-units that can be used for sampling. This helps in counting a mixture of fragments of rare and common prey types. For example, Stone Curlew *Burhinus oedicnemus* droppings often contain hundreds or thousands of earthworm chaetae, each of

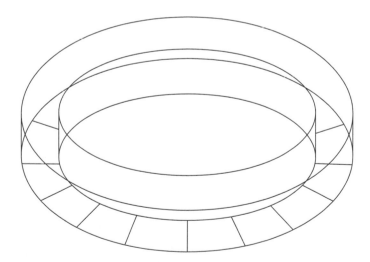

Fig. 10.1 A modified petri dish for counting fecal samples. An internal ring is glued in, so that the channel between the internal ring and the edge of the petri dish is the same width as the microscope field of view, which makes it much easier to search the entire sample without double counting or missing any (From Green and Tyler 1989).

which represents a small mass of digestible flesh, and a few large beetle mandibles, each of which is equivalent to much more food. By searching the whole of the channel for beetle mandibles, but only one in every six of the 10 degree sections for chaetae and scaling up, it was possible to increase the efficiency of sample processing (Green and Tyler 1989).

Some plant food items can be identified macroscopically, for example, from seed coats, but many other items cannot. The plant epidermis is reasonably resistant to herbivore digestion and so provides a more reliable measure. For example, it has been used to quantify the fruit, leaf, and seed diet of Madeira Laurel Pigeons *Columba trocaz* (Oliveira *et al.* 2002). The droppings can be dispersed in water (a couple of drops of sodium hypochlorite helps clear the sample) and examined under 10× or higher magnification. Phase contrast illumination microscopes are also useful for identifying plant epidermis because they allow cell walls to be seen without having to stain the sample first.

A reference collection of potential prey is essential. Ideally watch the species feeding and then catch prey within that habitat. Slides are prepared by dissecting out the part, such as a mandible and placing within a drop of a mountant such as Faure-Berlese solution (or Canada Balsam if the reference collection is not permanent) on a slide and then placing a cover slip on top. It may be desirable to remove soft tissue with potassium hydroxide or a proteolytic enzyme solution.

Nestlings typically have less efficient digestive efficiency, which makes the identification of fragments in droppings easier. Nestling diet often differs from adult diet, and if both are being studied, it is easiest to start with nestling droppings to gain experience in identifying fragments.

A common mistake is to round up individuals per dropping and record one mandible as one individual. This will greatly increase the estimated abundance of rare prey. Five mandibles should thus be recorded as 2.5 individuals rather than 3.

A considerable problem is differential digestibility between prey species (e.g. Tigar and Osborne 2000). Few remains of a species may be found either because few were eaten or because their identifiable remains are usually digested. This is best overcome by conducting calibration trials with captive birds. Feed a captive animal on an identifiable food (e.g. poultry pellets, mealworms), then give a known number of prey items (perhaps a range of species), record the number uneaten, return to feeding with identifiable food, collect all droppings, and examine for prey remains. After a gap the experiment can be repeated. Stone Curlew voided all prey within 24 h (Green and Tyler 1989), geese voided all food within 2 h (Marrion and Forbes 1970), and Knot *Calidris canutus* voided all within 4 h (Dekinga and Piersma 1993). The recovery rate is then calculated as the number found divided by the number eaten. Similar experiments can determine if digestibility varies with prey size.

10.2.5 Pellet analysis

Many species regurgitate pellets comprising bones and other hard parts along with fur and feathers. These provide an excellent means of quantifying the diets of owls but are less useful for diurnal raptors which digest bone. Many other bird species also produce pellets, particularly skuas (jaegers), gulls and waders (shore-birds), and shrikes. Pellets differ from fresh mammal feces in that they neither smell nor compress readily. Dried feces may look similar to pellets but the contents of feces are more digested and they are normally longer with straighter edges. Pellets may be found at roosts and nesting locations, as well as on the feeding areas. Pellets can be stored by drying or freezing.

For analysis, the pellets are placed in water until they are easy to tease apart and assessment is made the same way as for droppings (see 10.2.4). For larger prey the head may not be eaten and it is then necessary to identify other body parts. The problem of differential digestions of different species (see 10.2.4) can be even greater for pellets than droppings. For example, Green and Tyler (1989) showed that Stone Curlew pellets contained small mammal bones and hard parts of large insects, but that remains of small arthropods and earthworms occurred only in trace amounts, though remains of all prey were abundantly detectable in droppings. Calibration trials with captive birds (10.2.4) is a solution.

10.2.6 Stomach analysis

Birds may be found dead and the contents of the stomach analyzed. This used to be the main technique for the purpose, but shooting birds solely to determine diet is nowadays usually considered unacceptable. The methods for analyzing stomach contents are similar to those for droppings (10.2.4). The stomach should be removed as soon as possible and placed in alcohol (high concentrations are preferable as water contents of stomachs may be high) because the contents deteriorate rapidly. There may be differential digestibilities of prey types and different passage times, with hard items persisting for longer (Rosenberg and Cooper 1990). Most of the principles that apply to dropping analysis (see 10.2.4) apply to stomach contents, remembering that the tendency of the stomach to retain large hard parts and allow small soft items to pass into the intestines can lead to bias.

In some species, esophageal contents can be quantified and do not suffer from differential digestibilities (Kundle 1982). Some granivorous species have gullets or crops in which food is stored before it enters the stomach and again the contents are unaffected by differential digestion.

10.2.7 Direct observations of crop

For some granivores species the seeds and some invertebrates within the gullet can be identified. The feathers on the dorsal surface are blown aside and the gullet contents identified through the translucent skin of the neck (Newton 1967). This works for both nestlings and adults.

10.2.8 Regurgitates

Some species regurgitate food if disturbed. For example, swifts and swiftlets can be caught when they return to the nest whereupon they often regurgitate boluses, which can be preserved in alcohol and identified (Lourie and Tompkins 2000). Young herons regurgitate whole fish and other items. The nestlings are then deprived of a meal and repeated collections from the same nest could affect nestling survival.

10.2.9 Cafeteria experiments

Cafeteria experiments consist of providing a range of prey items and seeing which are taken by which species. Cafeteria experiments may be carried out in the field or with captive animals. A typical experiment would be to place out identical patches of seed of different species on the ground (or on platforms to reduce loss to rodents) in such a way as to minimize the differences between patches, such as disturbance or distance to cover. The patches are then watched to measure the number of items taken by each species or the time spent in each patch. Other options would include providing bunches of fruit or containers with invertebrates. For captive birds, it is sufficient to count or weigh the food and recount/reweigh it later to measure relative consumption. However, in the field, observations are usually necessary to ascertain the bird species involved. Cafeteria experiments are useful for understanding some components of the choice, but need not reflect the actual choices in the real world. Seeds or invertebrates preferred in experiments may be inaccessible in natural conditions.

10.2.10 Morphology

The structure of the bird can give insights into the likely foraging behavior. In general, insectivores which pick items from foliage have fine bills, sallying insectivores have wide bills, which are not deep and they often have long and stiff rictal bristles, seed eaters have broad and deep conical bills, specialized fruit eaters have wide gapes and mouths, birds feeding in dark conditions have large eyes, birds having to break tough prey have thick bills and the length of the bills of wading birds indicates the maximum depth to which they can probe. The spacing of the bill lamellae of dabbling ducks indicates the likely size range of prey. Similarly sex differences in morphology are often related to sex differences in foraging behavior (Durell 2000).

10.2.11 Neck ligatures

Neck collars have been applied to chicks so they cannot swallow food. This food is then collected and the collar removed. It has the advantage that the prey is undigested. Collars may be made of pipe cleaners, thread, or copper wire. If feeds are infrequent, then it may be necessary to compensate by feeding the chick. There can be increased mortality of chicks, adult birds may remove prey from collared nestlings, and smaller items may be swallowed (Rosenberg and Cooper 1990). Less food was delivered to nestling Grey Catbirds *Dumatella caralensis* with ligatures thus underestimating prey intake and larger items were often disgorged thus biasing diet (Johnston *et al.* 1980). This method is now rarely used due to the welfare and conservation considerations from both the loss of food and the risk of damaging the chicks.

10.2.12 Emetics and flushing

Emetics and stomach flushing do not usually kill the bird, but occasionally they do and these methods are usually considered too invasive. As an emetic 0.8 cm^3 of 1–1.5% antimony potassium tartrate per 110 g of body mass is administered via a syringe usually through a vaseline-coated narrow flexible plastic tube pushed gently down the esophagus. The bird is then placed in a dark box with a carpet of absorbent paper and released 15–20 min later. Of 3419 birds of 82 species studied in Venezuela, 3033 diet samples were obtained, of which 2712 had recognizable food, but 70 birds died (Poulin *et al.* 1994). Nectar is difficult to detect but pollen grains are obvious.

Stomach flushing apparently has lower risks. A vaseline coated narrow plastic tube is inserted into the stomach and lukewarm (often weak saline) water pumped in through a syringe until the contents of the esophagus and stomach are voided. In many countries use of both emetics and flushing would be illegal without permits.

10.2.13 Isotope differences between habitats

Marine, freshwater, and terrestrial foods typically differ in isotope "signatures" (see Chapter 9 for description of methods). These can then be used to identify likely feeding habitat. Sampling feathers grown during particular periods can then indicate whether birds concerned were feeding on marine, freshwater, or terrestrial foods. Thus Klaassen *et al.* (2001) showed that for each of ten different arctic breeding wader species the eggs were produced from food originating from tundra habitats rather than from coastal habitats in temperate regions.

10.3 Determining prey size

10.3.1 Direct observation of prey size

Direct observations can be sometimes a reasonable method of estimating prey size, for example, in relation to bill length. The relative measure (i.e. 2.5 times bill length) can be converted to actual length by multiplying by mean bill length (there may be a sex difference in bill length). The accuracy of this method can be assessed by holding prey items next to a stuffed bird and having observers estimate prey length in similar conditions to those used in the field. If the feeding birds have been videoed or photographed then these measurements can be made from the images. Sitters (2000) placed canes marked at 10 cm intervals on mudflats so he was able to assess the bill length of individually marked Oystercatchers *Hematopus ostralegus* by comparing on a video screen. He then used these measures to estimate the size of prey taken.

10.3.2 Determining size from prey remains

Prey remains can sometimes be found, and used to determine the size and species of items taken. Thus parts of carcasses left by predators can be measured, as can shells of molluscs and hard parts of other prey. For herbivores the size of the leaf at the base may give a measure of the size of piece bitten off. Thus Summers and Atkins (1991) measured the petiole (leaf stem) widths of Sea Aster *Aster tripolium* and showed the petiole width correlated with length. They used this measure to estimate the size of leaves eaten by Brent Geese *Branta bernicla*.

10.3.3 Determining prey size from regurgitates

Regurgitated prey can be measured directly. Quinney and Ankney (1985) collected the boluses from parent Tree Swallows *Tachycineta bicolor* returning to the nest. The collected insects were placed in 70% alcohol and at a later date the lengths of the insects could be measured.

10.3.4 Measuring fragments in pellets, droppings, or stomach

Large items within droppings can be measured with calipers. Smaller items can be measured using a graticule eyepiece in a microscope, which has been calibrated using a slide with a known scale. It is necessary to decide which items are suitable for measurement. These should be those that can be measured consistently and with minimal error. Mouthparts of invertebrates are often hard and relatively easy to identify so are usually counted and measured, but if mouthparts are hard to identify, legs might be counted instead.

Since mass (often ash-free dry mass) is a function of volume, it will usually not be directly related to linear measurements; therefore mass and length are best logged before calculating a regression equation for this relationship. Calver and Wasler (1982) suggest which diagnostic parts should be measured for a range of Hymenoptera, Coleoptera, and Diptera families and provide the regression equations to link these measures to total length. Morris and Burgis (1988) give the relationship between fresh weight W of UK passerines and humerus length H:

$$\ln W = -3.8027 + 2.4221\,(\ln H)$$

This relationship also provided a good fit for non-passerines with similar body shape but not for others such as swifts, gulls, and waders, with different relative leg lengths.

10.4 Prey quality

10.4.1 Energy content

It is often useful to assess the energy content of prey taken. The usual method is to assess ash-free dry mass as an indication of the energy content. Standard practice is to dry at 90 °C until a constant mass is reached. The duration needs to be tested initially but could be up to several days for larger items, which are therefore best cut into pieces beforehand. The prey is then placed in a crucible in a muffle furnace at 550 °C until all the organic matter has been burnt off (2 h is sufficient for most groups) and left in a desiccator to cool before weighing the ash content. High temperatures cause some conversion of carbonate to carbon dioxide providing inaccurate estimates for calcium rich species, such as molluscs and crabs, so it is best to first remove the shell if possible, for example, by placing molluscs in boiling water for 10 s. The ash-free dry mass is the dry mass minus the ash content. Measuring the prey beforehand allows a regression of ash-free dry mass against size to be calculated, so that the measures of prey size captured can be converted to intake. This can be used to calculate the energy intake from different prey types, in different areas, or at different times.

To actually measure energy, it is necessary to use a bomb calorimeter to determine the energy content per ash-free dry mass. This is usually calculated separately for each prey species or category. Robel *et al.* (1995) tabulate the energy and nutrient values of a wide range of invertebrates.

A standard method of assessing fruit quality is "relative yield"; that is, the dry mass (or better still ash-free dry mass) as a proportion of the total mass (Snow and Snow 1988). If necessary, the fruit pulp can be analyzed for its carbohydrate, fat, and protein content. Seed mass as a proportion of the whole fruit mass is a useful measure of the inedible weight that has to be carried if the fruit is swallowed.

10.4.2 Prey digestibility

The digestibility of a given food type is usually determined using captive birds. It can be assessed by providing a known mass of food f, then collecting all the droppings and determining their dry mass b. The droppings can be washed off a plastic sheet or tray.

The digestibility d can then be calculated as

$$d = b/f \cdot a$$

where a is the fraction of the mass of the fresh food retained when dried.

Digestibility can also be assessed in the field in some circumstances using an indigestible marker. The amount of indigestible marker M_i and the amount of energy (measured by microbomb calorimetry), protein C_i or carbohydrate or any other component of the diet is measured in the food and in the droppings (M_d and C_d).

The digestibility d can then be calculated:

$$d = \frac{C_i M_d}{C_d M_i}$$

In the past cellulose was often used, but it appears to be digested (Buchsbaum *et al.* 1986). Lignin is less readily digested and is probably preferable, but large species digest a higher proportion of such fibers. Trace elements like magnesium can also be used if the bird can be assumed to be in balance for them.

The food analyzed should contain the same proportion of marker and protein as the samples eaten. Birds are often highly selective for samples high in protein but low in fiber, such as young leaves, so analyzing a mix of young and old leaves would underestimate digestibility.

10.5 Foraging behavior

10.5.1 Time budgets

It can be useful to assess the time spent on different activities. This may involve comparing the time budgets of different individuals, ages or sexes, or birds at different locations. The two main methods are focal sampling and scan sampling. Focal sampling consists of watching an individual for a fixed period (e.g. 10 min) and recording the activities, such as the number of items eaten, the number of pecks, or the total amount of time spent foraging (use a second stopwatch). This is repeated for other individuals. It is important to ensure that the samples are not biased in a manner that affects the conclusions, for example, do not select active individuals in preference to sleeping ones.

Scan sampling involves systematically scanning each individual in turn and categorizing its behavior (e.g. sleeping, walking, or alert) at the instant when first observed. Scan samples are repeated but usually leaving a time gap to increase the independence of the records. Data can be collected using a Dictaphone, computer, or tally counters. Individuals can also be classified at the same time into categories (e.g. ages, sexes, habitats) to show differences in behavior between categories. An example of a scan sample would be to compare the numbers of adult and juvenile geese feeding in a flock. The method might involve starting from one end of the flock and systematically scanning across the entire flock recording the behavior and age of each bird in turn. Again the sampling should minimize bias, for example, as a result of individuals at the edge of the flock differing in age, status, intake rate, or vigilance from those in the center. Sampling across the flock can minimize this source of bias.

Scan samples are useful for quickly determining the time budget of abundant or flocking species. For example, a simple scan count of a hundred geese can be done in 10 min but a 10-min focal watch of a single individual is clearly insufficient (the individual might sleep for the entire period). Focal watches are better for dispersed species. They are also necessary when data are required on particular individuals: for example, to relate interactions between age, sex, intake rate, vigilance, walking rate, and aggression.

Data loggers have been used on Brünnichs Guillemots *Uria lomvia* to determine the time spent underwater, swimming, flying, and on the nest (Falk *et al.* 2000). The loggers also provided data on diving duration and depth. Time budgets can also be determined from radio-tracking if an activity sensor is fitted (see Chapter 6).

10.5.2 Time spent feeding per day

This is measured by assessing the mean number of hours spent away from the roost per day and multiplying by the percentage of time spent feeding through the day. Measures of percentage of time feeding have to sample through the day to allow for diurnal patterns.

10.5.3 Night observations

Nocturnal observations of foraging behavior are difficult to obtain, although the technology is improving. Recent, but expensive, equipment using a photocathode rather than an anode cone eliminates distortion away from the image center and has light amplification in the range 20,000–30,000 rather than 150–400 times. Infrared telescopes, binoculars, and videos can be improved by using a searchlight (e.g. 1 million candlepower) with an infrared filter. A searchlight did not effect the foraging behavior of oystercatchers (Sitters 2000). The range of these illuminators usually does not exceed 100 m. A bracket can be attached to the illuminator so that it moves with the optical equipment.

10.5.4 Handling time

Handling time, the time spent catching and consuming prey, can be estimated directly by starting a stopwatch when the individual concentrates on a prey item to the exclusion of others and stopping it when the bird moves on to the next activity. This is only accurate for relatively long handling times (e.g. over 5 s) or if the behavior is videoed so that it can be replayed at a slower speed.

The method of Goss-Custard and Rothery (1976) can be used for measuring handling times or pecking durations and is especially useful if handling times are short (e.g. under 3 s). The time taken for a given number of paces (say 40) is measured along with the number of pecks made. The linear regression is calculated with the numbers of pecks on the horizontal axis and the time taken for the pecks and 40 paces on the vertical axis. Each additional peck increases the total time, so the slope is the peck time. The intercept is the time for 40 paces and no pecks, and so dividing the intercept by the number of paces (40) gives the time for one pace.

10.5.5 Intake rate and the functional response

The intake rate is the rate at which prey is acquired. It can be measured as: prey items taken/foraging time or biomass intake/foraging time. The biomass intake is obtained through multiplying by both prey size (10.3) and some component of prey quality (see 10.4) such as energy content. Intake rate is measured either by watching foraging birds (and stopping if the bird stops foraging) or by selecting birds to watch regardless of their activity.

The intake rate depends upon the prey density; the functional response describes the relationship between the intake rate (expressed as number of items or biomass) and prey density. Following Holling (1959), the number of prey eaten E during time T is related to the prey density N by:

$$\frac{E}{T} = \frac{a'N}{1 + a'NT_h},$$

where a' is the searching efficiency and T_h is the handling time (see 10.5.5) and their values can be derived from using a curve fitting procedure (available in most statistical packages) to fit this equation. However, this approach should be used with caution. If a food category includes a range of items of different profitability, the bird may become more selective within a category as the density of that category increases. If this heterogeneity is not recognized, foraging parameters estimated in this way will be biased.

Daily energy expenditure can usually be assumed to balance consumption. Bennet and Harvey (1987) showed that across bird species the active metabolic rate (AMR) in KJ per day can be estimated from

$$\text{Ln (AMR)} = 0.61 \text{ Ln (body mass in kg)} + 1.18.$$

10.5.6 Interference

Interference is the short-term reduction in intake rate resulting from the presence of others, including the effect of disturbing the prey. This is usually assessed by marking out an area for which bird density varies markedly over the study period but the prey availability and density do not. Count bird density and intake rate. The standard method is to plot $\log_{10}$ intake rate against $\log_{10}$ bird density (Yates *et al.* 2000) with the slope indicating the extent of interference. Fighting and kleptoparasitism (food stealing) often contribute to interference and their rates are usually determined by focal animal sampling. Susceptibility to interference is likely to vary between individuals, with juveniles suffering most. Interference cannot be assessed by combining data across sites as better quality sites tend to have both higher intake rates and more birds, resulting in a positive correlation between intake rates and density.

10.5.7 Depletion

Depletion is the removal of food items that would otherwise be available to others. It can be studied using exclosures (see 10.5.11), by estimating the total intake of all individuals or by relating the decrease in prey population to the initial prey density. The maximum number of individuals P that can be sustained in a site, assuming no replenishment or growth of food items, can be calculated from:

$$P = T_h \sum_{d_c}^{M} (j - d_c) f_j + 1/a' \sum_{d_c}^{M} f_j \log_e(j/d_c)$$

where f is the area with density of available prey j at the start of the study, M is the highest recorded prey density, a' is the searching efficiency, T_h is the prey handling time and d_c is the threshold prey density at which feeding is no longer possible (Sutherland and Anderson 1993). The values of a' and T_h can be determined from the functional response. The value of d_c can be estimated from combining the functional response (see 10.5.6) with the daily energy requirements to ascertain when the daily intake is insufficient, or by field measures of the prey density at which feeding ceases. This model provided a good description of the number of Black-tailed Godwits *Limosa limosa* using different areas (Gill *et al.* 2001).

10.5.8 Prey availability

A major problem in relating foraging behavior to prey density is that not all prey are available. Thus, prey may seem abundant to the observer, but is largely inaccessible to the bird. Availability is always difficult to quantify and sometimes impossible. The usual first step is to analyze the prey species, size classes, and locations in which birds feed and restrict the study to these prey. For long-billed shorebirds, the depths at which prey are taken can be assessed by comparing the probing depth to bill length.

Prey depth of slow moving species can be quantified by digging out soil/mud cores pushing out the contents (or having cores that open along their length) and then quickly slicing through at measured depths. The depth at which bivalves occur has been assessed by gluing thin threads of known length to bivalves and measuring the length above the surface after they have reburied themselves (Zwarts and Wanink 1996). Furthermore, by exposing these prey to predation (easiest in captivity), it is possible to relate predation risk to depth.

The depth from which immobile prey can be extracted can be determined within artificial feeding sites by experiments on captive or free living birds. Mark prey individuals at each depth with a different mark and allow birds to feed and record those taken or left. Thus Robinson (1997) marked seeds with felt tip pens and placed the seeds in different depths within trays of soil placed in the wild. He watched to see which species fed there. After birds had fed he sieved the soil and recorded the seeds left and could thus determine the proportion taken from different depths.

10.5.9 Exclosures

It is often useful to compare changes in food abundance within exclosures from which birds are excluded and control areas. The main issues are:

1. Excluding the birds without also reducing the use of nearby control patches. This is a risk especially if the materials flap in the wind or otherwise scare birds from a wide area. For timid ground-feeding species, four corner posts with bird-high wires around and across the posts is often sufficient. Netting may be necessary for more confident species.
2. Whether prey will move into the exclosures so that depletion will be underestimated. It is often unrealistic to consider using exclosures,

for example, on mudflats with active intertidal prey. Even soil invertebrates move laterally. Movement can sometimes be prevented by prey proof exclosures. One technique is to compare between the edge and center of the exclosure to give an indication of extent of movement.

3. Whether there is compensatory mortality. Excluding the bird species may increase the local food density, and so attracts another predators undeterred by the exclosure, such as rodents. This then underestimates the depletion caused by birds. This can be tested using other exclosures that exclude both birds and other predators.

4. Whether the exclosure alters the microclimate and thus affects the survival and growth of those animals and plants inside. Even netting can have an impact by reducing wind speed and can also influence sedimentation and water flow in aquatic environments.

10.5.10 Mate provisioning and brood provisioning rates

Such rates are often measured from a hide for set periods of time, varied to reduce the effect of time of day. Other methods are to use nest cameras with a clock adjacent to the nest (see 10.2.2) or a camera with a built in clock, perhaps placing the entire nest on a balance so that from the increase in mass after a feed, the meal mass can be estimated. This works best for species that consume irregular large meals at long intervals, such as albatrosses (Huin *et al.* 2000).

Acknowledgements

Thanks to Aldina Franco, Simon Gillings, Rhys Green, Ian Newton, Ian Sherman, and Ron Summers for useful suggestions.

References

Bennett, P.M. and Harvey, P.H. (1987). Active and resting metabolism in birds—allometry, phylogeny and ecology. *J. Zool.*, 213, 327–363.

Dekinga, A. and Piersma, T. (1993). Reconstructing diet composition on the basis of faeces in a mollusc-eating wader, the knot *Calidris canutus*. *Bird Stud.* 40, 144–156.

Durell, S.E.A. Le V. dit (2000). Individual feeding specialisation in shorebirds: population consequences and conservation implications. *Biol. Rev.*, 75, 503–518.

Falk, K., Benvenuti, S., Dall'antonia, L., Kampp, K., and Ribolini, A. (2000). Time allocation and foraging behaviour of chick-rearing Brünnicks guillemots *Uria lomvia* in high-arctic Greenland. *Ibis*, 142, 82–92.

Gill, J.A., Sutherland, W.J., and Norris, K. (2001) Depletion models can predict shorebird distribution at different spatial scales. *Proc. R. Soc. Series B*, 268, 369–376.

Goss-Custard, J.D. and Rothery, P.A. (1976). A method of measuring some components of foraging birds in the field. *Anim. Behav.*, 24, 545–550.

Green, R.E. and Tyler, G.A. (1989) Determination of the diet of the stone curlew (*Burhinus oedicnemus*) by faecal analysis. *J. Zool., Lond.*, 217, 311–320.

Holling, C.S. (1959). Some characteristics of simple types of predation and parasitism. *Canadian Entomol.*, 91, 385–398.

Huin, N., Prince, P.A., and Briggs, D.R. (2000). Chick provisioning rates and growth in black-browed albatross *Diomedia melanophris* and grey-headed albatross *D. chrysostoma* at Bird Island, South Georgia. *Ibis*, 142, 550–565.

Lourie, S.A. and Tomkins, D.M. (2000). The diets of Malaysian Swiftlets. *Ibis*, 142, 596–602.

Klaassen, M., Lindstrom, A., Meltofte, H., and Piersma, T. (2001). Arctic waders are not capital breeders. *Nature*, 413, 794.

Morris, P.A. and Burgis, M.J. (1988). A method for estimating total body weight of avian prey items in the diet of owls. *Bird Stud.*, 35, 147–152.

Newton, I. (1967). The adaptive radiation and feeding ecology of some British finches. *Ibis*, 109, 33–98.

Oliviera, P., Marrero, P., and Nogales, M. (2002). Diet of the endemic Madeira Laurel Pigeon and fruit resource availability: a study using microhistological analyses. *Condor*, 104, 811–822.

Poulin, B., Lefebre, G., and McNeil, R. (1994). Effect and efficiency of tartar emetic in determining the diet of tropical land birds. *Condor*, 96, 98–104.

Quinney, T.E. and Ankney, C.D. (1985). Prey size selection by tree swallows. *Auk*, 102, 245–250.

Robel, J.R., Press, B.M., Henning, B.L., Johnson, K.W., Blocker, H.D., and Kemp, K.E. (1995). Nurient and energetic characteristics of sweepnet-collected invertebrates. *J. Field Ornithol.*, 66, 44–53.

Robinson, R.A. (1997). Ecology and Conservation of seed-eating birds on farmland. Unpublished PhD thesis, University of East Anglia.

Rosenberg, K.V. and Cooper, R.J. (1990) Approaches to avian diet analysis. *Stud. Avian Biol.*, 13, 80–90.

Sitters, H.P. (2000). The role of night feeding in shorebirds in an estuarine environment with special reference to oystercatchers. D.Phil thesis, Oxford.

Summers, R.W. and Atkins, C. (1991). Selection of brent geese *Branta bernicla* for different leaf lengths of *Aster trifolium* on saltmarsh. *Wildfowl*, 42, 33–36.

Sutherland, W.J. and Anderson, C.W. (1993). Predicting the distribution of individuals and the consequences of habitat loss: the role of prey depletion. *J. Theoret. Biol.*, 160, 223–230.

Tigar, B.J. and Osborne, P.E. (2000). Invertebrate diet of the houbara bustard *Chlamydotis [undulata] macqueenii* in Abu Dhabi from calibrated faecal analysis. *Ibis*, 142, 466–475.

Yates, M.G., Stillman, R.A., and Goss-Custard, J.D. (2000). Contrasting interference functions and foraging despersion in two species of shorebirds (Charadrii) *J. Anim. Ecol.*, 69, 314–322.

Zwarts, L. and Wanink, J. (1993). How the food supply harvestable by waders in the Wadden Sea depends upon the variation in energy content, body weight, biomass, burying depth and behaviour of tidal-flat invertebrates. *Netherlands J. Sea Res.*, 31, 441–476.

11

Habitat assessment

William J. Sutherland and Rhys E. Green

11.1 Introduction

Most field studies of birds incorporate measures of habitat extent and quality. By definition, an ecological study seeks to investigate trophic and other relationships among different species and the relationships of species with abiotic aspects of the environment. Autecological and behavioral studies of a focal bird species attempt to identify the environmental factors that influence population processes and behavior. Applied research directed at bird conservation usually attempts to improve understanding of habitat preferences and the relationships between demographic rates or population density and habitat area and quality. Habitat is usually assessed either to determine habitat associations or to document changes over time.

11.1.1 Habitat associations

Habitat associations relate bird distribution data (e.g. presence, abundance, or nest site location) to habitat data. One of the main methods, *area comparisons*, is to select a range of areas and relate abundance or presence to habitat. An example would be to select a number of blocks of mangrove and quantify both the habitat and the number of birds in each. Abundance or presence could then be related to habitat. Area comparisons are more likely to reveal the habitat associations if a wide diversity of sites are used. This approach can be carried out on a range of scales. At a patch scale the frequency with which different foraging patches are used could be quantified and related to habitat, while at a site scale the density of birds in different forest blocks or on different lakes could be related to habitat.

In other cases it is impractical to select sites beforehand and measure the habitat in them all, either because the area is not readily divisible, or because the species only occurs in few of the selected areas. The other main method, *presence–absence comparisons* comprises comparing areas used with either a selection of areas available or a selection of unused areas. If there are a large number of potential areas (e.g. pools or nesting trees) then the occupancy of the study species of these is determined and the habitat of occupied areas is compared with that of a random selection of unused areas. If the study area is not readily divisible then the study areas is searched uniformly and the habitat in places used by birds is compared with the habitat at points that are representative of the study area as a whole. As an example, Sutherland and Crockford (1993) located flocks of Red-breasted Geese *Branta ruficollis* and mapped the area that was visible from the transect. The map had a square grid. Representative points were taken at those intersections of the gridlines that lay within the observable area. Slope, aspect, altitude, distance from the roost, and distance from the road were compared between the observed locations and the representative points. In comparing occupancy it is necessary to ensure that the bird data is collected without bias. Thus if birds are more likely to be seen if close to roads, or on certain tree species, then this will bias the results.

11.1.2 Documenting changes over time

It is extremely useful to record long-term changes in variables such as habitat structure, water chemistry, fruit abundance, predator abundance, or disturbance levels. These can be used to explain changes in bird abundance or demography. However, in practice it is often extremely hard to relate changes in abundance to such variables because several variables may change at the same time. If this happens, a large sample of statistically independent study areas with different trends in abundance would be needed to assess the relative importance of the variables. Unless well planned, the methods used for collecting long-term data tend to change over time, especially if carried out by different observers. It is tempting to improve the methods but the data set is then broken unless the old method is also carried out for at least 1 year to allow comparisons. Such changes weaken its usefulness. It is also important to write down the exact method used and ensure this is followed.

11.2 Protocols

The habitat variable selected for measurement should relate to the ecology of the study species. Thus knowledge of nesting location, foraging habitat, foraging behavior, and diet are important for devising suitable measures.

When quantifying the habitat in a site, the objective is to make unbiased estimates of the habitat of the entire site. Though it may sometimes be possible to make a complete inventory of the features of interest in a study area, for example, a complete land cover map from satellite imagery, it is often necessary to measure habitats at sample points. To make these measurements representative they should be made at random locations, or better still, locations placed on a regular grid covering the study area. One frequent error is to intensively sample one or a few small plots as being a measure of the entire area. However, when relating habitat to point counts it is usual and sensible to collect data just around the point.

Much of the skill in measuring habitat considers of devising an efficient and effective protocol that can easily be repeated by different observers or by yourself after a long interval. For example, the protocol for habitat sampling around a point might be to count the number of trees within a 50-m radius circle around the point, to measure the soil pH at the central location and to measure the sward height at five locations at 20-m intervals along a north–south transect through the central location.

We know of many examples in which overenthusiastic habitat measurement has weakened an otherwise perfectly good field study. A major problem is balancing effort against precision. If habitat assessment is too time consuming then this may result in too few sites being visited, so estimates of habitat attributes have low precision because of low sample size. However, if the assessment is too quick and crude then the essential variables relevant to the birds may not be measured accurately, though precision is good because of the large sample size.

There are a number of possible means of planning a sensible programme including:

1. Estimate how long the habitat measures will take. This can even be done before setting out to the field site. One approach is to try them out in the most similar habitat nearby.
2. Estimate the tradeoff between the number and quality of data points. For example, how many more sites could be visited if the trees were not identified to species?
3. Consider the balance between the time and accuracy of the bird and habitat data. For example, if the habitat measures are time consuming (e.g. 10-min point counts followed by 3 h measuring habitat) then it might be better to repeat each bird count twice to make the point count more accurate at the cost of slightly reducing the sample size.
4. Consider when time is most limited. One useful trick to maximize the use of the early morning, when birds are most active, is to collect bird data along transects or points while walking out and when returning collect the habitat data from locations that have been marked.

Think carefully about the questions and plan the research design that is most likely to answer these. If there is insufficient time to answer the question then do something else.

Some judgment is involved with many habitat measures, so thorough training is needed before the survey work begins. Ideally all the data should be collected by the same researcher and, if that is not possible, a pilot study should be carried out, after initial training is complete, in which all researchers collect at least 30 data points independently in the same area. Analysis of these data can identify how strongly the measures by different researchers are intercorrelated and estimate correction factors to make the results comparable.

11.3 Physical environment

Abiotic aspects of a bird's environment can be important influences on distribution, abundance, reproductive success, and behavior. The technology for measuring abiotic variables is continually improving. In particular it has become routine to place sensors in the environment linked to dataloggers so that measurements can be taken and stored automatically at specified intervals for later transfer to a computer. Examining catalogs or websites of distributors is often a good way to get ideas for appropriate methods for measuring abiotic variables of interest. Distributors include Alana (alanaecology.com) in the United Kingdom, Ben Meadows (benmeadows.com) and Forestry Suppliers, (forestry-suppiers.com) in the United States. If ordering American equipment ensure it does not use imperial measures or Fahrenheit! Sophisticated instruments and automatic measuring and logging devices are usually expensive, so there may be a trade-off between measuring variables of interest in great detail at a few sites and doing this more crudely with simple equipment at many more sites.

11.3.1 Temperature and thermoregulation

Thermoregulation is an important cost for birds and depends upon ambient temperature and exposure to wind. Temperature also affects birds indirectly via their food supply, especially for insectivorous species. A study may require the placement of portable meteorological recording equipment in a site using small sensors to record the microhabitats of particular importance to birds, such as nest or roost sites. While some studies may require recording temperature at frequent intervals with a thermistor and datalogger, in others it may be sufficient to measure an integrated average temperature over a longer period with much cheaper equipment. For example, glucose in solution changes into fructose at a rate determined by temperature. By placing containers of glucose solution in nest boxes for ten-day periods and measuring the amount converted to fructose with a polarimeter, O'Connor (1978) was able to measure the mean temperature within the boxes with a precision of 0.1°C. Instrumented models of birds can be used to measure the combined effect of ambient temperature, microclimate, and wind on their rate of heat loss (Bakken *et al.* 1981; Wiersma and Piersma 1994).

11.3.2 Rainfall and soil wetness

As with temperature, rainfall can affect birds directly by effects on thermoregulation or flooding of nest sites or habitats or indirectly via effects on animal prey or the growth and seeding of plants. Analysis of bird population size and demographic parameters in relation to long-term rain-gauge records from networks of meteorological stations has yielded many insights into the factors affecting populations (e.g. Peach *et al.* 1991). Records of water levels in seasonally flooded wetlands can be used to estimate direct and indirect effects on bird populations and may usefully be combined with measurements of flood extent from satellite imagery (Nott *et al.* 1998). The water content of soil affects birds indirectly by influencing the abundance, activity, and depth distribution of soil invertebrates. The depth of the water table below the soil surface can be measured by reading the water level in permanent dipwells (pipes 6–50 mm diameter drilled with say 4 mm holes at least every 10 cm) or temporary dipwells using a 2–5 cm soil auger and recording at 30-min intervals until the level has stabilized. The water level can be measured using a ruler and torch or an electronic dipmeter. Water content can be measured by weighing, drying, and then reweighing samples of soil, but this is time consuming and rapid measurements of soil wetness in relation to depth can now be made using a theta probe (Gaskin and Miller 1996). Approximate estimates of soil moisture content can be obtained using daily rainfall records and a water balance model and these have been found to represent the availability of earthworms to foraging birds reasonably well (Chamberlain *et al.* 1999; Green *et al.* 2000).

11.3.3 Slope, aspect, elevation, and topography

Slope, aspect, and topography affect birds via influences on the local climate, including the exposure of an area to winds and can be quantified using contour maps. Topography and wind direction also affect the availability of updrafts to soaring birds. For direct measurements in the field a clinometer and compass can be used to measure slope and aspect while elevation can be measured using a GPS or altitude meter.

11.3.4 Type, chemistry, and penetrability of soils

Soil characteristics can influence the distribution and abundance of birds by influencing the effectiveness of their camouflage or that of their eggs and chicks and by effects on vegetation or invertebrate prey. Detailed soil maps are available for some regions and there are sometimes strong associations between bird distribution and soil type (e.g. Green *et al.* 2000). Maps that combine information on soils, topography, and climate in order to assess the suitability of land for arable agriculture or forestry may also provide useful information. The density of breeding Sparrowhawks *Accipiter nisus* in British woodland is positively associated with an index of the suitability of the landscape for agriculture. The mechanism of this effect is that the small bird prey of Sparrowhawks is more abundant on more productive land (Newton 1986).

Soil invertebrate abundance varies with soil characteristics, especially acidity, which can be measured using a pH meter. Mix soil with twice the volume of distilled water (pH 7) and wait for 10 min before taking the reading. Soil pH depends partly on the type of soil, but it is not readily predictable from soil maps because of the effects of agricultural management and the accumulation of leaf litter. Earthworms tend to be less abundant in acid soil (low pH) and measurements of both pH and earthworm abundance have been found to be good predictors of habitat preferences of earthworm predators such as the Woodcock *Scolopax rusticola* (Hirons and Johnson 1987). Some birds that feed on invertebrates in soil or intertidal substrates do so by probing with the bill. For these species the ease with which the substrate can be penetrated may influence the suitability of habitats for foraging. For example, the penetrability of wet grassland soils, was a more important determinant of the duration of the breeding season and foraging site selection in the Common Snipe *Gallinago gallinago* than prey abundance (Green 1988; Green *et al.* 1990). Penetrability can be quantified using commercially available penetrometers. A much cruder approach is to drop a graduated pointed stick from a constant height and record the depth to which it penetrates.

11.3.5 Water chemistry

The chemical composition of freshwaters influences birds mainly by affecting the animals and plants on which they feed. The population density and breeding success of Dippers *Cinclus cinclus* along freshwater streams in Britain is strongly correlated with water acidity because stream acidity affects the aquatic invertebrates upon which the birds feed (Ormerod *et al.* 1991). The pH is measured with a pH meter or, less accurately, indicator paper. In coastal lagoons the salinity of water, and fluctuations in it over time, has strong effects on plants and animals that may be important to birds. Salinity can be measured using a portable conductivity meter.

Eutrophication of rivers, lakes, and shelf seas by discharges of water contaminated with nitrogen and phosphorus from agricultural fertilizers, sewage, and wastes from livestock, also affects the food supply of birds by influencing the growth of algae or macrophytes and the abundance of fish and invertebrates. Chemical analysis of water samples allows the level of eutrophication of different water bodies to be compared (see Jones and Reynolds 1996 for the main methods). The Secchi disk measures water clarity, which can be important for birds feeding underwater and is a measure of eutrophication. This is a disk 30 cm across with alternate black and white quarters that can be either bought or made. It is submerged using a calibrated line with the depth recorded at which the white and black can no longer be distinguished. It is then submerged slightly more and raised until the quarters can be distinguished. The mean of the two measures is used.

11.4 Vegetation

The type and structure of vegetation is important to birds in providing nest sites, roost locations, refuge from predators, acting as food for herbivorous birds or providing herbivores for carnivorous birds, and, by its structure, enabling or constraining foraging.

11.4.1 Mapping of broad habitat types

Mapped data on the distribution of vegetation communities is valuable for many types of studies. Maps are usually required for large areas, so detailed descriptions of small quadrats over the entire area are not practical. Instead, the researcher may make detailed descriptions of a few representative samples of particular habitat types in order to identify their defining characteristics, such as tree density or species composition. It is essential to devise and document precise definitions, for example, when does savannah become grassland or woodland. Without precise definitions it is impossible to relate to other studies or repeat to document changes. With this information it is then often possible to walk around the study area mapping the boundaries of patches of particular habitats. Recording habitat edges using a GPS makes this much easier. Having an aerial photograph or high resolution satellite image of the study area is also useful. If a print of the image is taken into the field it is then often the case that habitat type boundaries visible on the ground can also be identified on the image and this can save time in mapping. This is a valuable approach even when differences in the appearance of habitats on the image are too subtle or affected by topography and lighting to allow habitat mapping using the image alone. If rigorous comparison between ground surveys with aerial photographs or satellite images indicates that the latter can be reliably interpreted, then it may be possible to map habitat types over a huge area using remote sensing.

11.4.2 Species composition of vegetation

Detailed recording of all plant species present in quadrats can provide useful measurements of vegetation as habitat for birds (Rotenberry 1985). One method is to remove all the plants from each of a series of quadrats, sort them by species, dry them, and measure their above-ground dry weight. However, although having this level of detailed information for large numbers of quadrats can yield useful insights into the ecology of the focal bird species, collecting it can be so time consuming that it would only be available for a miniscule sample of the area to be evaluated. Furthermore, although sometimes certain plant species are critically important for nesting or providing fruit, it is usually difficult to relate the bird abundance to the abundances of a long list of species. Furthermore, in practice the habitat structure is usually more important than species composition. Instead, bird habitat studies usually involve more rapid measures of the cover of dominant species and broad taxonomic groups or morphospecies (e.g. grasses) or plant species or higher taxa that are known to provide the focal birds with important resources that are specific to them, such as palatable leaves, seeds, or nest sites.

To gather data on vegetation composition, quadrats may be placed on a regular grid or in random locations within the area to be assessed. Estimates of percentage cover can be sufficiently accurate, especially if carried out by a single individual. For species feeding on the ground it is often useful to measure bare ground. The standard method is using quadrats (often 0.5 × 0.5 m) but larger quadrats (1 × 1, 2 × 2, or 5 × 5 m) have the advantage of reducing the local variation. Even more rapid, but rougher, assessments can be made using a sighting tube. A researcher looking down into a 50-mm length of 30-mm diameter plastic pipe fixed vertically to a holder on a belt around the waist height sees a circular area of about 10 cm diameter on the ground. The vegetation cover of this circle can be rapidly assigned to a category and the appropriate box ticked on a recording form. It is important to score at least 10 (and preferably about 30) circles at each sampling place, but this approach often yields more accurate measurements per unit fieldwork time than making a detailed assessment of a single quadrat. The researcher can walk rapidly between sampling points on a regular grid or transect with the sampling places being located by pacing or use of a GPS.

11.4.3 Vegetation architecture

The height, structure, and density of vegetation often affects birds by providing perches or cover and by limiting the bird's field of view and ability to run or fly to capture prey. The height of ground layer vegetation is a useful measure, but the maximum height is often unsatisfactory, because a single flowering grass stalk may be completely unrepresentative of the surrounding area. The use of a sward stick overcomes this problem by providing a measure of the height of the bulk of the vegetation within a defined area. Sward sticks often consist of a circular disc of thin wood or plastic with a hole in the center, which slides up and down a vertical rod. The rod is placed vertically with its end on the soil surface and the disc is allowed to rest on the vegetation. The height of the disc is then read off from graduations marked on the rod (easiest if graduations allow for disc thickness). The weight and diameter of the disc are chosen to suit the aims of the study and we recommend the use of standard discs. We suggest a 20-cm diameter disc weighing 144 g (to give 4.6 kg m^{-2}) for grass swards. A heavier disc could be used for denser and more rigid vegetation. The bulk of the vegetation should be compressed but not flattened.

The concealment provided by ground vegetation or its density can be assessed using a vertical board with a chequerboard or square grid pattern. The researcher looks horizontally at the board from the height of its center at a distance of 1 m and records the number of grid intersections that are visible through the vegetation. This gives measures of vegetation density and can be carried out at different heights, such as ground level and 1.5 m. Some use the chequerboard in different ways, such as the number of squares without any cover. These are, however, difficult to standardize.

The heights of trees and bushes can be estimated using a clinometer to measure the sighting angle to the top of the tree, when the researcher is at a measured distance from its trunk. Measurements of the diameter or of the stem at a 1.3-m height (diameter at breast height), is the standard measure of tree size, which is usually determined by measuring circumference. If the tree has a buttress then the diameter is taken just above the buttress. If the tree has multiple stems then each stem should be counted and measured.

A thin rod marked with 10-cm graduations can be pushed through bushes or hedges to measure their horizontal extent. It can also be held vertically and the number of contacts with vegetation in each 10-cm band counted (Weins 1973). The quantity of foliage at different heights in a tree can be assessed by lying under the tree at a sample of points and looking upwards into the canopy through binoculars on which the focusing wheel has been calibrated so that the distance between the observer and the object can be estimated. By focusing the binoculars on foliage or branches at different levels an assessment of the amount of vegetation at different heights can be made.

Canopy cover can be estimated using a sighting tube made simply by adding a cross wire to any tube or using commercially available sighting tubes with mirror and levels to ensure it is vertical. It is used to look upwards at points placed on a regular grid or along a transect and record whether the cross piece is covered by vegetation or sky is recorded. Thus, if 37 out of 100 points have vegetation, then the cover is 37%. This method is unbiased but requires a considerable number of sample points for a reasonable estimate of cover. Using a spherical densiometer involves looking at the reflection of the canopy in a curved mirror and recording the number of points on a grid that are covered by vegetation. It has the advantage that a number of data points can be obtained from one location thus speeding up data collection but as many of the points are not vertical they are thus more likely to hit some vegetation and thus overestimate cover.

Recent developments in airborne remote sensing make the rapid measurement of vegetation architecture possible over large areas. Airborne laser scanning (also called LIDAR) uses laser reflections to make a detailed map of the height profile of woodland and scrub which provides valuable habitat measures for woodland birds (Hinsley *et al.* 2002). The method can also be used to estimate the height of agricultural crops less than a meter high (Davenport *et al.* 2000). More subtle attributes of woodland vegetation architecture can also be estimated using this technique (Lefsky *et al.* 2002).

Summary statistics can be used to describe the diversity of the vertical distribution of foliage. The Shannon–Weiner information statistic, which is frequently used as a measure of species diversity, can be used to estimate foliage height diversity.

11.5 Quantifying habitat selection

Habitat selection can be studied in several ways depending on the type of data available. Commonly used approaches include the following.

11.5.1 Comparing the relative abundance of birds or records of tracked birds in each of several habitats with the relative areas of the habitats available

A key feature of this type of analysis is that every bird record can be attributed to a habitat. If this is possible then the selection or preference shown can be assessed. Underlying the concept of habitat selection is the idea that the number of birds or records per unit area, that is, density, varies among the habitats present within a particular region in a way that reflects the birds' preferences for using some habitats over others. The number of birds in each habitat reflects habitat utilization. However, differences in utilization do not imply differences in selection or preference because a rare preferred habitat may not be utilized as much as an abundant, less preferred habitat. Hence, measures of habitat selection have to take into account how much of each kind of habitat is available to the birds. Similar principles apply when analyzing the number of records obtained in different habitats for a radio-tagged or otherwise individually identifiable bird, which has been followed and its location recorded at intervals.

Deciding which and how much of each habitat is available is an important part of any analysis. For example, foraging habitats that are a long way from the nearest roosting or nesting sites may not really be available, or at least not as available as those near those habitats, and this can lead to bias if the habitat also varies with distance. A solution to this problem is to look at selection within areas that are carefully defined so that they can be regarded as equally available (e.g. within a certain distance). Alternatively the distance to the nest site or roost can be considered in the analysis. This approach allows different kinds of habitat preference to be examined separately.

A simple way to analyze this type of data is to calculate the density of records (numbers of birds or bird locations) in each habitat. If the records can be regarded as statistically independent from one another, a chi-squared test can be used to compare the observed distribution of records across habitats with that expected if record density was the same in all habitats. However, records are often not mutually independent and assessing preference becomes more complicated if there are several study areas, several survey dates or if a radio-tracking study covers several individually identifiable birds. Pooling bird count and habitat area data from several study areas or time periods can give misleading results if the relative areas of the different habitats vary amongst study areas or over time. It is also the case that the values of often-used measures of preference or selection, such as the forage ratio, Ivlev's index and Jacob's index, change if the relative areas of habitats available differ, even if the ratios of the densities of bird records in the habitats remains the same. This is clearly unsatisfactory and these indices are not recommended when it is envisaged that results for several areas or time periods with varying areas of habitats will be analyzed.

A measure of preference that does not suffer from this defect is the B_{i1} index of Manly *et al.* (1972). The calculation of this index involves first dividing the proportion of bird records in a habitat by the proportion of the available area in the habitat. This is done separately for each study area or time period. This is the forage ratio. The forage ratio for any particular habitat is then divided by the sum of the forage ratios for all habitats to give an index that will have the same value in different areas or surveys provided that the birds maintain the ratios of their densities across habitats at the same values.

Statistical testing could be done on the values of Manly's index from independent areas or surveys to examine the degree to which habitat preference was consistent across areas, or survey times. However, a more satisfactory approach to the analysis of data from many individual birds, study areas, or times is to use compositional analysis (Aebischer *et al.* 1993) or log-linear modeling (Heisey 1985) modified to include a randomization test (Green *et al.* 2000). Both methods analyze all the data at once and yield a ranking of relative density of use. Log-linear modeling gives values for relative density that are appropriately weighted for sample size. Both methods also have the advantage that they regard the data for each study area, survey data, or individual bird (for tracking data) as statistically independent. This is desirable because it is clearly unsatisfactory to regard multiple records of the same animal on the same habitat patch as being independent. This defect was present in some earlier widely used methods for testing the significance of habitat selection such as that of Neu *et al.* (1974). Manly *et al.* (1993) provide a useful account of the problems of measuring and testing for selection.

11.5.2 Relating numbers or densities of individuals or records of tracked birds in spatial units to the habitat composition of those units

If the data are counts of birds or records of tracked birds in areas such as transect sections, circles around point counts, grid squares, fields, or woods that each contains several habitats, then it may not be possible to attribute all the bird records to a particular habitat. This might be because birds were detected by their calls and not seen. A suitable analytical approach is then to carry out multiple regression with the density of birds or records as the dependent variable and the proportions of habitat types in each spatial unit as the independent variables. Preferred habitats will tend to have statistically significant positive regression coefficients. If the size of these sampled areas varies then this can be taken into account in the analysis, for example, by converting the counts to densities (numbers per unit area).

11.5.3 Comparison of habitat at places used by birds with that at places that are representative of the study area or known to be unused

This approach is useful when it is not feasible to map and measure habitats over large areas. Instead habitat is recorded at a sample of small sites chosen at random or on a regular grid to be representative of a much larger area available to the birds. Habitat is also recorded at places where birds are seen. Data of this type can be analyzed by multiple logistic regression with used or representative places being scored as a binary dependant variable (1 or 0) and the habitat measures as independent variables (Manly *et al.* 1993). As with the other analyses of selection, it is important to think about the availability of the random or representative places to the birds. For example, in the case of selection of foraging habitat by individually marked breeding birds, it might be appropriate to pair each foraging location with a site selected at random on the circumference of a circle centered on the nest with radius equal to the distance of the foraging site from the nest.

11.6 Food abundance and availability

Birds feed on a wide variety of organisms and methods to measure the density of all of them are beyond the scope of this chapter. Practical methods for estimating density or an index of density for many taxonomic groups can be found in Sutherland (1996). Rough measures of density may often be sufficient to answer the required questions. We want to emphasize here that studies of the food of birds require information not only on the density of food item, but also on their availability to birds, which is affected by prey activity, protective attributes (such as thorns, camouflage or poisonous compounds), depth in the substrate or height above ground in vegetation. The researcher should find out enough about the basic ecology of the bird to make choices about how to combine the measurement of food abundance with measures of habitat that influence food availability. One simple precaution is to ensure that the sampling of food abundance is being done in the type of habitat in which the birds can forage. It is often the case that birds have strong preferences for particular types of vegetation architecture in the areas where they forage. Samples taken in places that have vegetation cover that prevents birds from foraging their may give a misleading picture of food availability. Hence, pilot studies of the bird's foraging behavior and diet are recommended before large-scale sampling of food abundance and availability begins.

In some cases it may be appropriate to use a method that measures a combination of abundance and activity. For example, the catch per trap per day of ground-living invertebrates in pitfall traps is influenced by the activity of the animals as well as their abundance. This would be a disadvantage in a study of invertebrate population dynamics, but in a study of the availability of food for birds that locate prey when it is active on the surface, pitfall trap catches could be a useful measure. However, the researcher should be careful in deciding which invertebrates from those obtained in pitfall to include as being available to the study's focal bird species. A high proportion of arthropods caught in pitfall traps are nocturnally active and thought should be given to whether these are available to a diurnally foraging bird. This will depend on the foraging behavior of the bird and the resting location by day of the prey.

The availability of flying insects to birds that catch them on the wing can be assessed by powered suction nets with intakes placed high above the ground (Woiwod and Harrington 1994). Catches from even a single trap have been shown to be a good predictor of breeding parameters of local aerial feeding birds.

Suction traps often yield large quantities of insects, which would be difficult to sort and identify, but the daily total volume of the catch can be a useful measure (Bryant 1975). Insects flying near the ground can be sampled by flight interception traps (Ausden 1996) or by counting insects seen through binoculars during a watch of a standard volume of air for a fixed time (Flaspohler 1998). Visual counts of the moths and dipteran flies active during the night that are the prey of Nightjars *Caprimulgus europaeus* have be made using a vertically oriented spotlamp that is switched on briefly to minimize the attraction of insects to the light (Bowden and Green 1994).

Sweep nets can provide a useful quick way of assessing relative invertebrate abundance in dense vegetation. It is essential they are standardized, for example, 10 sweeps of constant strength and then the invertebrates counted.

The abundance of soil invertebrates can be assessed by hand sorting or otherwise separating them from the soil cores. However, this may include animals that are inactive or too deep in the soil to be available to the birds. Chemical extraction of soil invertebrates by applying a solution of an irritant chemical such as mustard to a quadrat in the field (Ausden 1996), appears to measure a combination of abundance and activity or proximity to the surface because the number of earthworms extracted per unit area shows short-term variations that are correlated with soil moisture levels that affect earthworm behavior (Green *et al.* 2000).

The availability of plant foods such as leaves and seeds can be measured by counting plants or their parts in quadrats or along transects. This approach is appropriate for seed-eating bird species that mainly take seed while it is still on the parent plant, but for species that take seed from the soil surface it is necessary to scrape a layer of soil from a measured area and separate the seeds by sieving. The seeds can then be counted and weighed. The contents of many seeds in the soil seed bank may have rotted away or have been removed by soil invertebrates and been replaced by soil. Hence, each seed (or a sample) should be crushed to check that it contains endosperm.

For ground-layer plants that are the food of grazing birds, counts of leaves or seedling cotyledons can be done in quadrats or along transects. It may be useful also to score plants for signs of damage from grazing birds to give a measure of utilization as well as availability. This is especially useful when the plant structures being eaten vary little in their number per plant, size, and shape so that the researcher can easily judge what is missing. For example, by carefully scoring damage to the paired cotyledons of seedlings being grazed by Skylarks *Alauda arvensis*, it was possible to estimate the species composition of the diet and the dry weight of cotyledon material of each species being eaten per day. The diet species composition results agreed closely with an independent assessment based upon the identification of fragments of cotyledon epidermis in Skylark droppings (Green 1980).

The main approaches for measuring fruit abundance are

1. Visit the same plants regularly and either use a quantitative measure of fruit abundance and ripeness or mark individual branches and count fruits at different stages. This shows differences in phenology between species, individuals, and years (seasonal patterns are reasonably constant in temperate regions but show considerable annual variation in the tropics).
2. Assess fruit density on the ground. This will pick up broad differences between sites and indicate timings but is obviously crude and the persistence on the ground clearly depends upon the abundance of ground dwelling frugivores.
3. Place fruit traps (e.g. suspended bags) under the canopy and count the fruit (usually ignoring aborted fruit) that has fallen. This can provide qualitative data even where the fruiting trees cannot be seen. However, to obtain sufficient data for a forest, a large number of traps are needed (75–300) (Blake *et al.* 1990). A smaller number of traps are needed if under individual trees. The abundance of fallen fruit is not the same as the abundance of fruit available to tree-dwelling birds.
4. Using transects or point counts (applying methods for bird censuses in Chapter 2) to assess fruit abundance of a range of plants (Blake *et al.* 1990).

The availability of food for nectar-feeding birds can be assessed by a combination of counts of flowers and measurements of the volume and concentration of nectar in a sample of flowers. Nectar is removed from the flower by probing it with a microcapillary tube. The volume of nectar is estimated by measuring the length of the column of liquid in the tube and its sugar concentration can be measured with a refractometer (Prys-Jones and Corbet 1987).

11.7 Predator abundance

Predator abundance may be important in determining habitat preferences. Fecal counts are often best for nocturnal mammals or for assessing the abundance of pet dogs. These may be carried out along a transect of measured length and the number within a set distance recorded. If the survey is being repeated then the feces are cleared away to prevent double counting. Creating patches of raked sand and counting the density of footprints provides a relative measure of mammal density. For most diurnal bird predators counts along transects (see Chapter 2) are usually the best approach. Eggs of domestic hens or models of eggs made from modeling clay (see Chapter 3) can be used to provide some measure of the relative nest predation risk.

11.8 Disturbance

If studying disturbance is necessary to relate the level of disturbance to the ecology of the species. A major difficulty is that there are numerous categories, for example, bird of prey, car, tractor, hunter, person on horse, person walking, or person walking with dog. With numerous categories it is then difficult to relate the impact on the birds to any one category. One approach is to measure each separately but combine similar groups in the analysis. Disturbance is usually measured by either scan counts or focal counts. In scan counts the number of disturbers is counted in each of a series of the observer's fields of view as if a series of non-overlapping photographs had been taken. For example, by moving from field to field and recording the disturbance at the first instance at which the entire field can be seen. If individuals then arrive or leave they are ignored. Scan counts can be repeated, for example, by measuring the number of disturbers at snapshots every 10 min. In focal counts an area is watched for a given period (e.g. an hour) and the number of potential disturbers counted. It is then often common to assess the impact on the birds by measuring some of number disturbed, distance at which disturbed, time disturbed for, and whether the birds resettle in the same area or elsewhere.

Acknowledgements

We thank Aldina Franco and Ian Newton for useful suggestions.

References

Aebischer, N.J., Robertson, P.A., and Kenward, R.E. (1993). Compositional analysis of habitat use from radio tracking data. *Ecology*, 74, 1313–1325.

Ausden, M. (1996). Invertebrates. In Ecological Census Techniques, ed. W.J.Sutherland, pp. 139–177. Cambridge University press, Cambridge.

Bakken, G.S., Buttemer, W.A., Dawson, W.R., and Gates, D.M. (1981). Heated taxidermic mounts: a means of measuring the standard operative temperature affecting small birds. *Ecology*, 62, 311–318.

Bowden, C.G.R. and Green, R.E. (1994). *The Ecology of Nightjars on Pine Plantations in Thetford Forest*. RSPB, Sandy.

Bryant, D.M. (1975). Breeding biology of the house martin *Delichon urbica* in relation to aerial insect abundance. *Ibis*, 117, 180–215.

Chamberlain, D.E., Hatchwell, B.J., and Perrins, C.M. (1999). Importance of feeding ecology to the reproductive success of Blackbirds *Turdus merula* nesting in rural habitats. *Ibis*, 141, 415–427.

Davenport, I.J., Bradbury, R.B., Anderson, G.Q.A., Hayman, G.R.F., Krebs, J.R., Mason, D.C., Wilson, J.D., and Veck, N.J. (2000). Improving bird population models using airborne remote sensing. *Int. J. Remote Sens.*, 21, 2705–2717.

Flaspohler, D.J. (1998). A technique for sampling flying insects. *J. Field Ornithol.*, 69, 201–208.

Gaskin, G.J. and Miller, J.D. (1996). Measurement of soil water content using a simplified impedance measuring technique. *J. Agric. Eng. Res.*, 63, 153–160.

Green, R.E. (1980). Food selection by skylarks and grazing damage to sugar beet seedlings. *J. Appl. Ecol.*, 17, 613–630.

Green, R.E. (1988). Effects of environmental factors on the timing and success of breeding of common snipe *Gallinago gallinago* (Aves: Scolopacidae). *J. Appl. Ecol.*, 25, 79–93.

Green, R.E., Hirons, G.J.M., and Cresswell, B.H. (1990). Foraging habitats of female common snipe *Gallinago gallinago* during the incubation period. *J. Appl. Ecol.*, 27, 325–335.

Green, R.E., Tyler, G.A., and Bowden C.G.R. (2000). Habitat selection, ranging behaviour and diet of the stone curlew (*Burhinus oedicnemus*) in southern England. *J. Zool., Lon.*, 250, 161–183.

Heisey, D.M. (1985). Analysing selection experiments with log-linear models. *Ecology*, 66, 1744–1748.

Hinsley, S.A., Hill, R.A., Gaveau, D.L.A., and Bellamy, P.E. (2002). Quantifying woodland structure and habitat quality for birds using airborne laser scanning. *Functional Ecol.*, 16, 851–857.

Jones, J.C. and Reynolds, J.D. (1996). Environmental variables. In Ecological Census Techniques, ed. W.J.Sutherland, pp. 281–316, Cambridge University Press, Cambridge.

Lefsky, M.A., Cohen, W.B., Parker, G.G., and Harding, D.J. (2002). Lidar remote sensing for ecosystem studies. *BioScience*, 52, 19–30.

Manly, B.F.J., Miller, P., and Cook, L.M. (1972). Analysis of a selective predation experiment. *Am. Natur.*, 106, 719–736.

Manly, B.F.J., McDonald, L.L., and Thomas, D.L. (1993). *Resource Selection by Animals*. Chapman and Hall, London.

Neu, C.W., Byers, C.R., and Peek, J.M. (1974). A technique for analysis of utilization-availability data. *J. Wildlife Manage.*, 38, 541–545.

Newton, I. (1986). *The Sparrowhawk*. Poyser, Calton.

Nott, M.P., Bass, O.L., Fleming, M., Killefer, S.E., Fraley, N., Manne, L., Curnutt, J.L., Brooks, T.M., Powell, R., and Pimm, S.L. (1998). Water levels, rapid vegetational changes and the endangered Cape Sable seaside-sparrow. *Anim. Conserv.*, 1, 23–32.

O'Connor, R.J. (1978). Nest-box insulation and the timing of laying in the Wytham Woods population of Great Tits *Parus major*. *Ibis*, 120, 534–537.

Ormerod, S.J., O'Halloran, J., Griffin, D.D., and Tyler, S.J. (1991). The ecology of Dippers (*Cinclus cinclus* (L.)) in relation to stream acidity in upland Wales: breeding performance, calcium physiology and nestling growth. *J. Appl. Ecol.*, 28, 419–433.

Peach, W.J., Baillie, S.R., and Underhill, L. (1991). Survival of British Sedge Warblers *Acrocephalus schoenobaenus* in relation to west Africa rainfall. *Ibis*, 133, 300–305.

Prys-Jones, O.E. and Corbet, S.A. (1987). *Bumblebees*. Cambridge University Press, Cambridge.

Rotenberry, J. (1985). The role of habitat in avian community composition: physiognomy or floristics? *Oecologia*, 67, 213–217.

Sutherland, W.J. ed. (1996). *Ecological Census Techniques*. Cambridge University press, Cambridge.

Sutherland, W.J. and Crockford, N.J. (1993). Factors affecting the feeding distribution of red-breasted geese *Branta ruficollis* wintering in Romania. *Biol. Conserv.* 63, 61–65.

Weins, J.A. (1973). Pattern and process in grassland bird communities. *Ecol. Monogr.*, 43, 237–270.

Wiersma, P. and Piersma, T. (1994). Effects of microhabitat, flocking, climate and migratory goal on energy expenditure in the annual cycle of red knots. *Condor*, 96, 257–279.

Woiwod, I.P. and Harrington, R. (1994). Flying in the Face of Change: The Rothamsted Insect Survey. In *Long term Experiments in Agriculture and Ecological Sciences*, eds. R.A. Leigh and A.E. Johnston, pp. 321–342. CAB International.

12

Conservation management of endangered birds

Carl G. Jones

12.1 Introduction

There is a long history of managing endangered birds. Techniques were first developed for game bird management and later adapted from falconry and aviculture to a wide range of species. Endangered birds have usually been managed at the population level by enhancing habitats, providing artificial nest sites or food, or controlling predators and pathogens. Manipulating the productivity of breeding birds has a more recent history and techniques are still being developed, especially in North America, New Zealand, and Mauritius.

In Mauritius and New Zealand, work on endangered birds on the mainland and on small offshore islands has involved habitat restoration and whole ecosystem management. This has led to integrated restoration programs addressing the ultimate environmental (e.g. habitat destruction and degradation), and the proximate demographic factors (poor survival and reproduction) that cause endangerment.

12.2 Process in the restoration of endangered species

The restoration of an endangered bird population usually starts with a synthesis of existing knowledge of the species, its life history and numbers, followed by an evaluation of the problems it faces. Research is often necessary to fill important gaps. The goal of the conservation effort is to alleviate the factors that prevent the population's recovery. With Critically Endangered and Endangered species, that by definition have small populations, it is important to increase the population as rapidly as possible and hence address the proximate limiting factors while at the same time working toward rectifying the ultimate causes of the species rarity.

The level of intervention and management is dictated by the rarity of the species. The IUCN criteria for threatened species; Critically Endangered, Endangered, and Vulnerable (IUCN 1994) provide a guide to the degree and intensity of management required. The process of restoration goes through several stages when the emphasis and priorities may change. Five broad overlapping stages provide a conceptual framework in which to develop restoration work.

12.2.1 Stage one: know your species

For many endangered species, we still know only cursory details of their life history and biology. Several early attempts to restore populations failed because not enough was known about their ecology to address, in an effective manner, the problems they were encountering. Thus the first stage is to know the life history, ecology, distribution, and numbers of the species concerned. A study of a small number of pairs will answer questions about the diet, habitat needs, and nest success. Studies of captive individuals have often been used to supplement studies in the wild, if necessary using related species to develop techniques and train staff.

12.2.2 Stage two: diagnose causes of population decline and test remedial action

There are several approaches to this problem (see Green 1995, 2002; Sutherland 2000 for useful background). Collation of existing knowledge is essential to assess previous distribution and population trends, especially information on mortality, productivity, causes of breeding failure, age structure, survival in different habitats, the impacts of weather, and other factors of possible relevance. Review any ecological changes that may have impacted upon the species, especially those brought about by recent human action.

From these exercises and information learnt in stage one, it is possible to list all the possible reasons for decline and to propose hypotheses on causes of rarity that can be tested in the field. For species where nest-sites might be limiting, this possibility can be tested by providing artificial sites or enhancing natural ones. Where food might be limiting, the provision of supplemental food, and monitoring the response of the population can give an indication of the extent and nature of the problem. (For details of previous experiments involving food and nest manipulations see Newton 1994, 1998).

For many threatened bird species on islands, it can be assumed *a priori* that known exotic mammal predators (often rats and feral cats) are likely to be affecting bird populations. On the basis of past experience, these species can be considered guilty until proven innocent, but data must be collected during any control program to evaluate its effect, and management should be modified accordingly. All management needs to be based on evidence.

It is often useful to keep a close watch on wild pairs, monitoring their behavior, anticipating and reacting to problems that may affect them. Such activities help to provide insight into the problems wild pairs face, and may enable the rescue of eggs and young from failing nests. They may also give evidence on how factors such as food, weather, and parasites affect the birds involved.

This is the stage when hypotheses are proposed and tested empirically, in order to identify the threats and to evaluate different approaches to improving productivity and survival. Staff are trained in the techniques of intensive management. These are the preliminaries to the next two stages.

12.2.3 Stage three: intensive management

Usually only applied to critically endangered populations, this stage is aimed at addressing limiting factors identified in stage two. The focus is on maximizing the productivity and survival of each individual, in order to increase numbers as rapidly as possible, and at the same time maintain as much genetic diversity, and where possible to avoid inbreeding. Intensive management may involve captive breeding and release, translocating birds onto predator free islands (where they can be carefully managed), and egg and brood manipulations. Intensive management requires great attention to detail, and may need help from avian pediatricians, veterinarians, reintroduction specialists, and other experienced support personnel (climbers, trappers, predator control, and captive-breeding personnel).

12.2.4 Stage four: population management

Populations that have not reached critically low levels can be managed without resorting to any form of intensive care, assuming that management actions enable numbers to recover to safe and sustainable levels. Management is under-taken at the population level, and is aimed at increasing a population's growth by addressing previously identified limiting factors. Typical approaches would include protection against human persecution, provision of habitat or secure nest-sites, supplemental feeding, predator or disease control, translocations to more suitable areas (but without the intensive management of Stage three). This is the stage when numbers are sufficient for detailed research to identify the most important factors that limit the population. Management is driven by these findings, and for many bird species may have to be long-term. The staff need to include researchers.

12.2.5 Stage five: monitoring

It is important to carefully monitor populations of conservation concern, both during and after restoration, so the impacts of the management can be evaluated. Consistent long-term population monitoring requires, at the least, continual assessment of numbers and distribution, and if possible also of productivity and survival.

These various components of species restoration may seem self-evident, but a surprisingly large number of restoration projects have proceeded without a clear and coherent knowledge of the problems and a plan of how to address them. Often the ultimate goal is clear but the intermediate steps are less evident. A clear step-by-step approach to species conservation allows managers to plan the work as a series of short-term achievable goals, where the roles of managers, technicians, consultants, and scientists can be clearly defined.

12.3 Broad population management approaches

Within their particular habitats, most bird populations are naturally limited by a relatively few variables, of which availability of food and safe nest-sites, predation, competition, and disease are among the most important (Lack 1954, 1966; Newton 1998). In the absence of human impact, it can be assumed that one, or an interaction of several such factors, will be limiting the size of most bird populations. The species may respond to a broad approach simultaneously addressing several of the more likely limiting factors. In declining or very small populations, productivity or survival may be enhanced by management without fully understanding the causes of population decline. This approach was applied to the Chatham Island Black Robin *Petroica traversi* and Echo Parakeet *Psittacula eques* restoration programs. The species' extreme rarity was addressed by providing some supplemental food, enhancing and protecting nest-sites, controlling/excluding predators, competitors, and parasites around nest-sites (Butler and Merton 1992; Jones and Duffy 1993). These management actions were implemented, even though it was not known at the time what was limiting Black Robin and Echo Parakeet populations, and which actions were the most important. In very small populations, empirical evaluation of the factors affecting numbers can often be the most efficient approach for understanding the causes of decline or rarity, reacting to problems as they are identified. The Chatham Island Black Robin, Mauritius Kestrel *Falco punctatus*, Pink Pigeon *Nesoenas mayeri*, and Echo Parakeet had declined to such low numbers that there was no other option available.

When attempting to restore a population, the temptation to focus exclusively upon the causes of past decline may not always be necessary (Goss-Custard 1993; Green 1995). This might in any case not be correctable in the short-term, as with habitat loss. But a species that is declining due to high adult mortality may be saved by boosting its productivity, and this may be achieved by improving its food supply or nest-sites.

In restoring critically endangered species, the initial aim is to address the proximate causes of population decline to prevent extinction and to boost numbers to a more viable level, while the long-term goal is to address the ultimate cause of decline and rarity, such as habitat destruction. Many projects flounder by failing to differentiate between the proximate and ultimate limiting factors, because the remedial actions required to address them are different.

12.3.1 Supplemental feeding

Supplementary feeding has shown a range of effects upon wild bird populations, including:

- Increasing the percentage of birds breeding
- Improving the productivity of individuals by inducing earlier laying, increased clutch size, or increased chick survival
- Improving juvenile and adult survival.

Such responses clearly demonstrate the importance of food supply in influencing individual performance, and can in turn lead to increased numbers (Newton 1998). Not surprisingly, supplemental feeding has been a main component in many bird restoration projects, often implemented alongside other measures. In particular, it may help to support other forms of management. Some Mauritius Kestrel pairs, that had been given foster young to rear, were provided with extra food (dead passerines), which allowed them to rear larger broods than normal. At Peregrine Falcon *Falco peregrinus* nests, when one of the pair was killed, the partner still managed to rear the young when given dead *Coturnix* quail (Craig *et al.* 1988; Walton and Thelander 1988).

The recovery of the central North America population of the Trumpeter Swan *Cygnus buccinator* is attributed to supplemental feeding during the winter. This population declined to about 130 birds in the 1930s. The swans fed on submerged aquatic vegetation and in late winter, when most ponds froze over, many swans died from food shortages. Annual supplemental feeding of grain was started in 1936. The level of winter mortality dropped dramatically and within 20 years the population had increased to about 600 birds (Archibald 1978a).

Winter feeding programs have also greatly benefited populations of cranes. A nonmigratory population of Red-crowned Cranes *Grus japonensis* in southeast Hokkaido was stable at about 30 birds. In the winter these cranes used to feed along streams, but in the unusually cold winter of 1952 the streams froze and so the cranes were given grain to prevent them from starving. This winter-feeding became a tradition and within 15 years the population had increased to about 200 birds, and has stayed around that level ever since. Similar population increases have been recorded in response to winter feeding programs in Hooded Cranes *G. monacha* and White-napped Cranes *G. vipio* (Archibald 1978*b*).

Some species, such as the Trumpeter Swan and cranes may be easy to feed, because they readily take grain. However, others are more difficult: attempts to feed wild Echo Parakeets, for example, were surprisingly largely unsuccessful. After trials with a range of food types offered in various ways, the best that was achieved was to feed a small number of individuals for a few weeks only (Jones and Duffy 1993). Captive-reared and released Echo Parakeets proved easier to feed and took a pelleted diet from hoppers. Young ones reared by released birds learned to use the hoppers, and thereafter some wild birds also started using the hopper, presumably through social facilitation. Birds feeding at the hoppers reared larger broods than the other wild birds, and readily accepted and reared fostered young. Even lone female parakeets successfully reared young.

Supplementary feeding has proved important in the restoration of the New Zealand Kakapo *Strigops habroptilus*. The females fed from hoppers placed in their home ranges, each female being individually managed. It is hoped that the provision of supplementary foods will promote and sustain regular breeding (James *et al.* 1991; Powlesland *et al.* 1992), and so far it has increased breeding frequency and enhanced chick survival (Elliot *et al.* 2001).

12.3.2 Enhancing nest-sites and the provision of nest-boxes

In many bird populations, nest-sites are limiting, especially for species that nest in tree cavities or on cliff ledges (Newton 1998). In addition many birds have only poor sites, which do not protect them against predators or adverse weather. Many species have increased in numbers after the enhancement of existing nest-sites or creation of new ones, thus demonstrating the limiting effect of nest-sites on breeding density and success (Newton 1998). Many different artificial nest-sites have been successful, such as nest ledges and cavities for cliff nesting species, platforms, and artificial stick nests for tree nesting species, artificial burrows for terrestrial hole nesters, rafts for wetland birds, and nest-boxes for a whole range of cavity nesting species.

In the absence of high quality nest cavities, Echo Parakeets and Mauritius Kestrels tried to nest in sites prone to predation, flooding, or overheating. Consequently, it became policy to improve nest cavities that were considered suboptimal. At cavities frequented by kestrels any debris, such as old nest material or loose rocks lying on the cavity floor, was replaced with washed gravel. If necessary, the cavity entrance was modified by placing rocks to provide landing spots or perches for the kestrels, and in exposed sites rocks were arranged to provide shade and shelter. Pairs subsequently raised young in many of the modified

cavities. Sites that were unsuitable because they were easily accessible to predators were permanently blocked (Jones *et al.* 1991). The use of nest boxes has
greatly increased the numbers of breeding kestrels on Mauritius by providing
nest sites in areas previously lacking them. In one subpopulation exposed to a
shortage of natural cavities in the 2002–03 season, 27 (63%) out of the 43
known pairs used nest-boxes.

Nest cavity modification has been a major feature of the Echo Parakeet
restoration work. Cavities were modified according to the characters of the most
secure and successful sites, with changes to every occupied cavity, and others
where parakeets were seen prospecting. In the 2002–03 season, there were
21 breeding pairs, 17 in modified tree cavities and 4 in nest-boxes.

Any nest-sites in rotten trees that were in danger of falling were destroyed,
while others were reinforced and repaired. In sites that flooded in wet weather,
drainage holes were inserted, or weather guards were placed around the entrance
to keep out driving rain. Shallow cavities are deepened to at least 70 cm. An
entrance door was built into the side of every cavity, so that field-workers could
gain access to eggs and young. The substrate was changed in all cavities before the
breeding season, and again every week or two during the nesting period, in order
to maintain hygiene and to prevent the build up of nest parasites.

In some cavities, the size of the entrance hole was reduced to exclude White-
tailed Tropic Birds *Phaethon lepturus*, (which have plenty of other sites), in order
to increase the numbers available to parakeets. At some cavities a network of
branches was placed near the hole, to prevent long-winged, non-perching tropic
birds from gaining access to the cavity, but to provide perches for adult and newly
fledged parakeets. In all nest trees, predator guards in the form of smooth plastic
sheeting wrapped around the trunk were fixed and any interlocking branches
from neighboring trees were pruned off to discourage monkeys and rats. Between
1987 and 2002, 45 modified cavities were used between one and seven times.

Nest-site enhancement has been an important component in the restoration
of the Puerto Rican Parrot *Amazona vittata*. For this species nest-sites were found
to be in short supply and accessible to predatory Pearly-eyed Thrashers *Margarops
fuscatus*, a recent colonist of Puerto Rican forests (Snyder 1978; Wiley 1985;
Snyder *et al.* 1987). Some cavities were used for 20 years or more, as good sites
that were both secure and of suitable size were scarce. The modification of nest
cavities was also a component of the California Condor *Gymnogyps californianus*
restoration project, where cliff cavity floors were leveled and rock baffles built for
protection (Snyder and Snyder 2000).

The Echo Parakeets on Mauritius would readily accept modified cavities, but
for many years refused to use nest-boxes, until at last a design was found that was

acceptable to some wild birds. Released parakeets, and wild males paired to released females, have readily accepted nest-boxes. The reluctance with which the wild Echo Parakeets have accepted nest-boxes is mirrored by the experience of others working with wild parrots. For example, efforts with three *Amazona* parrot species in Mexico, with St Lucia Parrots *Amazona versicolor* and Puerto Rican parrots have almost all failed (N. Snyder personal communication). However, by contrast, Blue and Gold Macaws *Ara ararauna* readily accepted nest-boxes (Munn 1992) as did Green-rumped Parrotlets *Forpus passerinus* (Beissinger and Bucher 1992). Nest-boxes increased the number of breeding pairs of the Green-rumped Parrotlet, and were more secure than natural holes. Birds nesting in boxes had more frequent and larger broods. This is a common finding with nest-boxes where predation rates are often lower. In addition cavity size may influence clutch and brood size.

Artificial ledges and cavities have been successfully constructed for many cliff nesting bird species, including Northern Bald Ibis *Geronticus eremita* (Hirsch 1978) and various raptors. In Germany artificial sites suitable for Peregrine Falcons *Falco peregrinus* have been made in quarries and cavities have been blasted in cliff faces (Hepp 1988) so that about 80% of eyries in the Black Forest area were in artificial sites, where breeding success was "distinctly higher than in natural nests" (Brucher and Wegner 1988).

Competition over cavities can be severe. The size of the entrance hole is often important, and for most species the smallest hole through which they can enter is the safest, since this excludes larger species. Minimizing the entrance hole was used to exclude White-tailed Tropic Birds, which were competing for nest-sites with the smaller and much rarer Bermuda Petrel *Pterodroma cahow* (Wingate 1978). The exclusion of the tropic birds led to improved breeding success and numbers of petrels, with pairs increasing from 18 in 1962, when management first started, to 26 in 1977 (Wingate 1978) and an estimated 180 birds (53 breeding pairs) by 1997 (Stattersfield and Capper 2000).

Nest-boxes are widely used in Europe and have resulted in increases in populations of many hole-nesting birds, including Pied Flycatcher *Ficedula hypoleuca*, Collared Flycatcher *F. albicollis*, Redstart *Phoenicurus phoenicurus*, various tits *Parus* spp., Tree Sparrow *Passer montanus*, and Starling *Sturnus vulgaris* (Newton 1994). In North America nest-boxes have resulted in a steady increase in the populations of Bluebirds *Sialia* spp. (Zeleny 1978, Newton 1998). Nest-boxes represent an alternative to natural hollows, but are not always an adequate replacement because they do not reflect the diversity of natural hollows (Gibbons and Lindemayer 2002). Some species prefer natural cavities to nest-boxes (e.g. Treecreeper *Certhia familiaris*), and for these more research is needed on nest-box design.

12.3.3 Disease control

Parasitic disease was once considered to have little impact on most bird species, only in exceptional circumstances being a major cause of mortality (Lack 1954, 1966). We now know that disease is an important component in the population limitation of many birds (Newton 1998), while the role of pathogens in threatened bird populations has been reviewed by Cooper (1989). Introduced diseases may have profound impacts on native hosts, a well known example being the introduced avian malaria and pox in Hawaii, which limits the endemic honeycreepers to upland areas where mosquito densities are low (Van Riper *et al.* 1986). The Pink Pigeon on Mauritius suffers high nestling mortality from trichomoniasis, caused by a flagellate protozoan believed to have been introduced to Mauritius with exotic doves (Swinnerton 2001).

A knowledge of the disease profile of the focal species is often useful so that:

- The likely effect of disease upon the survival and breeding of the species can be understood.
- The disease can be combated as a cause of poor breeding or survival.
- New diseases can be excluded by quarantine measures.

A health audit of a wild population of a managed critically endangered species needs to be implemented during the early stages of the project (Stage 2). Surveys of disease, and knowledge from similar surveys on related species, provide useful indicators (Joyner *et al.* 1992; Gilardi *et al.* 1995). It is important to find what diseases are present, how these may be influencing survival and productivity, and how they can be managed to minimize their impact.

The species should be screened for diseases known to be important to closely related species (where such information is available). For example, pigeons are prone to trichomoniasis and parrots to several viral infections such as psittacine beak and feather disease and poliomavirus. There also needs to be a more general screening for parasitic diseases to look for ecto-parasites, blood parasites, and endo-parasites. Fecal samples should be screened for pathogenic bacteria, with selective culture for fungi, yersinia, and chlamydia. All dead adults, chicks, and eggs should be postmortemed in an attempt to understand the causes, and also the patterns of mortality (Greenwood 1996).

A careful health audit of both wild and captive birds enables measures to be taken to avoid transmission of disease from captive to wild populations and *vice versa*, or from one species to another. Where there are *in situ* captive facilities ideally these should be for single species only, and where other species are held in the same place, they should be screened to avoid transmission of disease to the focal species. Young captive Pink Pigeons reared by domestic pigeons *Columba livia* died when they contracted pigeon herpesvirus from their foster parents (Snyder *et al.* 1985).

Birds intended for reintroduction should be raised and kept away from unnecessary contact with other captive birds, which may be carrying a disease to which the population is naive. Concerns about the risks of introducing disease into the wild by restocking with captive bred birds need to be set against a background of disease in the existing population. If a disease is already present, we may be less concerned about introducing that disease from captivity (Greenwood 1996).

There are disease risks associated with many management procedures, although these risks are often small. Fieldworkers may transmit parrot viruses on their clothes. Many of the problems can be minimized by good hygiene. Supplemental feeding stations need to be kept clean and only good quality food used. The spread of salmonellosis among British wild birds concentrated at garden feeding tables is well recognized (Wilson and Macdonald 1967). At the supplementary feeding stations used by free-living Echo Parakeets and Pink Pigeons, the hoppers have been designed to exclude most other species. A high incidence of trichomoniasis in young Pink Pigeons was believed to have been exacerbated by exotic doves drinking and feeding from the same hoppers (Greenwood 1996). The incidence of trichomoniasis decreased when the exotic doves were excluded.

Disease management can reduce mortality and some parasitic diseases can be treated in the field, especially at nest-sites. Tropical Nest Fly *Passeromyia heterochaeta* larvae feed on nestling Echo Parakeets and can be major source of mortality (Jones and Duffy 1993). This problem was eliminated by the addition of an insecticide dust (5% carbaryl) to the nest substrate at the beginning of the season, and every 2 weeks during the nestling period. Similarly, insecticide powder was applied to Black Robin nests to reduce the build-up of mites that, if unchecked, could cause brood desertion and mortality (Butler and Merton 1992). At Echo Parakeet nest cavities, the substrate was also treated with the abendazole to control fungal diseases such as aspergillosis.

12.3.4 Predator control

Predator control is a component of many bird restoration projects. When causes of predation are unknown, the approach should be protective, to minimize the possible impact of predators, and reactive, responding to any detected predation. The indiscriminate killing of even common predators is not recommended, as it may lead to unforeseen problems. An exception is when dealing with exotic predators known to be a problem elsewhere. In restoring native bird species on islands, introduced rats *Rattus* spp., feral cats *Felis catus*, Small Indian Mongoose *Herpestes auropunctatus*, Stoat *Mustela erminea*, Mink *M. vison*, foxes *Vulpes*, and *Alopex* all are potential problem species (Merton 1978).

On mainland or continental areas, predator control is usually a localized or a short-term option where it is best focused around areas where the focal species is most vulnerable, that is, nest-sites, roost sites, feeding areas, supplementary feeding stations, and release sites. Long-term predator control over large areas is usually not sustainable. However, in New Zealand biologists are experimenting with managed areas of up to 6000 ha in which smaller core areas are intensively managed. These areas are termed "Mainland Islands." Within these areas large herbivores are shot from helicopters and the exotic rats and Brushtail Possums *Trichosurus vulpecula* are controlled by the aerial distribution of toxic baits, and in the intensively managed areas trapping grids are set for exotic mammals including mustelids and feral cats. Mainland Islands have been successful in providing secure habitat for a range of native species and numbers of kiwi *Apteryx* spp. and Kokakos *Callaeas cinerea* have increased (Innes *et al.* 1999). A more sustainable long-term option is to eradicate exotic predators on islands, which can then be used as reintroduction sites for endangered bird species (see *Translocations*) or to enclose areas in predator proof fences.

Approaches available include close guarding, the provision of safe nest-sites, predator guards around nest trees, and the placement of supplemental feeding, and release sites in safe fenced locations. An advantage of close guarding is that it may reveal unknown predation problems. The first young Mauritius Kestrels released spent time on the ground, where they were susceptible to mongoose and cat predation, which explained losses of up to 25% of young at some sites. A close guarding and trapping program around some nest-sites and most release sites reduced the losses to these predators (Jones *et al.* 1991, 1995; Cade and Jones 1994).

The removal of introduced predators from islands may allow native birds that still exist to recover rapidly. On Little Barrier Island, New Zealand, fewer than 500 Stitchbirds *Notiomystis cincta* remained, but after the eradication of cats the population recovered within a few years to 3000 (Veitch 1985). Similarly on Raratonga in the Cook Islands, the endemic flycatcher *Pomarea dimidiata* declined to 29 birds in 1989, due primarily to nest predation by Black Rats *Rattus rattus*. Rat control with poison laid out on a grid system throughout the habitat, together with rat-proofing of nest-trees with predator guards, resulted in an increase in the population to 189 over the next 10 years (Bell and Merton 2002).

The Aleutian Canada Goose *Branta canadensis leucopereia* declined after Arctic Foxes and Red Foxes *Vulpes vulpes* were introduced to the islands where they bred. The breeding geese were reduced to just one fox-free island.

Foxes were removed from several islands and the goose population recovered to its former densities, helped by some reintroductions (Springer *et al.* 1978; Byrd *et al.* 1994).

The restoration of islands that can be used as refuges for endangered birds is a well-proven technique and is successful because harmful predators can be completely eradicated. On mainland areas, all that is usually possible is localized control or the fencing out of some problem species. Fencing technology is becoming increasingly sophisticated and the New Zealand "super fence" keeps out all mammals including rats and mice. The Karori Wildlife Sanctuary, Wellington has been surrounded by 8.6 km of fencing and threatened species that may formerly have occurred there such as the Little Spotted Kiwi *Apteryx owenii* are being re-introduced within the fenced area (Bell and Merton 2002; J. Mallam personal communication).

Predator control raises issues of ethics and welfare and should always be carried out to the highest standards. Although the science of predator control and eradication is well established, a great deal of experience is usually needed before trappers become efficient. Good trappers approach the subject with meticulous detail and a keen intuition and consequently may catch many more animals than a novice.

12.4 Intensive management of focal pairs

12.4.1 Close guarding and monitoring of nests

The main purposes of close guarding are to:

- Monitor the progress of the focal pairs and to build a body of knowledge on the biology and behavior of the species.
- Assess the suitability of the watched pair for possible clutch and brood manipulations.
- Monitor the results and progress of any manipulations.
- React to problems that threaten the pair or their nesting attempt (e.g. nest-site enhancement, supplemental feeding, control of parasites, predators, and competitors).
- Rescue clutches and broods from failing nesting attempts, or if necessary hand-feed and re-hydrate ailing chicks.

In the most critically endangered species 24-h guarding and monitoring has sometimes been undertaken. Some are monitored by video systems (Kakapo project) or by teams of volunteers (California Condor, Pink Pigeon, Echo Parakeet projects), but there must be clear guidelines on procedures if the nest shows signs of failing. Close guarding has been an important component in restoration programs for the Kakapo, Chatham Island Black Robin, Californian Condor, and Echo Parakeet. It has enhanced the productivity of focal pairs (e g. Butler and Merton 1992; Jones and Duffy 1993; Jones et al. 1998; Merton et al. 1999; Snyder and Snyder 2000; Elliott et al. 2001).

12.4.2 Clutch and brood manipulations

The purposes of clutch and brood manipulations are to increase the productivity of focal pairs, providing the birds concerned will tolerate the intrusion. In most species of birds, the number of fertile eggs laid is considerably greater than the number of young that leave the nest. There are losses during incubation and rearing that can often be minimized by careful management, and the eggs or young can be harvested without increasing the overall loss.

In some species, if eggs are harvested one at a time or as whole clutches, replacement eggs, or clutches are laid, thereby increasing the number of viable eggs produced. The harvested eggs can then be hatched in other ways, and the young reared by hand or fostered in other nests. Brood manipulations increase or decrease the number of young in the nest, but can also involve cross-fostering, fostering, or supportive care to the chicks and parents.

Harvesting and rescuing eggs

These techniques have been applied to many species to minimize the loss of viable eggs. The Whooping Crane *Grus americana* normally lays two eggs but only one young usually survives. The "surplus" eggs were harvested for captive rearing. Of 50 eggs harvested from the wild, 41 (82%) hatched and 23 (56%) of the chicks were reared to at least 6 months old (Kepler 1978). These were used to establish a captive population to provide eggs and young for reintroduction.

An important egg harvesting study involved Peregrine Falcons in North America. The falcons had poor breeding success due to DDE contamination (from the insecticide DDT) that was causing the females to lay thin-shelled eggs. Most pairs failed because the incubating adults accidentally smashed the eggs, and this caused populations to decline. In one study, the hatch rate of thin-shelled eggs under the wild birds was only 7% (Craig et al. 1988). In a sample of 661 harvested eggs, 536 were apparently fertile and alive when harvested, 386 (72%) hatched, and 356 (92%) chicks were reared to fledging. The majority of these were released by fostering and hacking (Burnham et al. 1988; Walton and Thelander 1988). The latter is a procedure that allows young to fly naturally from an artificial nest-site, to which they can return for food until they have learned to hunt for themselves about a month later.

Fostering of eggs

Eggs may be fostered to nests to add to those already there or to replace non-viable eggs. Fostered eggs should be at the same stage of development as the rest of the clutch.

Clutch augmentation is used with captive birds where pairs may incubate and hatch larger clutches than normal, but is limited by the number of eggs that the incubating bird can effectively cover. Wild birds on clutches larger than normal are likely to succeed in rearing larger broods only when natural food is not limiting or when extra food is provided.

Egg augmentation and replacement is an easy way to ensure that all nests in a population have the possibility to hatch young, and is an useful technique to introduce captive–produced eggs into a wild population.

Sequential egg removal

If an egg is removed soon after it is laid, the bird keeps laying further eggs in an attempt to complete a clutch, sometimes producing more eggs than the usual. This technique only works on birds that have an indeterminate clutch size. It is a technique most often used on captive birds, where the laying of eggs can be carefully monitored. Captive Sandhill Cranes *Grus canadensis* have exceptionally laid 18 and 19 eggs in succession, yet the normal clutch size does not exceed four. Single egg removal resulted in an average of 6.4 eggs per bird per year compared with 5.3 if the birds were "double clutched" (i.e. removal of a complete clutch to stimulate the laying of another). These egg removal studies did not have any marked effect upon egg viability (Ellis *et al.* 1996).

Multiple and double clutching

Many species of birds that normally produce only one clutch per season can lay a replacement clutch if the first nesting attempt fails, and some species can lay several clutches in a season. This ability to recycle can be exploited by removing the first clutch, and sometimes successive clutches, for artificial rearing (or fostering) and then leaving the pair with a final clutch to incubate and to rear themselves.

This technique was used for California Condors in the wild prior to the last birds being brought into captivity. Californian Condors lay single egg clutches and if they successfully rear a chick it is dependent on its parents for so long that they do not breed the following year; hence successful wild Condors can produce only a single independent young every other year. Between 1983 and 1986, 16 eggs were taken for artificial incubation from five different pairs. Thirteen (81%) of the eggs produced surviving chicks, far exceeding the 40–50% fledging success of wild pairs. Of ten pairs whose first clutch was removed, six relayed; and following removal of these second clutches, three pairs went on to lay a third time in the one season. The condors were retained to establish the captive-breeding program for the species (Snyder and Hamber 1985; Snyder 1986; Toone and Wallace 1994; Snyder and Snyder 2000).

Double clutching of wild Peregrine Falcons became routine in the Western United States (see *Harvesting and rescuing eggs*) (e.g. Burnham *et al.* 1988, Walton and Thelander 1988). The clutch was typically removed 7–10 days after completion. The delay in removal was to allow some natural incubation, which increased the subsequent hatchability in incubators, compared to eggs in incubators throughout. In one sample of 13 removed first clutches, all pairs laid a second clutch, usually after about 12 days. Second clutches were sometimes smaller and averaged 3.2 eggs, compared to 3.5 eggs in first clutches (Craig *et al.* 1988).

Double clutching was tried on Echo Parakeets, but with poor success. First clutches were removed, or lost, from wild pairs on 18 occasions, and in twelve (67%) of these a second clutch was laid. The second clutch was started 19–21 days (once about 30 days) after the loss of the first clutch. It was difficult to predict in Echo Parakeets if a pair would renest, and repeat nesting attempts were not as successful as first ones. Only 23% of eggs from second clutches resulted in fledged young (Jones and Duffy 1993; Jones *et al.* 1998). In view of this poor success, double clutching was stopped.

Many species respond to loss of a clutch by moving to another nest-site, so clutch removal has been used to move Peregrine Falcons from unsuitable nest-sites to secure ones (Craig *et al.* 1988) and to move Mauritius Kestrels from cavities that were accessible to predators to predator-proof nest-boxes.

Because egg quality often declines in replacement clutches, there are tradeoffs in management that have to be considered. A protocol for the harvesting of eggs from wild pairs of Mauritius Kestrels was developed, in which no pair was made to lay more than one extra clutch in a season. Harvested eggs of many species of wild birds have a better hatchability in artificial incubators if they have received some natural incubation, yet the birds recycle more readily if the eggs are harvested soon after the first clutch has been laid. In Mauritius Kestrels, the eggs were harvested about 5–7 days after clutch completion. Following the removal of the clutch, Mauritius Kestrels would usually move nest-site, so alternative nest-boxes were provided. First time breeders were left with their first clutch and not encouraged to lay additional eggs. It was considered important that young birds should rear young if they were to become good breeders in future years (Jones *et al.* 1991). Similar protocols have been applied to Peregrine Falcons (Walton and Thelander 1988).

Fostering

Three main types of fostering can be distinguished:

- **Augmentation**. The addition of young, thus increasing the size of the brood.
- **Replacement**. The replacement of a clutch of eggs with a brood of young, or the replacement of one brood with another.
- **Swapping**. The swapping of young between broods, so that all the young are about the same size, thus reducing the risk of mortality.

Fostering has been widely practiced in captivity in a range of species from many different orders. Work on wild birds has been limited, and the most detailed and successful studies have involved birds of prey.

In general, the more experienced the pair, the more liberties can be taken. Some pairs are poor at rearing and can never be trusted with their own or fostered young. The young to be fostered should not have developed fear reactions or they may refuse to accept food from the adults. In species that produce altricial young, fear reactions do not usually develop until the second half of the nestling period. Fostering attempts with species that produce precocial young are usually done as eggs since the young form attachments to their parents soon after hatching.

Augmentation fostering. The candidates for augmentation fostering are usually birds with a smaller than normal broods. The fostered young should be close to the age and size of the young that the adults are rearing.

The enlargement of normal brood size by adding extra chicks to the nests of altricial species has given variable results. In 11 out of 40 brood enlargement experiments reviewed by Dijkstra *et al.* (1990), enlarged broods suffered greater mortality and yielded fewer fledglings than control broods, suggesting that in these cases food was limiting. In the remaining 29 experiments, the enlarged broods produced more fledglings, on average, than control ones, showing that many species were able to raise larger than normal broods. If birds are to be given extra young, extra food provision is a good precaution.

Replacement fostering. The replacement of whole clutches of eggs with young is usually applied to birds that have been incubating non-viable eggs or whose eggs are needed for other purposes. The young do better if they are several days old and hence are stronger and easier to feed than newly hatched chicks. Large falcons accept young up to about 3 weeks old (Fyfe *et al.* 1978). In Echo Parakeets the optimal age for fostering is 4–7 days, although experienced females will accept and rear younger or older chicks.

Experience with wild Mauritius Kestrels and Echo Parakeets has shown that, if birds fail in breeding, but are going to be required to foster young later, then they can be given dummy eggs, even for up to 5 days after the young have hatched. They will incubate these for up to 5 days and when given the foster young will look after them.

Swapping. In some species, such as raptors and parrots, where hatching is asynchronous, the smaller young have poorer survival. The swapping of young between broods, so that all in each brood are about the same size, can enhance the survival of the compromised young and increase brood size at fledging. This has worked with several species including Kakapo, Echo Parakeet, Mauritius Kestrel, Pink Pigeon, and Spanish Imperial Eagle *Aquila adalberti* (Meyburg 1978).

Cross-fostering of eggs and young. Cross-fostering the young (or eggs) from one species to another has been tried in many taxa. Usually a common species is used to rear the young of a rarer species, freeing up the rarer species to lay additional clutches. Sometimes, however, the young of a common species have been fostered to a rarer species to test parental abilities and to provide rearing experience before a fostering attempt with a conspecific.

For centuries strains of domestic chickens have been used to incubate the eggs and rear the chicks of captive game birds and waterfowl. Domestic Bengalese Finches *Lonchuria striata* have been used to rear rarer estrildid finches, especially Gouldian Finches *Chloebia gouldiae*. But most of the conservation-orientated cross-fostering studies on wild birds have involved diurnal birds of prey and the Chatham Island Black Robin, where cross-fostering has been attempted using a common species to rear the young of a rarer species. Olendorff *et al.* (1980) and Barclay (1987) review cross-fostering studies in raptors involving 12 different species.

Intra-generic cross-fostering. McIlhenny (1934) pioneered cross-fostering as a successful conservation technique on wild birds. Snowy Egret *Egretta thula* eggs were harvested and cross-fostered to the nests of the commoner Little Blue Herons *E. caerulea* and Tricolored Herons *E. tricolor*. The Snowy Egrets recycled and were left to incubate their second clutches and rear the young. The cross-fostering was successful and the Snowy Egret population rapidly increased.

Subsequently the cross-fostering of wild birds was attempted in ethological studies. Schutz (1940 quoted by Cade 1978) placed the eggs from a tree-nesting colony of Common Gulls *Larus canus* in the nests of Black-headed Gulls *L. ridibundus* among reeds. On reaching sexual maturity, the Common Gulls returned to their hatching and rearing location and formed a small colony within the Black-headed Gull colony. The Common Gulls adopted a new breeding location and apparently paired preferentially with their own species rather than with Black-headed Gulls.

Harris (1970) swapped the eggs of the Lesser Black-backed Gulls *Larus fuscus* with Herring Gulls *Larus argentatus* on Skokholm Island, Wales. Some 496 Lesser Black-backed Gull eggs were placed in the nests of Herring Gulls and 389 Herring Gull eggs were placed in the nests of Lesser Black-backed Gulls. Cross-fostered gulls were subsequently found breeding on Skokholm and, of these, 71 were in mixed species pairs and 44 were breeding with their own species. Harris found that cross-fostered females usually mated with the males of their foster parent while the males mated with either species.

In an attempt to reintroduce Peregrine Falcons into their former range, young were fostered in several areas into the nests of Prairie Falcon *Falco mexicanus*. In California 113 nestling Peregrine Falcons were fostered into nests of wild Prairie Falcons, and all or most fledged. About ten of these cross-fostered Peregrines were later found breeding normally in the wild and none was seen mated to a Prairie Falcon (Walton and Thelander 1988; Cade and Temple 1995).

In the Mauritius Kestrel, inexperienced wild pairs were, on four occasions, given pipping eggs of Common Kestrels *Falco tinnunculus* to gain experience of hatching and rearing. One pair proved competent enough for the Common Kestrels to be replaced with Mauritius Kestrels (Jones *et al.* 1992). Similarly, Chatham Island Tits *Petroica macrocephala* were fostered under the rarer Chatham Island Black Robins to give the robins rearing experience (Butler and Merton 1992).

The cross-fostering of Chatham Island Black Robins has been the most successful use of this technique for the conservation of a critically endangered species. Black Robin eggs and young were cross-fostered under Chatham Island Tits and Chatham Island Warblers *Gerygone albofrontata*, a procedure which encouraged Robin pairs to lay replacement clutches. Although the warblers hatched the eggs and reared the young robins successfully for the first week, they could not bring sufficient food to rear them beyond 10 days old. Subsequently, if warblers were used as foster parents, the young were moved back to robins after a week. The tits could rear the Black Robins successfully to fledging, but these young imprinted to, and attempted to breed with tits. This was overcome by: (1) swapping the cross-fostered young back to robins before fledging so that they developed the appropriate species fixation; and (2) translocating robins that fledged under tits to an island lacking tits so that the robins had no option but to breed with each other, which they did with some initial reluctance. Of 180 black robin eggs incubated by tits, 156 (87%) hatched—a figure comparable to that from eggs incubated entirely by robins (Butler and Merton 1992).

Cross-fostering should proceed with care because of the very real problems of sexual imprinting to the foster species (Immelman 1972). Due to the taxonomic closeness of species within the same genus there is the possibility that they would produce fertile hybrids. For example, Scarlet Ibises *Eodocimus ruber* were introduced to Florida by cross-fostering the eggs to White Ibises *E. albus*. This procedure resulted in considerable hybridization, producing "pink" ibises (Long 1981).

Inter-generic cross-fostering. After a series of cross-fostering experiments in captivity, Fyfe *et al.* (1978) experimented by placing broods of Prairie Falcons under three species of *Buteo* hawks. This was done to test the suitability of these different *Buteo* species as foster parents, but also to see if the procedure could be used to extend the Prairie Falcon's distribution. Two pairs of Ferruginous Hawks *Buteo regalis* raised four out of five young, three pairs of Red-tailed Hawks *B. jamaicensis* raised ten out of eleven young and a pair of Swainson's Hawks *B. swainsoni* raised four out of five young in a single brood. Subsequently three pairs of Prairie Falcons were found breeding near to the cross-fostering sites but outside of the natural range of Prairie Falcons; they were assumed to have been the young from the cross-fostering experiments. One pair even nested for two seasons in an old *Buteo* nest, the first time this behavior had been recorded in Prairie Falcons. Clearly this trial, although modest in terms of sample size, demonstrates the potential of the technique. Similarly, in Germany Common Buzzards *Buteo buteo*, Goshawks *Accipiter gentilis*, and Common Kestrels nesting in trees have been used to rear young Peregrine Falcons in an attempt to re-establish a tree nesting tradition in the latter species (Saar 1988). Several pairs of Peregrine Falcons have subsequently nested in trees in the area concerned (Cade 2000).

Where possible, cross-fostering should be of whole broods so that the siblings can socialize with each other. Where single young are raised by foster parents of another species, then the only possibilities of early socializing and sexual imprinting are with the foster parents, as indeed happened when Whooping Cranes were fostered under Sandhill Cranes *Grus canadensis* (Ellis *et al.* 1996). Cross-fostering has worked well in some inter-generic attempts, where hybridization between the species was highly unlikely, as demonstrated by the successful rearing of young falcons by *Buteo* and *Accipiter* hawks.

There are clear behavioral constraints on which species will tolerate the young of another species. In small passerines, neither domestic Bengalese Finches (Estrildidae) nor domestic Canaries (Fringillidae) reared the young of Madagascar Fodies *Foudia madagascariensis* (Ploceidae) even though they incubated and hatched the eggs. Neither species responded to the begging cries and gaping of the young fodies. The Bengalese Finches did, however, rear the young of the congeneric Spice Fince *Lonchura punctulata* even if they were not themselves either incubating or rearing young at the time (C.G. Jones, unpublished).

Chick rescue

Some species of birds hatch more young than they successfully rear, and in any restoration project opportunities may arise to rescue failing chicks. Rescued chicks are likely to be under-weight and dehydrated, with reduced survival chances, but some survive for release to the wild.

The young of several species of eagles engage in siblicide or "cainism," where one chick may kill its siblings. Studies on Lesser Spotted Eagles *Aquila pomarina*, Spanish Imperial Eagles and Madagascar Fish Eagles *Haliaeetus albicilla* have all demonstrated that the productivity of these eagles can be increased by removing the weakest chick for hand-rearing, fostering or cross-fostering (Meyburg 1978; Watson *et al.* 1996; Cade 2000). This approach has potential to boost the productivity of rare eagles.

Rescuing chicks that are not growing well has been a valuable technique in the management of the California Condor (Snyder and Snyder 2000) and Echo Parakeet. Wild parakeets lay clutches of 2–4 eggs but usually only rear one chick, sometimes two, apparently because insufficient food is available. To avoid this loss, wild parakeets are allowed to hatch their eggs and keep the young for the first 5–8 days. Young were then removed and either hand reared or given to foster pairs that had failed to hatch their own young. Of 38 chicks rescued when starving, 29 (78%) fledged, while of 14 chicks removed earlier, all subsequently fledged.

Supportive care of young birds in the nest

Providing supportive care to young birds in the nest during periods of temporary food shortage, when the adults are having difficulty feeding them, may improve their survival chances. Young birds fostered under inexperienced adults, in replacement of a clutch of eggs, may need some hand-feeding while the parents learn how to feed them adequately. During inclement weather, when adults have difficulty foraging, it has sometimes proven necessary to feed and re-hydrate young Mauritius Kestrels and Echo Parakeets in the nest to help them through a brief period of food shortage. However, it is usually more efficient to feed the adults, if they will accept supplemental food, and let them pass it on to the young.

12.5 Reintroduction and translocations

12.5.1 Reintroduction

Reintroduction is usually defined as the release of captive-bred or captive-reared birds into an area which was once part of their range but from which they have become extirpated (IUCN 1998), but can also include the addition of individuals to an existing population. This latter category of reintroductions is often treated separately as re-enforcement or supplementation (IUCN 1998), but since both types of reintroduction are often used in the same project, they are here treated together.

Reintroductions have received a great deal of attention due to their high profile nature (Fyfe 1978; Cade 2000). The reintroduction of some species works well (birds of prey), and others have proven problematic (parrots, hornbills, and some passerines). Successes are becoming more frequent, as we learn more about the needs of different species. There are several different release techniques, of which fostering and cross-fostering of eggs and young have already been described. Other release processes can be divided into hard or soft releases. The hard release (also termed abrupt release) is when the bird is released without any preliminary conditioning to the area and is not given any support thereafter, on the assumption that it will be able to look after itself. Many early reintroductions were of this type and were characterized by a high failure rate. As a rule, hard releases are best avoided.

In soft release (also called gentle release), the birds are habituated to the area before release and are provided with some form of support during the release process. Usually the birds are provisioned after release with food and water, so that they can become independent of human care gradually. Soft releases are more successful than hard-releases. For example, post release survival to 1 year was 12% of 51 Sandhill Cranes that were hard-released, and 68% of 238 cranes that were soft-released (Nagendran et al. 1996).

The process of a soft release falls into three stages: (1) pre-release training and conditioning; (2) the release process; and (3) post-release support. Birds intended for release have to be correctly socialized, with due care taken to ensure that the stimuli during the early learning stages are appropriate. For most releases, parent reared, or foster raised birds are to be desired because they will have experienced normal imprinting and socialization. If hand-raised birds are being used, it is important that they are raised with siblings to ensure socialization with conspecifics or, if they are being raised alone, that they are fed with the aid of a puppet that mimics the adult so that the young birds imprint upon an appropriate image (Wallace 2000). Puppet rearing has been used with Californian Condors, and Takahe *Porphyrio mantelli*. For cranes, the rearer wears a full body crane suit. Attention also needs to be given to early learning of nest-site characteristics, and all individuals should be provided with opportunities to develop physical and survival skills (Wallace 2000).

Birds of prey are usually released using a soft release technique called "hacking" (Sherrod et al. 1981). The birds are placed in an artificial nest-site once they are homeothermic and are old enough to feed themselves from provisioned food.

The birds fledge from the artificial nest-site at the usual fledging age and develop their flying and hunting skills gradually at ages comparable to wild-reared birds. During this learning period, food is provided on or near the nest until it is no longer taken.

The hacking technique is not suitable for some other groups of birds that are incapable of feeding themselves from provisioned food until after fledging. Hence, they are usually kept in captivity until they have been fully weaned, and released.

Released birds are unlikely to do as well as young reared in the wild. There are exceptions, however, for survival of released birds of prey is very high and in Mauritius Kestrels is comparable to that of wild fledged birds. Survival in reintroduced captive-reared Takahe was also equal to that of wild reared birds (Maxwell and Jamieson 1997), as was that of Pink Pigeons, although for the pigeons additional food was provided and predators were controlled (Swinnerton 2001).

For most species, the earlier in the post-fledging period they can be released, the better their subsequent survival. In the Pink Pigeon, 68% of 196 birds released before reaching 150 days survived for at least a year post-release, compared to 56% of 52 birds released at a greater age (Swinnerton 2001). Griffon Vultures *Gyps fulvus* were unusual in that birds released as adults survived better than those released as juveniles (Sarrazin *et al*. 1994). In this social species, the released vultures joined previously released birds from which they presumably learned their survival skills. These vultures were artificially provisioned with food (Terrase *et al*. 1994). In all social species, releases seemed more successful if there were other birds from previous releases nearby from which newcomers could learn.

Once the birds have been free for a designated period, food and water are gradually reduced and the birds are left to fend for themselves. The degree of post-release care is variable between projects and some also provide close guarding, individual monitoring, veterinary backup, and predator control. Some species need to be supported long-term following release, especially if the release environment is suboptimal in some way. Some social species, such as parrots and hornbills may, in some cases, have to be supported for at least a generation post-release, especially if there are no wild or previously released conspecifics established in the area to pass on appropriate social and survival skills.

12.5.2 Translocations

Translocations involve the movement of wild birds from one area of habitat to another. The most appropriate birds to move are usually juveniles, using the same release and post-release management as for captive-raised birds. Adults are more likely to have the necessary survival skills, but are also more likely to leave the release area and return to their site of origin. An early successful translocation was of 3100 Snowy Egrets that were moved from Louisiana to Florida in the United States in 1909. They were held captive for several months and then released. These birds helped to re-establish the species in Florida (McIlhenny 1934).

Translocations that have worked well include island endemics moved onto other islands from which introduced mammalian predators have been eradicated. The New Zealanders are the pioneers in this type of management and have successfully translocated the Eastern Weka *Gallirallus australis*, North Island Weka *G. a. greyi*, the two races of saddleback *Philesturnus carunculatus* (Merton 1975), Chatham Islands Snipe *Coenocorypha aucklandica pusilla*, Black Robin, Brown Teal *Anas aucklandica chorotis*, Kakapo, Kokako *Callaeus cineria*, North Island Brown Kiwi *Apteryx mantelli* and Little Spotted Kiwi, Stitchbird *Notiomystis cincta* and Takahe (Bell and Merton 2001). In the Seychelles, the Magpie Robin *Copsychus sechellarum* (Watson *et al.* 1992) and Seychelles Brush Warbler *Acrocephalus sechellensis* (Komdeur 1994) have been successfully moved to other islands. In western Australia, the Noisy Scrub-bird *Atrichornis clamosus* has been successfully translocated from its last natural stronghold in the southwest of the state to a number of mainland sites and one island (Bell and Merton 2002).

Some of these translocations have probably ensured the survival of the species involved. In New Zealand, many of the endemic birds cannot coexist with the introduced predators that now exist on the mainland, and the two races of saddleback, Eastern Weka and Kakapo all now exist on islands beyond their natural range (Bell and Merton 2002).

12.6 Supportive management for bird restoration projects

12.6.1 Role of captive facilities

Captive breeding projects that are established near to wild populations have the advantage that the movements of birds and eggs from the wild to captivity and vice versa is relatively easy. In addition, skilled personnel can be readily moved from the captive-breeding program into the field for the application of avicultural techniques to the wild birds, while field researchers can be used in the captive-breeding program.

Captive breeding has played an important role in the restoration of several critically endangered species and populations. The restoration of Peregrine Falcon populations in North America, Sweden, Germany, and elsewhere relied almost exclusively on captive-produced birds (Cade *et al.* 1988), as did the restoration of the Hawaiian Goose *Branta sandvicensis* (Black and Banko 1994), Whooping Crane (Ellis *et al.* 1996), and Pink Pigeon (Jones *et al.* 1992). However, captive breeding is not essential if the free-living populations can be closely managed, as in the Kakapo and Black Robin (Bell and Merton 2002). In the Kakapo, harvested and rescued eggs and young were brought into captivity for artificial incubation and hand-rearing, with subsequent reintroduction to the wild. Avian pediatric medicine and care are proving to be important for most intensively managed bird populations.

Captive-breeding facilities have a role in the development of techniques, and training personnel for use in future bird restoration projects. Some of the intensive management techniques can best be learnt on captive birds. We need to know which clutch and brood manipulations work for which species, and what are the costs and benefits of each technique in terms of lifetime reproductive output.

12.6.2 Model or surrogate species

Closely related surrogate species, with similar ecology to the target species may serve a number of functions.

1. Development of techniques before being applied to the rarer species. These may include captive breeding, artificial incubation, hand-rearing, and release techniques. For example, on Mauritius, Ring-necked Parakeets *Psittacula krameri* were released to develop techniques for the Echo Parakeet releases, and in California, Andean Condors *Vultur gryphus* were released (and later recaptured) to test release techniques for Californian Condors (Wallace and Temple 1987).
2. Staff training. Staff can learn handling and management techniques on a commoner species before they are applied to the focal species.
3. Foster parents for cross-fostering in captivity (e.g. Ring-necked Parakeets for Echo Parakeets and Barbary Doves *Streptopelia risoria* for Pink Pigeons).

The use of surrogate species, both in the wild and captivity, has been extensive in North America. For example, the Patuxent Wildlife Research Center worked first on Sandhill Cranes in order to develop captive breeding and management techniques applicable to the rarer Whooping Crane (Kepler 1978). Similarly Prairie Falcons were experimentally manipulated in the wild and kept in captivity by both The Peregrine Fund and the Canadian Wildlife Service to train personnel and to develop management techniques for application to the rarer Peregrine (Fyfe 1976).

12.6.3 Artificial incubation and hand-rearing

Avian pediatrics has primarily been developed in captive-breeding facilities and is most advanced in groups of birds of greatest commercial value: ratites, raptors, waterfowl, parrots, and passerines, although the number of experienced personnel is small.

Artificial incubation and hand-rearing provide supportive captive management for clutch and brood manipulations. In established projects with experienced personnel and good facilities, hatchability of fertile eggs is likely to reach 80%, and rearing success 90% in many groups of birds.

12.7 Integrated management

Some of the most marked recoveries of critically endangered species entailed a range of management practices, some of which were applied simultaneously. In Table 12.1 the management procedures that were applied to the Mauritius Kestrel, Pink Pigeon, Echo Parakeet, and Black Robin are all listed. All these species recovered from tiny populations and management involved the whole population.

The Mauritius Kestrel recovered from a wild population of four known birds in 1974 to between 600 and 800 in early 2003; the Pink Pigeon from 10 wild birds in 1990 to about 350 free-living birds in 2003, the Echo Parakeet from 8 to 11 known birds in 1987 to 175–200 free-living birds in 2003; and the Black Robin from 5 birds in 1980 to about 300 in 2001. In all these species, the genetic variance in the tiny remnant populations must have been small, yet they recovered to give large free-living populations (Groombridge *et al.* 2000). For some of these, however, continued management may be necessary in the future.

12.8 Discussion

In effect, species are rare or declining because of poor productivity and/or reduced survival, whatever the ultimate cause. The application of intensive management techniques to small and declining populations offers high chances of a rapid increase. However, because these techniques are intensive they are less appropriate for use on widespread populations, and many of the most successful examples are from relatively tame island species. Moreover, they are unlikely to succeed long-term unless the ultimate causes of poor status are addressed, whether these are loss of habitat and food supply, predation from people or introduced predators, new diseases or other factors.

The application of techniques, such as fostering, and cross fostering of eggs and young, works best with species that have high nest success (e.g. raptors, cranes, and parrots). With many other species the levels of nest failure in the wild are too high to justify the investment of time and energy. A broader approach to population management is often more appropriate, including, for example,

Table 12.1 *The management techniques used in the restoration of four Critically Endangered island endemics*

Management technique	Mauritius Kestrel	Pink pigeon	Echo parakeet	Black Robin
Supplemental feeding of free-living adults	*	**	**	**
Supplemental feeding of released birds	**	**	**	N/a
Supplemental feeding of dependent young	*	—	*	**
Nest site enhancement	**	—	**	**
Provision of artificial nest sites	**	*	**	**
Egg manipulations	**	**	**	**
Fostering of young	**	*	**	**
Cross fostering of young	*	*	*	**
Nest guarding	**	**	**	**
Rescue of eggs and young from failing nests	**	**	**	**
Captive breeding	**	**	**	N/a
Release of captive bred/reared young	**	**	**	N/a
Translocation of free-living birds	*	**	—	**
Predator control in breeding areas	*	**	**	—
Predator control at supplemental feeding sites	—	**	**	—
Predator control at release sites	**	**	**	N/a
Predator exclusion (fencing)	—	—	—	**
Control of nest competitors	*	—	**	**
Disease control	*	**	**	**
Control of disease vectors	—	*	—	—
Genetic management	**	**	**	**
Habitat restoration	—	**	**	**

* = Management technique used experimentally or on small scale and did not have a significant effect on the population.
** = Technique used extensively and is thought to, or known to have had a beneficial effect on the population.
N/a = Not applicable.

For details of these techniques and their application see Cade and Jones (1994), Jones and Duffy (1993), Jones *et al.* (1991, 1992, 1995, 1998, 1999), Jones and Hartley (1995) and Jones and Swinnerton (1997), Butler and Merton (1992).

predator control, supplemental feeding, nest-site management, together with reintroductions, and translocations.

To be effective, endangered species management has to be focused. It is possible only with teams of dedicated personnel, long-term commitment from supporting organizations, and access to skilled technicians.

Programs for the four species considered in Table 12.1 followed the steps discussed in Section 12.2. From the first conservation orientated field studies (Stage 1) to the start of intensive management (Stage 3) took between 8 years for the Black Robin and 20 years for the Echo Parakeet. The intensive management stage took about 9 years for the Black Robin (Butler and Merton 1992), 10 years for the Mauritius Kestrel (Jones et al. 1995) and 10 years for the Pink Pigeon (Swinnerton 2001). The Echo Parakeet is still at Stage Three and is likely to be intensively managed for a total of 10–12 years. Hence, these data suggest that it can take from 17 to 30 years to restore a population from being Critically Endangered (and poorly studied) through to the stage at which it requires minimal further management. For large, long lived, slow breeding species (Kakapo, Californian Condor, Whooping Crane), this time is likely to be much longer.

As we learn more about bird management, it should be possible to compress Stages 1 and 2 to a few years. But it seems likely that the restoration of Critically Endangered bird populations will always be a relatively long-term commitment. It is also likely that, with the increasing loss and degradation of habitat, it will be necessary to manage some bird populations in perpetuity if they are to survive, providing safe nest-sites, and food and managing predators.

A basic premise of intensive management is that it addresses proximate rather than ultimate causes of endangerment (Temple 1978). A population may be rescued in the short term by intensive management but long-term survival is best guaranteed if this management is coupled with efforts to address the ultimate problems, often related to habitat loss or degradation (Cade and Temple 1995). Intensive management often helps to clarify which environmental problems are causing the species' rarity, and it is recovery work that may drive the efforts to address the ultimate problems.

On Mauritius, species restoration has driven habitat restoration. The political will to establish a National Park arose as a direct result of restoration work on the endemic birds. Similarly, in New Zealand the restoration of many offshore islands has been done primarily to provide refuges for endangered birds.

Most critically endangered species would probably respond favorably to intensive management. Gurney's Pitta Pitta gurneyi had a known population of nine pairs in 1997 due primarily to habitat destruction (Stattersfield and Capper 2000). In June 2003 a survey revealed 31 birds of which 18 were males and an estimate of

15–20 pairs (A. Owen personal communication). The species was also rediscovered at four sites in neighboring Myanmar where there were 10–12 pairs at one site, although none of these sites are officially protected. While there may be possibilities for habitat restoration or to establish additional populations, in the long-term the immediate concern is to improve productivity and to use "surplus" birds derived from clutch and brood manipulations, to establish managed and or captive populations. Captive pittas of other species readily lay repeat clutches and there is every likelihood that Gurney's Pitta would also do so. The clutch size is usually 3–4, but nest predation is frequent and nest success low, with brood size at fledging usually being one or sometimes two (Lambert and Woodcock 1996). With a high level of natural egg and chick mortality, this productivity would be improved by close guarding and the application of clutch and brood manipulations.

Released animals can be managed at liberty and this offers opportunities for re-establishing species that may otherwise be difficult to reintroduce. The Spix Macaw *Cyanopsitta spixii* is now extinct in the wild, but there are about 70 birds in captivity. It has been proposed to release captive bred birds but this may prove difficult, because large parrots may rely to some extent on cultural transmission of information across generations, and with no wild birds left such learning will not be possible. Intensive management of released birds, with close guarding and provision of food, nest-sites and predator control may help their survival and breeding success. It is likely that the high level of management that would be necessary to establish Spix Macaws at liberty could be reduced as successive generations become more self sufficient.

Management of released birds also allows the possibility of maintaining populations in areas that would normally be unsuitable or marginal. This approach has resulted in the establishment of populations of formerly critically endangered species (Pink Pigeon, Hawaiian Goose) or species extinct in the wild (Kakapo). It opens up possibilities for the management at liberty of species that have critically endangered wild populations but have thriving captive populations (Northern Bald Ibis and the Bali Mynah *Leucopsar rothschildi*). Managed reintroduced populations of Bali Mynah and Northern Bald Ibis would provide data that could help in understanding the needs of the wild birds and increase the public profile of these species. In the long-term, having free-living birds with a low level of management would be more desirable than having the species existing only in captivity.

Applied population management offers potent possibilities for the restoration of most species of endangered birds, but is time consuming and may be expensive.

Acknowledgements

I thank the staff of the Mauritian Wildlife Foundation for help with this chapter in particular Kirsty Swinnerton, Jason Mallam, Steve Cranwell, Tom Bodie, and

Nancy Bunbury. Drs Andrew Greenwood and Diana Bell commented and greatly improved the manuscript.

References

Bell, B.D. and Merton, D.V. (2002). Critically endangered bird populations and their management, pp. 105–138. In K. Norris and D.J. Pain (eds.). Conserving Bird Biodiversity. Cambridge University Press, Cambridge.

Beissinger, S.R. and Bucher, E.H. (1992). Sustainable harvesting of parrots for conservation, pp. 73–117. In S.R. Beissinger and N.F.R. Snyder (eds.). New World Parrots in crisis. Smithsonian Institution Press, Washington and London.

Black, J.M. and Banko, P.C. (1994). Is the Hawaiian Goose saved from extinction? pp. 394–410. In P.J. Olney, G. Mace, and A. Feistner (eds.). *Creative Conservation: Interactive management of wild* and captive animals. Chapman and Hall: London.

Brucher, H. and Wegner, P. (1988). Artificial Eyrie Management and the protection of the Peregrine Falcon in West Germany, pp. 637–641. In T.J. Cade, J.H. Enderson, C.G. Thelander, and C.M. White (eds.). *Peregrine Falcon Populations, their management and recovery*. The Peregrine Fund, Inc.: Boise, Idaho.

Butler, D. and Merton, D. (1992). The Black Robin, Saving the world's most endangered bird. Oxford University Press, Oxford.

Burnham, W.A., Heinrich, W., Sanfort, G., Levine, E., O'Brien, D., and Konkel, D., (1988). Recovery effort for the Peregrine Falcon in the Rocky Mountains, pp. 565–575. In T.J. Cade, J.H. Enderson, C.G. Thelander, and C.M. White (eds.). Peregrine Falcon Populations, their management and recovery. The Peregrine Fund, Inc., Boise, Idaho.

Cade, T.J. (2000). Progress in translocation of diurnal raptors. In R.D. Chancellor and B.-U. Meyburg (eds.). *Raptors at Risk*. WWGBP/Hancock House.

Cade, T.J. and Jones, C.G. (1994). Progress in restoration of the Mauritius Kestrel. *Conserv. Biol.*, 7(1), 169–175.

Cade, T.J. and Temple, S.A. (1995). Management of threatened bird species: evaluation of the hands-on approach. *Ibis* (Suppl.), 137, s161–s172.

Cade, T.J., Enderson, J.H., Thelander, C.G., and White, C.M. (1988). Peregrine Falcon Populations, their management and recovery. The Peregrine Fund, Inc. Boise, Idaho.

Cooper, J.E. (1989). The role of pathogens in threatened populations: an historic review, pp. 51–61. In J.E. Cooper (ed.). Disease and threatened birds. International Council for Bird Preservation Cambridge, U.K.

Craig, G.R., Berger, D.D., and Enderson, J.H. (1988). Peregrine management in Colorado, pp. 575–585. In: eds. T.J. Cade, J.H. Enderson, C.G. Thelander, and C.M. White (eds.). Peregrine Falcon Populations, their management and recovery. The Peregrine Fund, Inc.: Boise, Idaho.

Dijkstra, C., Bult, A., Bijlsma, S., Daan, S., Meyer, T., and Zylstra, M., (1990). Brood size manipulations in the kestrel (*Falco tinnunculus*): effects on offspring and parent survival. *J. Anim. Ecol.*, 59, 269–285.

Elliot, G.P., Merton, D.V., and Jansen, P.W. (2001). Intensive management of a critically endangered species: the kakapo. *Biol. Conserv.*, 99, 121–133.

Ellis, D.H., Gee, G.F., and Mirande, C.M. (eds.) (1996). Cranes: their biology husbandry and conservation. Hancock House Publishers, Blaine, WA., USA.

Fyfe, R.W. (1976). Rational and success of the Canadian Wildlife Service Peregrine breeding project. *Canadian Field Natur.*, 90, 308–319.

Fyfe, R.W. (1978). Reintroducing endangered birds to the wild—a review, pp. 323–329. In: *Endangered Birds*, (ed.) S.A. Temple. Management techniques for preserving threatened species. University of Wisconsin Press: Wisconsin.

Fyfe, R.W., Armbruster, H., Banasch, U., and Beaver, L.J. (1978). Fostering and cross fostering in birds of prey, pp. 183–193. In *Endangered Birds. Management techniques for preserving threatened species*, (ed.), S.A. Temple, University of Wisconsin Press, Wisconsin.

Gibbons, P. and Lindenmayer, (2002). *Tree Hollows and wildlife conservation in Australia*. CSIRO publishing, Collingwood, Victoria, Australia.

Gilardi, V.K., Lowenstine, L.J., Gilardi, J., and Munn, C.A. (1995). A survey for selected viral, chlamydial, and parasitic diseases in wild Dusky-headed Parakeets *(Aratinga weddellii)* and Tui Parakeets *(Brotogeris sanctithomae)* in Peru. *J. Wildlife Dis.*, 31, 523–528.

Goss-Custard, J.D. (1993). The effect of migration and scale on the study of bird populations: 1991 Witherby Lecture. *Bird Stud.*, 40, 81–96.

Green, R.E. (1995). Diagnosing causes of bird population declines. *Ibis*, 137(Suppl.), s47–s55.

Green, R.E. (2002). Diagnosing causes of population declines and selecting remedial actions, pp. 139–156. In *Conserving Bird Biodiversity*, eds. K. Norris, and D.J., Pain, Cambridge University Press: Cambridge.

Greenwood, A.G. (1996). Veterinary support for *in situ* avian conservation programmes. *Bird Conserv. Int.*, 6, 285–292.

Hepp, K. (1988). Contributions towards the recovery of Peregrine Falcons in West Germany, pp. 643–648. In *Peregrine Falcon Populations, their management and recovery*. T.J. Cade, J.H. Enderson, C.G. Thelander, and C.M. White, eds. The Peregrine Fund, Inc., Boise, Idaho.

Hirsch, U. (1978). Artificial nest-ledges for Bald Ibises, pp. 61–69. In: *Endangered Birds*, ed., S.A. Temple, *Management techniques for preserving threatened species*. Wisconsin: University of Wisconsin Press.

Immelmann, K. (1972). The influence of early experience upon the development of social behaviour in estrildine finches. *Proceedings of the XV International Ornithological Congress*, vol. 15, pp. 316–338.

Innes, J., Hay, R., Flux, I., Bradfield, P., Speed, H., and Jansen, P. (1999). Recovery of North Island Kokako *Callaeas cinerea wilsoni* populations, by adaptive management. *Biol. Conserv.*, 87, 201–214.

IUCN (1994). IUCN Red List Categories. IUCN: Cambridge.

IUCN (1998). IUCN Guidelines for re-introductions. IUCN: Gland, Switzerland.

James, K.A.C., Waghorn, G.C., Powlesland, R.G., and Lloyd, B.D. (1991): Supplementary feeding of kakapo on Little Barrier Island. *Proc. Nutrit. Congr. N.Z.*, 16, 93–102.

Jones, C.G. and Duffy, K. (1993). The conservation management of the echo parakeet *Psittacula eques echo. Dodo, J. Wildlife Preserv. Trusts*, 29, 126–148.

Jones, C.G. and Hartley, J. (1995). A conservation project on Mauritius and Rodrigues: An overview and bibliography. *Dodo, J. Wildlife Preserv. Trusts*, 31, 40–65.

Jones, C.G. and Swinnerton, K.J. (1997). Conservation status and research for the Mauritius Kestrel *Falco punctatus*, Pink Pigeon *Columba mayeri* and Echo Parakeet *Psittacula eques. Dodo, J. Wildlife Preserv. Trusts*, 33, 72–75.

Jones, C.G., Heck, W., Lewis, R.E., Mungroo, Y., and Cade, T.J. (1991). A summary of the conservation management of the Mauritius Kestrel *Falco punctatus* 1973–1991. *Dodo, J. Jersey Wildlife Preserv. Trust*, 27, 81–99.

Jones, C.G., Swinnerton, K.J., Taylor, C.J., and Mungroo, Y. (1992). The release of captive-bred Pink Pigeons *Columba mayeri* in native forest on Mauritius. A progress report July 1987–June 1992. *Dodo, J. Jersey Wildlife Preserv. Trust*, 28, 92–125.

Jones, C.G., Heck, W., Lewis, R.E., Mungroo, Y., Slade, G., and Cade. T.J. (1995). The restoration of the Mauritius Kestrel *Falco punctatus* population. *Ibis*, 137, s173–s190.

Jones, C.G., Swinnerton, K.J., Thorsen, M., and Greenwood, A. (1998). The biology and conservation of the Echo Parakeet *Psittacula eques* of Mauritius, pp. 110–123. Fourth International Parrot Convention. Loro Parque, Tenerife, Spain.

Jones, C.G., Swinnerton, K.J., Hartley, J. and Mungroo, Y. (1999). The restoration of the free-living populations of the Mauritius kestrel (*Falco punctatus*), Pink pigeon (*Columba mayeri*) and Echo parakeet (*Psittacula eques*), pp. 77–86. In *Linking Zoo and Field Research to Advance Conservation, the 7th World Conference on Breeding Endangered Species*. Cincinnati, Ohio, USA. May 22–26, 1999. Cincinnati Zoo.

Joyner, K.L., DeBerger, N., and Lopez, E.H. (1992). Health parameters of wild psittacines in Guatemala. *Proceedings of the Association of Avian Veterinarians*, 287–303.

Kepler, C.B. (1978). Captive propagation of Whooping Cranes: A behavioural approach, pp. 231–241. In: *Endangered Birds. Management techniques for preserving threatened species*, ed. S.A. Temple. University of Wisconsin Press, Wisconsin.

Komdeur, J. (1994). Conserving the Seychelles Warbler *Acrocephalus sechellensis* by translocation from Cousin Island to Aride and Cousine. *Biol. Conserv.*, 67, 143–152.

Lack, D. (1954). *The Natural Regulation of Animal Numbers*. Clarendon Press, Oxford.

Lack, D. (1966). *Population Studies of Birds*. Clarendon Press, Oxford.

Lambert, F. and Woodcock, M. (1996). *Pittas, Broadbills and Asities*. Pica Press, Sussex.

Long, J.L. (1981). *Introduced Birds of the world*. David and Charles, Newton Abbott and London.

Maxwell, J. and Jamieson, I.G. (1997). Survival and recruitment of captive-reared and wild-reared takahe in Fiordland, New Zealand. *Conserv. Biol.*, 2, 28–30.

McIlhenny, E.A. (1934). *Bird City*. The Christopher publishing House, Boston, USA.

Merton, D.V. (1975). The Saddleback: its status and conservation, pp. 61–74. In *Breeding Endangered Species in Captivity*, ed. Martin, R.D. Academic Press, London.

Merton, D.V. (1978). Controlling Introduced predators and competitors on islands, pp. 121–128. In *Endangered Birds. Management Techniques for Preserving Threatened Species*, ed. Temple, S.A. University of Wisconsin Press, Wisconsin.

Merton, D., Reed, C., and Crouchley, D. (1999). Recovery strategies and techniques for three free-living, critically-endangered New Zealand birds: Kakapo (*Strigops habroptilus)*, Black Stilt (*Himantopus novaezelandiae)* and Takahe (*Porphyrio mantelli)*, pp. 151–162. In *Linking Zoo and Field Research to Advance Conservation, the 7th World Conference on Breeding Endangered Species*. Cincinnati, Ohio, USA. May 22–26, 1999. Cincinnati Zoo.

Meyburg, B.-U. (1978). Sibling aggression and cross-fostering of eagles, pp. 195–200. In: *Endangered Birds. Management Techniques for Preserving Threatened Species*, ed. S.A. Temple, University of Wisconsin Press, Wisconsin.

Munn, C. (1992). Macaw biology and ecotourism, or "when a bird in the bush is worth two in the hand", pp. 47–72. In *New World parrots in crisis*, eds. S. R. Beissinger and N. F. R. Snyder. Smithsonian Institution Press. Washington, U.S.A.

Nagendran, M., Urbanek, R.P., and Ellis, D.H. (1996). Reintroduction Techniques, pp. 231–240. In: *Cranes: Their biology Husbandry and Conservation*, eds. D.H. Ellis, G.F. Gee, and C.M. Mirande. Hancock House Publishers, Blaine, WA, USA.

Newton, I. (1994). Experiments on the limitation of bird breeding densities: a review. *Ibis*, 136, 397–411.

Newton, I. (1998). *Population Limitation in Birds*. Academic Press, London.

Olendorff, R.R., Montroni, R.S., and Call, M.W. (1980). Raptor management-the state of the art in 1980, pp. 468–523. Workshop proceedings, Management of Western Forests and Grasslands for nongame birds. USAD Forest Service General Technical Report INT-86.

Powlesland, R.G., Lloyd, B.D., Best, H.A., and Merton, D.V. (1992). Breeding biology of the kakapo Strigops habroptilus on Stewart Island, New Zealand. *Ibis*, 134, 361–373.

Saar, C. (1988). Reintroduction of the Peregrine Falcon to Germany, pp. 629–637. In *Peregrine Falcon Populations, Their Management and Recovery*, eds. T.J. Cade, J.H. Enderson, C.G. Thelander, and C.M. White. The Peregrine Fund, Inc., Boise, Idaho.

Sarrazin, F.C., Bagnolini, C., Danchin, E., and Clobert, J. (1994). High survival estimates of Griffon Vultures *Gyps fulvus fulvus* in a reintroduced population. *Ibis*, 111, 853–862.

Sherrod, S.K., Heinrich, W.R. Burnham, W.A., Barclay, J.H., and Cade, T.J. (1981). *Hacking: A Method for Releasing Peregrine Falcons and Other Birds of Prey*. The Peregrine Fund, Boise, Idaho.

Snyder, B., Thilstead, J., and Burgess, B. (1985). Pigeon herpesvirus mortalities in foster reared Mauritius Pink Pigeons. *Proceedings of the American Association of Zoo Veterinarians*, 69–70.

Snyder, N.F.R. (1978). Puerto-Rican Parrots and nest-site scarcity, pp. 47–53. In *Endangered Birds. Management Techniques for Preserving Threatened Species*, ed. S.A. Temple. University of Wisconsin Press: Wisconsin.

Snyder, N.F.R. (1986). California Condor recovery programme, pp. 56–71. In *Raptor Research Report No. 5: Raptor Conservation in the Next 50 years*. S.E. Senner, C.M. White, and J.R. Parish. Raptor Research Foundation, Provo, Utah.

Snyder, N.F.R. and Hamber, J.A. (1985). Replacement-clutching and annual nesting of California Condors. *Condor*, 87, 374–378.

Snyder, N. and Snyder, H. (2000). *The California Condor. A Saga of Natural History and Conservation*. Academic Press, London.

Snyder, N.F.R., Wiley, J.W., and Kepler, C.B. (1987). *The parrots of Luquillo: Natural History and Conservation of the Puerto Rican Parrot*. Western Foundation of Vertebrate Zoology, Los Angeles.

Stattersfield, A.J. and Capper, D.R. eds. (2000). *Threatened Birds of the World*. Lynx Edicions and BirdLife, Barcelona and Cambridge.

Sutherland, W.J. (2000). *The Conservation Handbook: Research, Management and Policy*. Blackwell Science Ltd., London.

Swinnerton, K.J. (2001). The Ecology and Conservation of the Pink Pigeon *Columba mayeri* in Mauritius. PhD thesis, University of Kent at Canterbury.

Temple, S.A. (1978). *Endangered Birds. Management techniques for preserving threatened species*. University of Wisconsin Press, Wisconsin.

Terrase, M., Bagnolini, C., Bonnet, J., Pinna J.-L., and Sarrazin, F., (1994). Reintroduction of the Griffon Vulture *Gyps fulvus* in the Massif Central, France, pp.479–491. In *Raptor Conservation Today*, eds. B.-U. Meyburg and R.D. Chancellor. WWGBP/The Pica Press, Berlin.

Veitch, C.R. (1985). Methods of eradicating feral cats from offshore islands in New Zealand. *ICBP Tech. Bull.*, 3, 125–141.

Van Riper, C., Van Riper, S.G., and Goff, M.L. (1986). The epizootiology and ecological significance of malaria in Hawaiian land birds. *Ecol. Monogr.*, 56(4), 327–344.

Wallace, M. (2000). Retaining natural behaviour in captivity for re-introduction programmes, pp. 300–313. In *Behaviour and Conservation*, eds. M. Gosling, and W.J. Sutherland, Cambridge University Press, Cambridge.

Walton, B.J. and Thelander, C.G. (1988). Peregrine Falcon Management Efforts in California, Oregon, Washington and Nevada pp. 587–597. In *Peregrine Falcon Populations, Their Management and Recovery*, eds. T.J. Cade, J.H. Enderson, C.G. Thelander, and C.M. White, The Peregrine Fund, Inc. Boise, Idaho.

Watson, J., Warman, C., Todd, D., and Laboudallon, V. (1992). The Seychelles magpie robin *Copsychus sechellarum*: ecology and conservation of an endangered species. *Biol. Conserv.*, 61, 93–106.

Watson, R.T., Thomsett, S., O'Daniel, D., and Lewis, R. (1996). Breeding, growth, development and management of the Madagascar Fish-Eagle (*Haliaeetus vociferoides*). *J. Raptor Res.*, 30, 21–27.

Wiley, J.W. (1985). The Puerto Rican parrot and competition for its nest sites. ICBP Technical Publication No. 3, 213–223.

Wilson, J.E. and Macdonald, J.W. (1967). Salmonella infection in wild birds. *Br Vet J*, 123, 212–219.

Wingate, D.B. (1978). Excluding competitors from Bermuda Petrel nesting burrows, pp. 93–102. In *Endangered Birds. Management techniques for preserving threatened species*, ed. S.A. Temple, University of Wisconsin Press, Wisconsin.

Zeleny, L. (1978). Nesting box programs for Bluebirds and other passerines, pp. 55–60. In: *Endangered Birds. Management Techniques for Preserving Threatened Species*, ed. S.A. Temple. University of Wisconsin Press, Wisconsin.

13

Exploitation

Michael C. Runge, William L. Kendall, and James D. Nichols

13.1 Introduction: assessment of exploitation

13.1.1 Taking a conservative approach

Many bird populations are exploited for human purposes, through subsistence or commercial harvest, or live collection, for food, recreation, medicine, ornaments, pets, or to reduce crop damage, or predation on game animals. While the motivations and methods are varied, they share one consequence—removal of birds from wild populations. Often, this removal and the consequences to the population are not quantified, but the outcome can be dire. Overkill is one of an "Evil Quartet" of causes of recent extinctions (Diamond 1989).

The conservation biologist or wildlife manager wishing to prevent loss of biodiversity and extinction due to exploitation is often faced with uncertainty regarding the status of the population, the level of harvest, and the dynamics of the population in question. How should one proceed? There are two kinds of error to guard against: doing nothing when exploitation is having a negative impact; and implementing unnecessary restrictions when exploitation is already sustainable. The focus of this chapter is to describe an assessment approach and monitoring tools to guard against the former error; that is, to take a conservative approach in the face of uncertainty.

Exploitation of bird and mammal populations has tremendous economic and cultural importance, and managing exploitation requires careful consideration of those forces (Bennett and Robinson 2000). The conservation biologist needs to bring robust data and defensible analyses to the discussion. The three general types of information required are: minimum estimates of population size, estimates of harvest levels, and an understanding of population dynamics.

13.1.2 Minimum estimates of population size

The effect of harvest on a population depends in large part on the magnitude of the harvest relative to the population size. In addition, the effect of harvest can be mediated by density-dependent dynamics. For these reasons, an estimate of the population size is an important element in the assessment of exploitation. Of course, the size of a bird population is rarely known with much precision. To guard against overexploitation due to this uncertainty, minimum population estimates can be used. Thus, abundance estimates based on the number of individuals actually counted or the lower end of a confidence interval around a population estimate might be used in computations of allowable harvest.

13.1.3 Estimates of harvest levels

To assess whether the current level of exploitation is sustainable, some measure of the harvest is needed. Harvest can be measured on an absolute scale, as the total number of animals removed, or on a relative scale, as a harvest rate relative to the population size. These two approaches are appropriate under different circumstances, but either can provide critical information for the assessment.

13.1.4 Population models and associated parameters

To assess whether a particular harvest level is sustainable, given a current population size estimate, an understanding of the underlying population dynamics is needed. Ideally, this would include age-, sex-, or size-specific estimates of survival and reproductive rates, a complete model of the life-history dynamics, and measures of the links between these life-history parameters, harvest rates, and environmental driving variables. A more minimal understanding of the population dynamics, however, can serve as a starting point that provides an initial, conservative assessment that can be revised as more information is gathered. Two parameters are valuable at the initial stage: the maximum potential growth rate for the population, r_{max}, and the carrying capacity, K (see Section 13.6 for details on estimating these quantities). As more information is gathered, it is useful to understand the density-dependent processes that regulate the population, and the nature of the density-independent forces that can affect it.

13.1.5 The use of trends

Is information about the trends in population size or harvest levels useful in assessing the impact of exploitation? For example, is an increasing trend in population size satisfactory evidence that the harvest is sustainable? It is tempting to conclude that it is, but there are several reasons why trends need to be interpreted with caution. (1) Harvest may be sustainable but still be too large. For example, in a species of conservation concern, it may be unacceptable to allocate a large portion of the net productivity to harvest and so delay recovery substantially. In such a case, the relevant comparison is between the observed growth rate and the growth rate expected in the absence of harvest, not simply whether the observed growth rate is positive. (2) A trend may be due to transient dynamics. For short periods, populations that are ultimately decreasing may show transient increases due, for instance, to shifts in age-structure. (3) A declining trend is ambiguous. Population size might decrease because the harvest is not sustainable; or it might decrease because a sustainable level of harvest is causing the population size to shift to a lower equilibrium point. Total harvest might decrease because the harvest rate is decreasing, or because the population is rapidly declining. (4) Estimates of trends can be spurious. For example, a trend in population size could be due to a change in survey effort, a shift in bird distribution relative to survey strata, or a change in the ability to detect the birds because of a shift in, say, vegetation structure. The methods described below provide an alternative to the use of trend information for assessing the effects of exploitation.

13.2 Theoretical basis for sustainable exploitation

The challenge in the management of exploitation is to find the right balance between allowing removal of animals for purposes sanctioned by society and preventing population declines, loss of biodiversity, and extinction; that is, to determine what level of exploitation is sustainable. There is no precise definition of sustainable. In the World Conservation Strategy (IUCN/UNEP/WWF 1980), use of a natural resource is considered sustainable when the effect on the wild population is not significant. But, exploitation can significantly affect the wild population (in particular, by reducing the population size) yet still be sustainable, if the removal does not exceed the net production (Bennett and Robinson 2000). This condition can be met, however, at many different combinations of population size and harvest rate, so ultimately, sustainability needs to be determined by the range of economic, cultural, and ecological values operating in a particular setting. Nevertheless, there is a formal theoretical basis for the *ecological* sustainability of exploitation, a basis derived from equilibrium population dynamics, and optimum sustained yield. For a more detailed treatment of this subject, see Reynolds *et al.* (2001).

13.2.1 Logistic growth model with perfect information

Consider the theoretical case where population size could be measured exactly, where there were no stochastic dynamics, where harvest rates could be precisely controlled, and where the population grew according to a logistic growth model. The population size at each time step is governed by:

$$N_{t+1} = N_t + r_{max} N_t(1 - N_t/K) - h_t N_t \qquad (13.1)$$

where N_t is the population size at time t, h_t is the harvest rate for the same time period, r_{max} is the maximum growth rate, and K is the carrying capacity. This model underlies the results of Caughley (1977:178–181) and is a simplified version (with $\theta = 1$) of the generalized logistic model used by Taylor and DeMaster (1993), Wade (1998), and Taylor *et al.* (2000).

Imagine harvesting at a fixed rate, $h_t = h$ (where the rate is relative to the population size), for an indefinite period of time. The population size will converge to and maintain an equilibrium value, $N_{eq}(h)$ that is a function of the fixed harvest rate (provided r_{max} is not too large, in which case cyclical or chaotic results can be obtained (May 1976)). The equilibrium population size is given by:

$$N_{eq}(h) = K\left(\frac{r_{max} - h}{r_{max}}\right) \qquad (13.2)$$

(see Runge and Johnson 2002 for calculation methods); that is, the equilibrium population size decreases linearly from K (when $h = 0$) to 0 (when $h \geq r_{max}$) (Figure 13.1a).

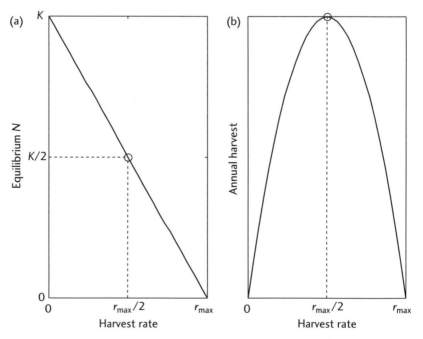

Fig. 13.1 Maximum sustained harvest from a logistic model. (a) Equilibrium population size as a function of a fixed harvest rate. (b) Annual sustained harvest as a function of a fixed harvest rate. The maximum sustained yield is achieved by harvest at a rate of $r_{max}/2$, which achieves an equilibrium population size of $K/2$.

Once the equilibrium population size (for a particular fixed harvest rate) is reached, the annual harvest is also constant and is given by

$$H_{eq}(h) = hN_{eq} = \frac{r_{max}Kh - Kh^2}{r_{max}},\qquad(13.3)$$

which is parabolic with respect to h (Figure 13.1b).

It is important to note that *any* fixed value of harvest rate less than r_{max} will produce an equilibrium population size and annual harvest that are both greater than zero. In this sense, any harvest rate less than r_{max} is sustainable. We can find the maximum sustainable harvest rate by differentiating equation 13.3 with respect to h, setting the result equal to zero, and solving for h (Runge and Johnson 2002). For the logistic growth model, the maximum sustainable harvest rate is $h^* = r_{max}/2$, which produces an equilibrium population size $N^* = K/2$, and an annual harvest of $H^* = r_{max}K/4$ (Caughley 1977).

13.2.2 Incorporating uncertainty: potential biological removal

For application to real populations, the preceding discussion forms the basis for sustainable harvest, but uncertainties complicate its implementation. First, real population dynamics are more complex than implied by equation 13.1, because of differences in life-history parameters by age, sex, reproductive status, etc. Second, the nature of density-dependence may be different from that expressed by equation 13.1. Third, most wild populations exhibit considerable stochasticity, due to random fluctuations in environmental driving factors. Fourth, sampling bias and error in measuring N_t and estimating r_{max} and K introduce uncertainty into the calculation of sustainable harvest levels. Finally, managers typically cannot control exploitation with much precision. This last point can be particularly important depending on whether total harvest or harvest rate is the management variable being controlled. When harvest *rate* can be managed to achieve a maximum sustained yield (by setting $h = h^*$), that equilibrium point is quite stable. But in many cases, it is easier to manage the total harvest (H), perhaps through a quota system. The maximum equilibrium point achieved by setting $H = H^*$ is dangerously unstable. If the harvest by chance is above H^*, or if the population size by chance drifts below N^*, continued extraction of the maximum sustained yield (H^*) will lead to extinction.

In a precisely managed system, with frequent and timely monitoring data, and the ability to adjust harvest or harvest rates on a regular basis in response to current conditions, these uncertainties can be accommodated. But how can a conservation biologist make an assessment in cases where the uncertainties are great? This question has been carefully considered by biologists worried about the impact of incidental take on populations of marine mammals, species whose status and dynamics are often poorly known. The approach that has emerged as a default assessment in the face of uncertainty is referred to as "Potential Biological Removal" (PBR, Wade 1998). The formula for PBR is:

$$\text{PBR} = \frac{1}{2}r_{max}N_{min}F_{R} \tag{13.4}$$

where r_{max} is the theoretical maximum growth rate at low population density (as above), N_{min} is the minimum population estimate, and F_{R} is a recovery factor between 0.1 and 1.0. The formula specifies the maximum harvest that can potentially be removed from the population while allowing the population to achieve or maintain its optimum sustainable population size (Wade 1998). Note that if recovery is not a concern, and F_{R} is set to 1.0, the formula can be rearranged,

$$h = \frac{\text{PBR}}{N_{min}} = \frac{1}{2}r_{max}, \tag{13.5}$$

showing that this method can be used to seek the maximum sustained yield by controlling harvest rate, an approach that produces a stable equilibrium. It is conservative because a minimum estimate of the population size is used, ensuring that the actual harvest rate is not greater than optimal harvest rate (h^{*}). The recovery factor, F_{R}, is used to seek further conservatism, particularly when the status and dynamics are very poorly known, and when there is a desire to allow a depleted population to quickly recover to an optimal level. The recovery factor is typically set to 0.1 for endangered species, and 0.5 for threatened species (Taylor et al. 2000). These levels have been shown to guard against the typical levels of uncertainty and potential bias associated with dynamics of marine mammals (Wade 1998; Taylor et al. 2000), but similar work has not been done with birds.

To apply the PBR formula, an estimate of population size, N_{min}, and an estimate of potential growth rate, r_{max}, are needed. The population size estimate should reflect confidence that the actual population size is greater than the estimate used. Where methods are available to obtain an unbiased estimate of N, the lower bound of the 95% confidence interval can be used (Taylor 1993). Where such data are not available, the minimum number known alive, or some similar measure, could be used. This value should be updated, and PBR recalculated, whenever new data are available. The maximum potential growth rate, r_{max}, is technically $\lambda_{max}-1$, where λ_{max} is the maximum discrete rate of population growth. If data are available about survival and reproductive rates for a species, there are methods for calculating λ_{max} (see Slade et al. 1998; Caswell 2001). Otherwise, biologists can rely on comparisons to related species for which such

Box 13.1 Case study: calculating the potential biological removal of Bearded Guans

---

The Bearded Guan (*Penelope barbata*) is a cracid endemic to Ecuador and northern Peru, and is considered vulnerable. Jacobs and Walker (1999) studied several unprotected forest blocks in Ecuador. In the 400-ha Selva Alegre forest, Bearded Guans were found at a density of 17.1 km^{-2} (95% CI: 10.4–27.9). Maximum potential growth rate for *P. barbata* has not been calculated, but using life-history data for the closely related Spix's Guan (*Penelope jaquacu*) of Peru (Begazo and Bodmer 1998), we can calculate $\lambda_{max} = 1.073$ (based on the formula for λ_L in Slade *et al.* [1998]). With $F_R = 0.5$ (because the species is considered vulnerable), the potential biological removal is

$$PBR = \frac{1}{2}\left(\lambda_{max}-1\right) N_{min} \ F_R = \frac{1}{2}(0.0733)(10.4 \text{ km}^{-2})(0.5) = 0.19 \text{ km}^{-2}$$

or 0.76 birds per year from the entire forest. Thus, to guard against overharvest of Bearded Guans from this forest tract, measures should be taken to ensure that no more than 3 birds are taken over the course of four years. By contrast, Robinson and Redford (1991) would calculate that a harvest greater than 6.7 birds per year would clearly be unsustainable.

measures are approximately known. (It is important to stress that estimates of growth rate from stable populations, where $\lambda \approx 1$, are not appropriate; the growth rates need to be maximum potential growth rates, see 13.6.1). The result of the calculation is a level of allowable removal, PBR, that can be used for assessment (by comparison to estimates of H) or management (as a target level for H) (Box 13.1).

13.2.3 A note about other methods of assessing sustainability

When used to assess whether a population is being overharvested, the PBR formula guards against concluding that exploitation is sustainable when in fact it is not. When used to set target harvest limits, the PBR formula is designed to allow a population to reach and stay above the maximum net productivity level (Wade 1998). A very similar method has been used fairly extensively to evaluate the sustainability of harvest, both in mammals and in birds (Robinson and Redford 1991; Robinson 2000). This method calculates the maximum possible production of a population and compares it to the actual harvest. Slade *et al.* (1998) point out that this method tends to overestimate growth rate and annual production. Thus, this method guards against the other type of error: determining that the exploitation is unsustainable when in fact it is. That is, if a harvest is determined to be unsustainable using Robinson and Redford's method, you can be assured of that conclusion; but the method cannot be used to determine whether the actual harvest is sustainable (Robinson 2000). It is critical that an assessment of sustainable exploitation acknowledges what sort of error it is guarding against.

13.3 Harvest control and management objectives

13.3.1 Harvest management strategies

The PBR equation provides a target level for total harvest, but the manager has to figure out what management actions can be implemented to achieve that harvest level. Newton (1998, Table 14.2) describes the advantages and disadvantages of 6 different harvesting systems: free-for-all, fixed quota, fixed effort, variable quota, fixed percentage, and fixed escapement. In a free-for-all system, there is no control of the harvest, which makes it easy to administer, but potentially a significant conservation concern. As the harvesting systems improve in their ability to conserve the resource, the administrative burdens and monitoring costs increase. In a fixed quota system, the same number of animals are allowed to be removed each year; this may be relatively easy to administer in small areas with good enforcement effort and results in a constant yield, but can be very unstable, leading to sharp population declines if the population size drops below a critical level. Stepping to a variable quota system, where the number harvested each year depends on the population size, greatly increases the ability to conserve the resource, at the cost of increased monitoring and enforcement. The PBR calculation implies either a variable quota system (equation 13.4) or a fixed percentage system (equation 13.5, with F_R retained). These systems are much more robust to uncertainty than the fixed quota system. The decision about what harvesting system to employ depends heavily on the specifics of the situation: local laws and customs, enforcement ability, avenues for communication with the users, monitoring potential, and other political pressures. Managers need to combine these considerations with the ecological properties of the harvesting system to identify the best course of action. Two suggestions can be offered: first, management actions that are intended to affect the harvest rate (relative to the population size) will be more protective of the resource (without sacrifice of harvest potential) than actions intended to affect the total annual harvest; and second, management actions that depend on the state of the system (especially population size or density) will be more protective, without long-term sacrifice of harvest, than actions that do not consider the population state.

In most situations, managers do not directly control either the total harvest (H) or the harvest rate (h). Instead, they implement and enforce some regulations that are thought to influence H or h. Understanding the link between the management actions (regulations and enforcement) and the population dynamics is important, and assumptions about this link should be examined carefully. In fact, the underlying endeavor of the manager is to understand how implementation of regulations affects the achievement of management objectives—all the intermediate effects do not necessarily need to be understood if the objectives are obtained. To this end, an experimental or adaptive harvest policy (see 13.7.2 and 13.7.3) can be useful, particularly when data are sparse.

13.3.2 Harvest management objectives

Clearly thought out and articulated objectives are crucial to proper harvest management. For a single species, the simplest objective is to maximize sustainable harvest. But as with any managed system, there are frequently many objectives for a harvested system. For instance, maximizing harvest opportunity, maintaining the population size above some target level, having regulations that are easy to understand and enforce, preventing decline or extinction, and avoiding having to forbid harvest of the resource at any time are all possible objectives. Any of these objectives can be included formally in an analysis of harvest. For example, a constraint could be added to the objective of maximizing harvest to provide a minimum number of birds for other purposes such as birdwatching (see Johnson *et al.* 1997). Finding the right balance between these objectives is a political decision, but it has a large impact on the harvest strategy that is implemented. In the PBR framework, two objectives are implied: allowing harvest, but constraining that harvest to guard against uncertainty and to allow recovery. These constraints are induced indirectly through the recovery factor (F_R).

The harvest of multiple species increases the complexity of the management problem, especially when the harvest processes or population dynamics for each species are not independent. In this case options include: managing for one species, and hoping for the best with regard to the others; managing primarily for one species while incorporating other species through constraints; or managing for a total yield across all species (either weighted or unweighted), perhaps constrained by minimum population sizes on each. Final management objectives are often the result of compromise among objectives of various stakeholders. This compromise might be significant if the objectives are conflicting. For example, as Newton (1998) discusses, imagine harvesting two species, an *r*-selected species and a K-selected species with a common regulatory framework. Managing only for the r-selected species might imperil the other species; while managing only for the K-selected species would require foregoing the significant harvest potential in the r-selected species. While there are biological implications of various objectives in a multi-species context, the articulation of objectives is a political endeavor.

13.4 Assessing harvest levels

13.4.1 Total harvest (H) versus harvest rate (h)

Assessments of harvest and harvest rate are related as

$$H = hN,\qquad(13.6)$$

where N is abundance. Harvest is the total number of individuals killed and retrieved by hunters in a specified time period. Harvest rate can be defined in different ways but is frequently the probability that an individual alive at the beginning of a hunting period is killed and retrieved by a hunter during this period. As is clear from the preceding equality, interpretation of an estimate of total harvest as a measure of magnitude of exploitation is difficult in the absence of an estimate of animal abundance. Harvest rate is thus more commonly used as a parameter in population-dynamic and management models. However, in the actual establishment of harvest regulations, it is frequently easier to specify a harvest quota or allowable number of individuals to be removed from the population.

In the models and discussion presented above, harvest has been referred to as the total number of individuals removed by exploitation, and harvest rate as the probability that an individual in the population prior to the hunting season dies as a result of hunting. For many forms of exploitation, animals may be killed as part of the exploitation process, yet not retrieved by hunters (e.g. Anderson and Burnham 1976; Pollock *et al.* 1994). For example, birds may be shot by hunters and their carcasses not located, or birds may be injured and escape the hunter only to die subsequently as a result of gunshot wounds. In such situations, the term "harvest" is typically reserved for the number of animals retrieved by hunters, whereas "kill" is sometimes used to refer to the number of animals that die as a result of exploitation.

13.4.2 Harvest estimation when harvest is legal and observable

Several methods exist for sampling hunters in order to estimate harvest, and selection of an appropriate method depends heavily on the geographic scale of harvest and the logistical ease with which hunters can be encountered either during or just following hunting activities. Private hunting clubs and estates are frequently relatively small in area and are under strict control of owners and their gamekeepers. In such situations, it may be possible to obtain exact counts of harvested birds, as for Red Grouse *Lagopus lagopus* on shooting moors in Great Britain (e.g. Potts *et al.* 1984; Hudson 1986). In many situations, however, such tight control is not possible and sample survey methods must be used to estimate harvest. On-site surveys involve efforts to contact hunters during or just following hunting activities, whereas off-site surveys involve other means of contacting hunters at times subsequent to hunting.

On-site survey methods generally fall into one of two categories: access-point surveys and roving surveys. Access-point surveys are appropriate for situations in which hunting takes place in a local area (or series of such areas) to which access is restricted to a relatively small number of entry points. Wildlife or conservation officers are then stationed at a sample of these access points during a sample of possible hunting times, and hunters exiting the area are stopped and their harvests recorded. Visits to access points by conservation officers can be selected randomly with known probability from the possible points and from the possible times during the hunting season, and harvest can be estimated. Although the statistical framework for such sampling has been best developed for fisheries (Robson 1960; Robson and Jones 1989; Pollock *et al.* 1994), surveys based on encounters at hunter check stations have been used successfully for birds (e.g. Mikula *et al.* 1972; Wright 1978).

Roving surveys may be useful in areas for which discrete access points do not exist or are too numerous to cover. The sampling frame for this design consists of the set of all possible times (e.g. days, part-days) available for hunting and all possible locations where hunting can take place (these might be woodlots, moors or portions of such habitat patches). Conservation officers travel over the selected sample locations counting, interviewing, and checking encountered hunters. The statistical framework for this design has again been developed primarily for fisheries surveys (Robson 1961, 1991; Pollock *et al.* 1994) but has general applicability for bird harvest as well.

Off-site surveys typically involve efforts to contact hunters before, during, or following the hunting season in an effort to obtain information about the harvest. The sample frame for such surveys is generally based on lists of hunters (e.g. purchasers of hunting licenses, members of hunting clubs). Questionnaires may then be mailed to hunters (e.g. Atwood 1956; Martin and Carney 1977; Wright 1978; Barker 1991; Barker *et al.* 1992; Dolton and Padding 2002) or conservation officers may contact and question selected hunters by telephone (Hayne and Geissler 1977; Barker 1991) or in person (door to door surveys,

Pollock *et al.* 1994). Another off-site approach involves contacting hunters before the hunting season and asking them to keep records of each hunt during the season (e.g. Caithness 1982; Pollock *et al.* 1994). Off-site surveys provide more opportunity for biased estimates resulting from such factors as poor memory, rounding error, false reporting associated with prestige or other motivations for deceit, question misinterpretation, species misidentification, and nonresponse.

13.4.3 Harvest estimation when harvest is illegal

Illegal harvest can be of two types that sometimes require different methods of estimation and investigation. One type of illegal harvest occurs during open hunting seasons and involves violations of bag limits (shooting more birds than legally permitted) and of species regulations (shooting protected species of birds in addition to birds for which the season is open). The other type of illegal harvest does not occur within the context of an open season and includes virtually any other type of illegal hunting activity, such as targeting protected species, illegal commercial harvest, and hunting at illegal times (e.g. at night).

Illegal harvest during the hunting season is frequently investigated by clandestine observation of hunters. North American studies of illegal activities in waterfowl hunting often employ a "spy blind" technique (e.g. Mikula *et al.* 1972; Martin and Carney 1977; Nieman and Caswell 1989). Roving hunter checks by conservation officers can also provide information on certain types of violations (e.g. harvested birds exceeding the bag limit). Mail questionnaire hunter surveys have been used to estimate illegal activity as well (Gray and Kaminski 1989, 1994), and may be useful when respondent anonymity is ensured.

Illegal harvest occurring outside the context of a legal hunting season typically cannot be investigated using hunter observation methods, as participants in illegal activities are secretive. One method for estimating the magnitude of illegal activities and harvest is sometimes termed "violation simulation" (Gray and Kaminski 1989). The approach uses investigators who pose as illegal hunters and harvest birds "illegally". The number of these simulated violators detected by conservation officers is used with the number of actual violators detected to estimate total violations under a capture-recapture approach. Another approach is for investigators to infiltrate the societal subset of interest (e.g. commercial poachers) in order to obtain information about the magnitude of harvest (Gray and Kaminski 1989). It is worth noting that in some countries, laws regarding hunting are so infrequently enforced that participants in illegal activities may not worry about being secretive. In such cases, the off-site surveys described above (13.4.2) may work.

13.4.4 Measuring harvest rate

We focus attention first on estimation of legal harvest rate. In cases where it is possible to estimate both harvest ($\hat{H}$, the hat denotes an estimate) and the size of the population from which the harvest is taken ($\hat{N}$), harvest rate may be estimated as:

$$\hat{h} = \hat{H}/\hat{N}, \qquad (13.7)$$

based on the relationship expressed in equation (13.6). Although this approach has seen some use (e.g. Anderson and Burnham 1976), resulting estimates are often imprecise, and the more common estimation approach involves use of marked birds (also see Chapter 5).

The most straightforward approach using marked birds requires a sample of n_t birds to be individually marked before the hunting season in year t. If r_t of these marked birds are shot and retrieved by hunters during the subsequent hunting season and reported to conservation officers, or to a national bird banding data repository, then we can estimate a new quantity, recovery rate (f_t), as:

$$\hat{f}_t = r_t/n_t. \qquad (13.8)$$

If multiple years of bandings and recoveries are available, then the band recovery models noted in Chapter 5 (e.g. Brownie *et al*. 1985) can be used to estimate recovery rate, as well as survival rate. Estimates based on these models make full use of recoveries obtained in all years following banding and are thus somewhat more precise, although the simple estimator in Equation (13.8) is fairly efficient.

Harvest rate can be estimated using estimates of recovery rate and band reporting rate, $\hat{\gamma}_t$, defined as the probability that a marked bird shot and retrieved by a hunter is reported to the conservation agency. Recovery rate is then related to harvest rate as $f_t = \gamma_t h_t$ so that harvest rate can be estimated from estimates of recovery rate and reporting rate as:

$$\hat{h}_t = \hat{f}_t / \hat{\gamma}_t. \qquad (13.9)$$

Reporting rate can be estimated from reward band studies in which some bands are clearly inscribed with a reward that can be obtained by reporting the band. When the reward is sufficiently large that reporting rate can be assumed to approach 1 (see Nichols *et al*. 1991), the relative recovery rates of standard bands and reward bands can be used to estimate reporting rate (e.g. Henny and Burnham 1976; Conroy and Blandin 1984).

Estimation of harvest rate associated with illegal kill is very difficult and can be accomplished using equation (13.7) if abundance and illegal harvest can both be estimated. Standard approaches using marked birds are not likely to be useful, because illegal hunters are not likely to report the kill of marked birds. Special studies using surgically implanted radio transmitters would provide a means of estimating harvest rate associated with illegal kill in local study areas, but this would be a relatively expensive endeavor.

Finally, we note that the translation of estimates of harvest rate (defined with respect to retrieved harvest) into estimates of kill rate (defined with respect to all hunting-caused death, regardless of hunter retrieval) requires additional information on the rate of retrieval of shot birds. Such information on "crippling rate" can be obtained from hunter observation studies and hunter questionnaire surveys (e.g. Martin and Carney 1977).

13.5 Assessing population size

13.5.1 Overview

The subject of estimating avian abundance and density was discussed in Chapter 2. The basic concept underlying all of the methods described in that chapter involves the issue of detectability. All estimation methods entail collection of some sort of count statistic, C. This may be the number of birds detected while walking a line transect, the number caught in a mist net, the number counted from an airplane, or the number shot and reported by hunters. This count can be viewed as a random variable, the expectation of which can be written as a function of the quantity of interest (N = abundance) and detection probability (β = probability that a member of N appears in the count, C): $E(C) = N\beta$. If detection probability associated with the count statistic can be estimated, then this relationship can be used to estimate abundance as:

$$\hat{N} = C/\hat{\beta}. \tag{13.10}$$

All of the methods available to estimate animal abundance require estimation of detection probability associated with the count statistic and then application of equation (13.10) (Chapter 2; also see Lancia *et al.* 1994; Williams *et al.* 2002). These approaches include such observation-based methods as distance sampling, multiple observers, sighting probability modeling, and temporal removal modeling, as well as capture-recapture and catch-effort approaches based on marked animals, and change-in-ratio approaches for harvested species (Seber 1982; Lancia *et al.* 1994; Williams *et al.* 2002).

13.5.2 Coping with uncertainty

Population size is seldom known with certainty, and this "partial observability" is an important source of uncertainty in the management process (e.g. Williams *et al.* 2002). There are two basic approaches to coping with this uncertainty. Decision-theoretic approaches such as adaptive management (see 13.7.3) incorporate uncertainty in a formal manner. In the case of partial observability, estimates of population size and their associated sampling variances are incorporated directly in the process of making decisions. As noted above, another *ad hoc* approach to dealing with uncertainty is to base harvest decisions on "minimum population sizes" that are typically smaller than true abundances. These minimum sizes might be obtained as the lower end of a 95% confidence interval for population size. This approach is intended to be conservative, in the sense that the usual error in determining allowable harvest will be to restrict take to a smaller harvest than could likely be sustained.

13.6 Assessing population dynamics

The assumption that underlies the justification of sustainable harvesting is that negative density-dependence creates surplus production at density levels below saturation. The certainty with which exploitation can be justified, and the precision with which it can be managed, depend on understanding the population dynamics of the species in question, especially with regard to density-dependence and the effects of harvest. The PBR formula requires minimal information because it makes many assumptions about the dynamics. Errors due to these assumptions are guarded against by taking a conservative approach. As more information is known about the specific population dynamics of a species, the management recommendations can be less restrictive.

13.6.1 Maximum growth rate (r_{max})

The maximum growth rate used in the PBR formula (equation (13.4)) is calculated by subtracting one from the discrete growth rate (λ_{max}) that would be experienced by the population in the absence of harvest, when the density was very low, and in the absence of Allee effects (see below, Section 13.6.2). It is important to note that this may *not* correspond to any observed growth rate for that population, since most populations will not be at such a low density. On the other hand, r_{max} should not be taken as a rate of population growth under unrealistic assumptions. Robinson and Redford (1991) provide a formula for calculating λ_{max} from minimal life-history information. They begin by solving Cole's (1954) formula:

$$1 = \lambda_C^{-1} + b\lambda_C^{-\alpha} - b\lambda_C^{-(\omega + 1)} \tag{13.11}$$

for λ_C, where α is the age at first reproduction, ω is the age at last reproduction, and b is the number of female offspring per reproductive female per time period. This formula assumes that both adult survival and survival to age at first reproduction are 1, and thus produces growth rates that are not realizable, even in the best of conditions. To account for these additional factors, Robinson and Redford (1991) recommended multiplying $(\lambda_C - 1)$ by 0.6, 0.4, or 0.2, depending on whether the maximum longevity of the species was <5 year, between 5 and 10 year, or >10 year. Slade *et al.* (1998) acknowledge these other life-history parameters more explicitly, by solving:

$$1 = p\lambda^{-1} + l_\alpha b\lambda^{-\alpha} - l_\alpha b p^{(\omega - \alpha + 1)} \lambda^{-(\omega + 1)} \tag{13.12}$$

for λ, where p is the adult survival rate and l_α is the survival rate from birth to age at first reproduction. This growth rate can be interpreted directly as λ_{max}, without any further adjustment. If estimates of p and l_α are not available, Slade *et al.* (1998) offer some alternatives for how to make reasonable guesses. In determining values for the life-history parameters (α, ω, p, and l_α), it is important to consider what these values might be in the absence of harvest and at low density.

Where data are unavailable to estimate r_{max} or the life-history parameters for the species in question, biologists can look to better-known closely related species with similar life histories. In establishing guidelines for the application of PBR to stocks of marine mammals, Wade (1998) recommended using default values for r_{max} of 0.04 for cetaceans and 0.12 for pinnipeds. To our knowledge, no such taxonomic generalizations of r_{max} for birds have been completed. Instead, biologists familiar with the species in question should consider the available knowledge for conspecific or congeneric birds.

If direct estimates of growth rates (i.e. from mark-recapture models, see Chapter 5 and Williams *et al.* 2002) were available for several populations at varying densities, one might consider estimating the maximum growth rate from the relationship between growth rate and density by finding the limit as density approached zero. However, this would require assuming that the populations were in equivalent habitat (i.e. that there were not source and sink areas) and that all populations had the same density at carrying capacity, and the same maximum growth rate.

13.6.2 Other aspects of density dependence

In the logistic growth model (equation 13.1), the ratio of population size, N, to carrying capacity, K, determines the reduction in growth rate due to density-dependence. Two approaches can be taken to estimate carrying capacity: (1) identifying the limiting resource, determining how much of it is there, then calculating the maximum population size it could sustain; or (2) observing an unexploited population at equilibrium, or calculating K from the relationship between growth and density at a number of replicates sites. Both of these approaches can pose significant challenges and are subject to considerable uncertainty. While determining the maximum sustained yield does require knowing

K, the PBR formula avoids this need by substituting an estimate of density for an estimate of K. Essentially, this sets a target harvest *rate*, rather than a total harvest quota, which, as discussed above, is a more conservative and stable strategy. Thus, for application of the PBR formula, an estimate of K is not required.

The logistic growth model assumes a quadratic relationship between N_{t+1} and N_t, and this results in a linear relationship between the equilibrium population size and the harvest rate (Figure 13.1(a)). A more general form for the logistic growth model includes another parameter, θ, to govern the shape of the density-dependent relationship:

$$N_{t+1} = N_t + r_{max} N_t \left[1 - \left[\frac{N_t}{K} \right]^{\theta} \right] - h_t N_t. \qquad (13.13)$$

When the shape parameter is greater than 1, the effects of density-dependence do not occur until the density is high, such as might be the case when the density-dependence is due to a limited number of territories (Gilpin *et al.* 1976). When the shape parameter is less than 1, the effects of density-dependence are apparent even at very low densities; this might be the case when resources are hetero-geneous and the first individuals consume the highest quality resources (Gilpin *et al.* 1976). The shape parameter affects the density at which yield is maximized— when $\theta > 1$, $N^* > K/2$; when $\theta < 1$, $N^* < K/2$ (Taylor and DeMaster 1993). Thus, knowledge about θ can increase confidence in the sustainability of harvest. The PBR formula assumes that $\theta = 1$. Estimation of θ requires measuring the growth rate of a population for a range of densities, ideally in the absence of harvest, and fitting those data to equation (13.13).

The models discussed above incorporate only negative density-dependence—the reduction in survival and/or reproductive rates, hence growth rates, with increases in density. But positive density-dependence at low density, known as an Allee effect (e.g. Dennis 1989), can also occur. For example, Allee effects can occur when poten-tial mates are too sparsely distributed to be able to locate each other. Allee effects can create instability in harvested systems—if the population drops below a critical density, an extinction vortex can result. That critical density, if it exists, is below the equilibrium population size that produces maximum harvest. Since the PBR method seeks to keep the population size above that optimal equilibrium size, it guards against potential Allee effects. Nevertheless, if there are aspects of the behav-ior or life-history of the organism that suggest a possible Allee effect, harvest strate-gies should guard against inadvertently lowering the density below the critical level.

Equations (13.1) and (13.13) are phenomenological, in that they do not ascribe a mechanism for density-dependence, but instead posit an empirical relationship that captures the phenomenon. A more mechanistic treatment of density-dependence would look at how individual life-history parameters (e.g. juvenile survival, adult survival, age at first reproduction, etc.) are affected by density. These relationships would then be combined in a structured model (see 13.6.4) to provide an alternative to equations (13.1) and (13.13) that captured more of the dynamics specific to the animal of interest.

13.6.3 Other driving forces

In the simple models discussed above, the only factors that affect population growth are density and harvest. But, in reality, many other factors can affect population growth, sometimes to such a degree that the density-dependent relationships are not discernable. For instance, reproduction, and hence population growth, of Mallard ducks (*Anas platyrhynchos*) in North America are strongly influenced by the availability of water on the prairie breeding grounds, so much so that annual harvest management decisions for this species are conditional on observed number of prairie ponds (Johnson *et al.* 1997). Increased understanding of the range of forces that affect the population dynamics can enhance the management of exploited populations.

13.6.4 Model structure

The models presented in equations (13.1) and (13.13) treat the population as a homogeneous group of individuals, all with the same life-history traits and vulnerability to harvest. The heterogeneity of real populations, due, for instance, to differences by sex, age, size, and reproductive status, can affect the growth rate and density-dependence of a population, and hence the impacts of harvest (Box 13.2). Further, some exploitation strategies target specific classes of

---

Box 13.2 Case study: assessing the population dynamics of the Maleo

Maleo (*M. maleo*) are burrow-nesting megapodes endemic to Sulawesi, Indonesia, that incubate their eggs in communal nesting sites on beaches and in soils heated by volcanic activity (Argeloo 1994). The eggs of the Maleo, like those of most other megapodes, have probably been harvested for millennia (Jones *et al.* 1995). In recent decades, traditional indigenous restrictions, which had served to make egg harvesting sustainable, have broken down, leading to overexploitation (Argeloo and Dekker 1996). Using a population dynamics approach to assess the potential growth rate would require knowledge of several life-history parameters: number of eggs laid per female per breeding season, hatching success rate, survival rate of hatchlings until sexual maturity, and adult survival rate. From those parameters, a structured population model could be built, and a growth rate could be estimated. One way to calculate a sustainable harvest for eggs would be to insert the maximum growth rate calculated from the structured population model into equation (13.4), using the minimum estimate of the number of eggs laid as N_{min}. This is a reasonable place to start, but may be unnecessarily conservative, since the PBR method ignores differences in reproductive value and density-dependence among classes within a population. To develop an assessment of sustainable harvest of eggs specific to this life-history would require additional information about the density-dependence of some of the life-history parameters, particularly hatching success and survival of hatchlings to maturity.

Unfortunately, the unique life-history of megapodes makes many of those parameters difficult to estimate. The hatchlings, which are completely independent of their parents from the moment they emerge, are highly vulnerable to predation, and are thus extremely secretive. The adults also are secretive, living in the forest except when they come to the communal nesting sites. The estimated number of eggs laid per female is 8–12 per breeding season (Dekker 1990). The hatching success rates in predator-proof hatcheries built at nesting grounds have been between 55% and 75% (Jones *et al.* 1995). Post-emergence mortality rates are not known. In the Australian Brush-Turkey (*Alectura lathami*), a "very rough estimate" of the mortality from emergence to sub-adult was 90–97% (Jones 1988). Maleo become sexually mature in their second or third year (Jones *et al.* 1995). There is no estimate of the adult survival rate. Thus, the data do not currently permit a quantitative determination of the potential growth rate or the sustainable harvest rate of eggs. To estimate growth rate, the two critical parameters that would have to be measured are: hatchling survival to maturity, perhaps by banding or otherwise marking hatchlings released from incubation programs; and adult survival, again by banding or marking, with subsequent recovery, recapture, or resighting of those marks (see Chapter 5). To estimate the sustainable harvest rate without the assumptions of the PBR calculation, additional information on the density-dependence of the life-history parameters would be needed.

Even if such monitoring programs can be initiated, it will take a number of years to obtain enough data to estimate reliably the parameters needed. As an alternative, an experimental or adaptive approach (see 13.7.2 and 13.7.3) could be taken, where portions of communal nesting grounds are completely protected from egg harvesting, and the number of pair-visits, the number of eggs harvested, and, if possible, the number of hatchlings emerged are monitored. Such an approach should be designed to assess what level of protection is needed, to produce an increase in population size, while still allowing some harvest. A good place to start is the level of protection afforded under traditional egg-harvesting methods (Argeloo and Dekker 1996).

individuals: e.g. parrots are preferentially captured as nestlings for the pet trade (Beissinger and Bucher 1992); eggs of the Maleo (*Macrocephalon maleo*) are collected for food, while the adults are left alone (Argeloo 1994). The effect of harvest of a particular class of birds on the population growth rate depends on the reproductive value of that class (Kokko *et al.* 2001). Thus, where life-history traits are strongly class-dependent (as in long-lived birds), or where exploitation targets a particular class of birds, a structured population model (Caswell 2001) is needed to assess the impact of exploitation. Field studies designed to estimate age-, size-, sex-, or stage-specific survival and reproductive rates are required to provide the parameters for this type of model (see Chapters 3 and 5).

13.7 Addressing uncertainty

13.7.1 Motivation

The advantage of an approach like the PBR formula is that it is simple to apply and requires minimal information about the species being exploited. The disadvantage is that it produces a very conservative estimate of sustainable harvest, in order to guard against the large amount of uncertainty about the life-history dynamics and effects of harvest. Where there is strong motivation for increased exploitation, managers and biologists can make more precise, less guarded, assessments of harvest potential by increasing their knowledge about the dynamics of the species in question.

Two critical uncertainties that need to be resolved to improve management of any exploited species are the effects of density-dependence and exploitation on life-history parameters. There are two approaches to resolving these uncertainties in the context of management: management experiments and adaptive management.

13.7.2 Management experiments

Management experiments involve applying different management treatments to multiple experimental units (e.g. separate populations) in a randomized and controlled statistical design (Walters and Holling 1990). The advantage of this approach is that results can be obtained fairly quickly, and the inference that can be made from the results is strong. That is, because of the random application of treatments to experimental units, conclusions about cause and effect can be made. Further, the methods for designing such studies and the techniques for analysis are standard. The disadvantage is that this approach is sometimes viewed as risky from a conservation standpoint: the range of treatments may need to include options that have not been previously tried, or which are greater in magnitude than previous experience; it may not be palatable to apply the more aggressive treatments to the more vulnerable population units; and the potential effects of some of the treatments might be irreversible. Further, it may be difficult to determine whether separate populations are truly independent—substantial and perhaps density-dependent movement of animals among the populations could invalidate the experiment.

For management experiments designed to estimate parameters needed to determine sustainable harvest levels, the key quantities to control are the population density and the harvest rate. The key quantities to measure will depend on the life-history of the particular species, but will likely include adult survival rate, reproductive rate, and juvenile survival rate; integrated measures of population dynamics, such as population growth rate and the annual harvest achieved, could also be measured directly.

13.7.3 Adaptive management

The goal of a management experiment is short-term learning, even if the learning comes at the expense of the management objectives, with the assumption that the knowledge acquired could then be applied to subsequent management decisions for the long-term benefit of the resource. Adaptive management is an alternative approach that seeks the reduction of uncertainty (i.e. learning) in the context of meeting the management objectives (Walters 1986) (Box 13.3). Thus, the primary goal is the management of the resource, with learning pursued only insofar as it will improve such management. There are several advantages of adaptive management: it can be applied to only one experimental unit, if necessary; concerns about conservation risk are built into the approach; and the appropriate balance between learning and management can be found. The disadvantage is that adaptive management may take longer to yield useful knowledge than experiments.

In its most formal application, adaptive management links a decision theoretic approach to resource management with an explicit method for tracking (and reducing) uncertainty. There are four elements required for this approach: explicit management objectives, a list of alternative management actions, multiple models that capture the uncertainty about the dynamics of the population, and a monitoring system to provide feedback (Nichols *et al.* 1995; Williams 1997). Each year (or whatever the time frame of management decisions is), a choice is made from the list of possible management actions that maximizes achievement of the management objectives, given the current state of the population, and the current state of knowledge, where this knowledge state is specified as a set of model "weights" reflecting the relative degrees of faith in the different models. The action is taken, and the consequences are monitored. The results of the monitoring are then used

Box 13.3 Case study: adaptive harvest management of Mallards

In North America, duck harvest is regulated by annually setting a common sport hunting season for a number of species, with the length of the season dependent on the current status of the duck populations and the conditions of the breeding habitat (Nichols *et al.* 1995; Williams and Johnson 1995; Johnson *et al.* 1997; Williams 1997). Presently, the status of Mallards (*Anas platyrhynchos*) is used as a surrogate for the status of most other duck species. An explicit adaptive management approach is used to choose the regulations each year (Johnson *et al.* 1997). The two fundamental uncertainties about population dynamics and responses to management that are being addressed by this approach are (1) whether the effect of harvest is additive to, or compensatory with, natural mortality; and (2) the degree to which reproduction depends on density; four alternative models of Mallard population dynamics capture this uncertainty. An optimal state-dependent harvest policy is calculated based on the alternative models and the objective to maximize cumulative harvest over an infinite time horizon, subject to a constraint that reflects a desired minimum spring population size. Each May, Mallard population size and the number of ponds in prairie Canada are estimated through aerial surveys. The particular set of regulations associated with values of these "state variables" is taken from the optimal harvest policy and implemented. The four models all predict the population size in the next year, the observed value for this quantity is used to adjust the weights associated with each model, and the process continues.

   The life-history information that went into building the models for management of Mallard harvest is substantial, and is based on a number of long-term, large-scale monitoring efforts: population size has been estimated with aerial surveys for over 50 years; annual survival rates, harvest rates, and differential vulnerabilities of adult and immature Mallards are estimated from band returns; and reproductive rates are estimated from the ratio of immature to adult birds in the harvest. Harvest of this species can be maximized, with little risk to the population because of this extensive knowledge of the population dynamics, because the monitoring is designed to specifically inform management, because relevant uncertainty is incorporated directly into the process and because management can respond annually to current conditions.

to inform the models of the population dynamics and to update their associated weights. The increased knowledge is used in the next time step to select the appropriate management action. Ideally, the optimization is made *actively*, that is, taking into account the potential effects of learning on future management decisions (Walters and Holling 1990).

It is critical that the monitoring system be tailored to the objectives and models particular to the management scenario in question. There are three purposes for the monitoring: to assess whether the objectives are being met, to measure quantities (such as population size) upon which the decisions will depend, and to measure response variables (such as survival rate or population size) that are predicted by the models to provide feedback about relative performance of the models. Ongoing monitoring also provides data for future refinements of the models. Because of the interrelatedness of objectives, models, actions, and monitoring, adaptive management requires careful articulation of all of these elements.

Acknowledgments

We wish to thank Bill Sutherland, Ian Newton, and Rhys Green for suggesting the scope of this chapter and for constructive comments during its development. Our work on this chapter was supported by the U.S. Geological Survey, Patuxent Wildlife Research Center.

References

Anderson, D.R. and Burnham, K.P. (1976). Population ecology of the mallard. VI. The effect of exploitation on survival. U.S. Fish and Wildlife Service Resource Publication 128.

Argeloo, M. 1994. The Maleo *Macrocephalon maleo*: new information on the distribution and status of Sulawesi's endemic megapode. *Bird Conserv. Int.*, 4, 383–393.

Argeloo, M. and Dekker, R.W.R.J. (1996). Exploitation of megapode eggs in Indonesia: the role of traditional methods in the conservation of megapodes. *Oryx*, 30, 59–64.

Atwood, E.L. (1956). Validity of mail survey data on bagged waterfowl. *J. Wildlife Manage.*, 20, 1–16.

Barker, R.J. (1991). Nonresponse bias in New Zealand waterfowl harvest surveys. *J. Wildlife Manage.*, 55, 126–131.

Barker, R.J., Geissler, P.H., and Hoover, B.A. (1992). Sources of nonresponse to the federal waterfowl hunter questionnaire survey. *J. Wildlife Manage.*, 56, 337–343.

Begazo, A.J. and Bodmer, R.E. (1998). Use and conservation of Cracidae (Aves: Galliformes) in the Peruvian Amazon. *Oryx*, 32, 301–309.

Beissinger, S.R. and Bucher, E.H. (1992). Can parrots be conserved through sustainable harvesting? *BioScience*, 42, 164–173.

Bennett, E.L. and Robinson, J.G. (2000). Hunting for the snark. Pages 1–9 In eds. *Hunting for Sustainability in Tropical Forests*, J.G. Robinson and E.L. Bennett, Columbia University Press, New York, USA.

Brownie, C., Anderson, D.R., Burnham, K.P., and Robson, D.R. (1985). Statistical inference from band recovery data—a handbook, 2nd edition. U. S. Fish and Wildlife Service Resource Publication 156.

Caithness, T. (1982). *Gamebird Hunting*. The Wetland Press, Wellington, NZ.

Caswell, H. (2001). *Matrix Population Models: Construction, Analysis, and Interpretation*, 2nd edn. Sinauer Associates, Inc. Publishers, Sunderland, Massachusetts, USA.

Caughley, G. (1977). *Analysis of Vertebrate Populations*. Wiley and Sons, New York, USA.

Cole, L.C. (1954). The population consequences of life history phenomena. *Quart. Rev. Biol.*, 29, 103–137.

Conroy, M.J. and Blandin, W.W. (1984). Geographic and temporal differences in band reporting rates for American Black Ducks. *J. Wildlife Manage.*, 48, 23–36.

Dekker, R.W.R.J. (1990). The distribution and status of nesting grounds of the Maleo *Macrocephalon maleo* in Sulawesi, Indonesia. *Biol. Conserv.*, 51, 139–150.

Dennis, B. (1989). Allee effects: population growth, critical density, and the chance of extinction. *Nat. Res. Model.*, 3, 481–538.

Diamond, J. (1989). Overview of recent extinctions. In *Conservation for the Twenty-first Century*, eds. D. Western and M.C. Pearl, pp. 37–41. Oxford University Press, Oxford, UK.

Dolton, D.D. and Padding, P.I. eds. (2002). Harvest Information Program: evaluation and recommendations. U.S. Fish and Wildlife Service, Washington, D.C.

Gilpin, M.E., Case, T.J., and Ayala, F.J. (1976). θ-Selection. *Math. Biosci.*, 32, 131–139.

Gray, B.T. and Kaminski, R.M. (1989). Strategies for estimating illegal waterfowl hunting and harvest. *Int. Waterfowl Symp.*, 6, 148–159.

Gray, B.T. and Kaminski, R.M. (1994). Illegal waterfowl hunting in the Mississippi Flyway and recommendations for alleviation. *Wildlife Monogr.*, 127.

Hayne, D.W. and Geissler, P.H. (1977). Hunted segments of the mourning dove population: movement and importance. Southeastern Association of Game and Fish Agencies, Technical Bulletin Number 3.

Henny, C.J. and Burnham, K.P. (1976). A reward band study of Mallards to estimate reporting rates. *J. Wildlife Manage.*, 40, 1–14.

Hudson, P. (1986). *Red Grouse: The Biology and Management of a Wild Gamebird*. The Game Conservancy Trust, Fordingbridge.

International Union for Conservation of Nature and Natural Resources, United Nations Environment Programme, and World Wildlife Fund [IUCN/UNEP/WWF]. (1980). World conservation strategy: living resources conservation for sustainable development. IUCN/UNEP/WWF, Gland, Switzerland.

Jacobs, M.D. and Walker, J.S. (1999). Density estimates of birds inhabiting fragments of cloud forest in southern Ecuador. *Bird Conserv. Int.*, 9, 73–79.

Johnson, F.A., Moore, C.T., Kendall, W.L., Dubovsky, J.A., Caithamer, D.F., Kelley, J.R., and Williams, B.K. (1997). Uncertainty and the management of Mallard harvests. *J. Wildlife Manage.*, 61, 202–216.

Jones, D.N. (1988). Hatching success of the Australian Brush-Turkey *Alectura lathami* in south-east Queensland. *Emu*, 88, 260–263.

Jones, D.N., Dekker, R.W.R.J., and Roselaar, C.S. (1995). *The Megapodes* (Megapodiidae). Oxford University Press, Oxford, UK.

Kokko, H., Lindström, J., and Ranta, E. (2001). Life histories and sustainable harvesting. In *Conservation of Exploited Species*, eds. J.D. Reynolds, G.M. Mace, K.H. Redford, and J.G. Robinson, pp. 301–322. Cambridge University Press, Cambridge, UK.

Lancia, R.A., Nichols, J.D., and Pollock, K.H. (1994). Estimating the number of animals in wildlife populations. In ed. Bookhout, T. pp. 215–253. *Research and Management Techniques for Wildlife and Habitats*. The Wildlife Society, Bethesda, MD, USA.

May, R.M. (1976). Simple mathematical models with very complicated dynamics. *Nature*, 261, 459–467.

Martin, E.M. and Carney, S.M. (1977). Population ecology of the Mallard: IV. A review of duck hunting regulations, activity and success, with special reference to the mallard. U.S. Fish and Wildlife Service Resource Publication 130.

Mikula, E.J., Martz, G.F., and Bennett, C.L. Jr. (1972). Field evaluation of three types of waterfowl hunting regulations. *J. Wildlife Manage.*, 36, 441–459.

Newton, I. (1998). *Population Limitation in Birds*. Academic Press, San Diego, California.

Nichols, J.D., Blohm, R.J., Reynolds, R.E., Trost, R.E., Hines, J.E., and Bladen, J.P. (1991). Band reporting rates for Mallards with reward bands of different dollar values. *J. Wildlife Manage.*, 55, 119–126.

Nichols, J.D., Johnson, F.A., and Williams, B.K. (1995). Managing North American waterfowl in the face of uncertainty. *Ann. Rev. Ecol. System.*, 26, 177–199.

Nieman, D.J. and Caswell, F.D. (1989). Field observations of hunters: the spy blind approach. *Int. Waterfowl Symp.*, 6, 160–165.

Pollock, K.H., Jones, C.M., and Brown, T.L. (1994). Angler survey methods and their applications in fisheries management. American Fisheries Society Special Publication 25.

Potts, G.R., Tapper, S.C., and Hudson, P.J. (1984). Population fluctuations in Red Grouse: analysis of bag records and a simulation model. *J. Anim. Ecol.*, 53, 21–36.

Reynolds, J.D., Mace, G.M., Redford, K.H., and Robinson, J.G. eds. (2001). *Conservation of Exploited Species*. Cambridge University Press, Cambridge, UK.

Robinson, J.G. (2000). Calculating maximum sustainable harvests and percentage offsets. Pages 521–524 In eds. J.G. Robinson and E.L. Bennett. *Hunting for Sustainability in Tropical Forests*. Columbia University Press, New York, USA.

Robinson, J.G. and Redford, K.H. (1991). Sustainable harvest of Neotropical forest mammals. Pages 415–429 In eds. J.G. Robinson and K.H. Redford. *Neotropical Wildlife Use and Conservation*. University of Chicago Press, Chicago, USA.

Robson, D.S. (1960). An unbiased sampling and estimation procedure for creel censuses of fishermen. *Biometrics*, 16, 261–277.

Robson, D.S. (1961). On the statistical theory of a roving creel census. *Biometrics*, 17, 415–437.

Robson, D.S. (1991). The roving creel survey. *Am. Fish. Soc. Symp.*, 12, 19–24.

Robson, D.S. and Jones, C.M. (1989). The theoretical basis of an access site angler survey design. *Biometrics*, 45, 83–98.

Runge, M.C. and Johnson, F.A. (2002). The importance of functional form in optimal control solutions of problems in population dynamics. *Ecology*, 83, 1357–1371.

Seber, G.A.F. (1982). *The Estimation of Animal Abundance and Related Parameters*. MacMillan, New York.

Slade, N.A., Gomulkiewicz, R., and Alexander, H.M. (1998). Alternatives to Robinson and Redford's method of assessing overharvest from incomplete demographic data. *Conserv. Biol.*, 12, 148–155.

Taylor, B.L. (1993). "Best" abundance estimates and best management: why they are not the same. U.S. Department of Commerce, NOAA Technical Memorandum NMFS-SWFSC-188. 20 pp.

Taylor, B.L. and DeMaster, D.P. (1993). Implications of non-linear density dependence. *Mar. Mam. Sci.*, 9, 360–371.

Taylor, B.L., Wade, P.R., DeMaster, D.P., and Barlow, J. (2000). Incorporating uncertainty into management models for marine mammals. *Conserv. Biol.*, 14, 1243–1252.

Wade, P.R. (1998). Calculating limits to the allowable human-caused mortality of cetaceans and pinnipeds. *Mar. Mam. Sci.*, 14, 1–37.

Walters, C.J. (1986). *Adaptive Management of Renewable Resources*. MacMillan, New York, NY.

Walters, C.J. and Holling, C.S. (1990). Large-scale management experiments and learning by doing. *Ecology*, 71, 2060–2068.

Williams, B.K. (1997). Approaches to the management of waterfowl under uncertainty. *Wildlife Soc. Bull.*, 25, 714–720.

Williams, B.K. and Johnson, F.A. (1995). Adaptive management and the regulation of waterfowl harvests. *Wildlife Soc. Bull.*, 23, 430–436.

Williams, B.K., Nichols, J.D., and Conroy, M.J. (2002). *Analysis and Management of Animal Populations*. Academic Press, San Diego, California, USA.

Wright, V.L. (1978). Causes and effects of biases on waterfowl harvest estimates. *J. Wildlife Manage.*, 42, 251–262.

14

Habitat management

Malcolm Ausden

14.1 Introduction

Habitat management is the manipulation of habitats to provide suitable conditions for species of interest, or in some cases to reduce the number of species considered as pests. Habitat management is most commonly used to:

- provide suitable conditions for species where these are no longer created by natural processes;
- maintain characteristic *assemblages* of species, where persistence of the *assemblage* is dependent on continuation or re-instatement of particular land-management;
- maximize the harvestable surplus of game species (Chapter 13).

Most habitat management involves preventing or reversing the direction of vegetation succession. In some cases it is also used to create suitable vegetation structure, increase food availability, and provide suitable nesting areas for birds.

The successional stage of an area, and hence its suitability for a given species, is influenced by processes such as grazing by large herbivores and disturbance by fire, floods, and storms. In many remaining fragments of semi-natural habitat, the key natural processes influencing succession either no longer operate, or if they do, they operate at an inappropriate scale or frequency to maintain suitable conditions for the species of interest. The absence of suitable natural processes is most acute in small and isolated patches of habitat. For example, small patches of semi-natural habitat are rarely grazed by large wild herbivores, let alone by their full complement of native species. Large-scale disturbances caused by fire or flood are largely prevented as a matter of policy. If they do occur, they are likely to affect the whole of the remaining small fragment of habitat, and so make it temporarily unsuitable for most of its associated species. If the fragment of habitat is isolated from sources of re-colonization, the less mobile species lost because of the disturbance are unlikely to re-colonize. In these situations, habitat management can be used to mimic the effects of these natural processes in a controlled manner to maintain a continuity of suitable conditions for desired species.

The main way of manipulating succession and creating suitable vegetation structure in terrestrial systems is by removing vegetation, either by cutting, grazing/browsing, burning, or herbicide use. Succession in wetlands can also be manipulated by controlling water levels and, in some cases, nutrient inputs.

When considering habitat management, it is useful to distinguish between phases of management aimed at restoring suitable conditions (restorative management) and those aimed at maintaining them (maintenance management). Wholesale habitat creation on land of little or no conservation value, such as arable land, usually involves increasing the rate of succession by optimizing conditions for the establishment and growth of desired vegetation, and in some cases also introducing seeds and plants. Techniques for habitat creation are outside the scope of this chapter.

Habitat management for birds has only been widely used in more intensively managed regions, such as Europe and parts of North America. Most of this chapter is therefore based on experience and research from these areas, although the principles involved are applicable elsewhere. An understanding of the general principles and effects of habitat management is fundamental to good site management, but the specific aims and details of any management should always be decided on a site by site basis. Habitats have been excluded where management is rarely, if ever, driven by the specific requirements of birds, such as deserts, mountain tops, sea cliffs, marine habitats, and rivers.

14.2 Deciding what to do

Habitat management has the potential to damage, as well as to benefit, important populations of plants and animals and ecological processes. Therefore, it is important to think out clearly what you want to achieve and how best to achieve it with minimal harm to other desired species. A good way to do this is first:

- To collate and summarize information relevant to the management of the site

and then to identify:

- The current and potential important features of the site. These can be individual species, groups of species, habitats, processes, or landscapes.
- The ideal "condition" of these features. This ideal condition may be a range of states with fluctuations influenced by semi-natural processes.

- The key factors influencing the "condition" of these features.
- Whether there are potential conflicts and constraints in achieving the ideal "condition" of these features, and if so, how these could be resolved or overcome.

Only after going through this process will it be possible to make informed decisions about:

- *What* you want to achieve (your objectives).
- *How* you intend to achieve it (your prescriptions).
- What *monitoring* you need to carry out to determine whether you are achieving your objectives.

One way to ensure that you have followed this process is to produce a management plan. There are a variety of formats for management plans (e.g. Hirons *et al.* 1995), but the ideal is short, simple, and focused on the key decisions. Those people who will implement the plan should be fully engaged in the process of producing it.

Most management plans include setting targets or upper and lower "limits of acceptable change" (LACs) for species and assemblages. Even though populations of many species fluctuate widely in response to factors unaffected by habitat management (such as the weather), setting targets and LACs is still a useful way of precisely defining objectives, and, subsequently, assessing success.

When considering objectives for a site with *existing* conservation value, it is best to start with the assumption that it is more important to maintain suitable conditions for key species and assemblages already present, than to create suitable conditions for those not currently there. In many parts of the world, virtually all areas of semi-natural habitat have survived only because they have been managed to provide something useful, such as grazing, hay, timber, or peat. Species present in these habitats may only have persisted because of this traditional management. The greatest chance of retaining suitable conditions for these species will therefore often be to continue existing management. However, alternative management might create even better conditions for the species already present. It may also provide suitable conditions for species unable to survive under the existing management regime. However, since many patches of semi-natural habitat are now isolated from areas of similar habitats, many of the potential colonists may never arrive (but see Section 14.4.4). This is likely to be less of a problem for many birds and flying insects than for less mobile species. Some forms of "traditional" land management may also be important in terms of their cultural heritage or aesthetic value, for example, flower-rich traditionally managed hay meadows. Over-reliance on traditional land-use practice can, though, have the disadvantage of fossilising potentially more dynamic systems. One option is to trial the modified management over only part of the site.

Management of newly created habitats is not constrained by previous land-use and offers greater opportunities to trial novel management. An example of an alternative approach to traditional management is that at Oostvaardersplassen, a wetland created during land reclamation in the Netherlands. Here, management has involved introducing ecological processes important in influencing habitat conditions (more or less naturally fluctuating water levels and year-round grazing by large herbivores), and then allowing these processes to operate with minimal further interference (Vulink and Van Eerden 1998) (Figure 14.1). This large site supports important populations of many wetland birds, including two breeding species rare or absent from the rest of western Europe, Spoonbill *Platalea leucorodia* and Great White Egret *Egretta alba*. In addition, greater reliance on natural processes is usually cheaper than more interventionist management.

Fig. 14.1 One approach to habitat management involves reintroducing ecological processes important in influencing habitat conditions, such as year-round grazing by large herbivores, and letting these processes operate with little further intervention. This is only practical at very large sites, such as here at Oostvardersplassen in the Netherlands, where free-ranging konik ponies (shown), heck cattle, and red deer have been introduced. (Malcolm Ausden)

14.3 Monitoring

Monitoring needs to be given a high priority. Without it, we have no way of knowing whether management is successful or needs altering. Thorough documentation is also essential if you want to inform others of the success or otherwise of your management. It is particularly important to record negative results, which can help to avoid the repetition of management that has already proved unsuccessful. Unfortunately, negative results are rarely disseminated and difficult to publish. The results of monitoring which re-confirms previous findings, but under different conditions at other sites, are also informative.

There are several levels at which monitoring can be used to determine the effects of habitat management on birds. The simplest is to record the numbers of birds using the area at different times (Chapter 2). The next level is also to monitor the effects of management on key factors likely to influence bird use, such as vegetation composition and/or structure or food supply (Figure 14.2). This will help assess possible reasons for the management's success or failure. If bird use declines and your management is not producing the desired habitat conditions, then you need to review the effects of your management on the habitat. If bird use declines, but your management is having the desired effect on habitat conditions, then you need to review the relationship between the desired habitat conditions and bird use.

A further level is to compare trends in bird use and key habitat features in both a managed and similar unmanaged area (a control) (Figure 14.3). This will help determine whether changes in the managed area are due to the management itself, or simply part of changes taking place over a wider area. It is still possible, though, that any differences in trends between managed and unmanaged areas are simply due to chance. The most rigorous level is to use a randomized, replicated experiment to determine the effects of management. It is rarely practical to set up such experiments on a large enough scale to investigate effects of management on bird use. However, it is, often feasible to use randomized, replicated experiments to determine the effect of management on biotic and physical factors thought to be important in influencing bird use. For example, many species of waders and other wetland birds feed on invertebrates in the mud (benthos) of shallow, brackish lagoons (Section 4.10.3). We do not know whether food supply is limiting use of a particular lagoon by these bird species. However, increasing the abundance of their prey is at worst likely to have no effect on bird use, and at best will increase bird use and possibly also survival and breeding success. The effects of organic matter on benthic invertebrate biomass in the mud can be investigated by marking out a number of plots and incorporating organic matter

Fig. 14.2 Monitoring the effects of habitat management is fundamental to good site management. Here, densities of invertebrates in the water column and mud are being monitored to determine food supply for waders, such as avocets. (Geoff Welch and RSPB images)

to half of each plot. The half to which organic matter is incorporated should be selected randomly to prevent any introduction of bias. Benthic invertebrate biomass can then be compared between the treated and untreated halves of each plot.

14.4 General principles of managing habitats for birds

14.4.1 Factors influencing habitat use by birds

It has long been recognized that habitat structure is of fundamental importance in influencing its use by birds (e.g. MacArthur and MacArthur 1961). However, factors other than the physical condition of the habitat can also be important. In particular, there will be no point in creating suitable conditions for particular birds if the site is too small, too disturbed, or in the wrong geographical location to attract them. Other factors influencing bird use at a site might include availability of food and nest sites, predation, brood parasitism, disease and pollutant, and pesticide levels. Use of an area by migratory and dispersive birds may also be influenced by population processes taking place on a much wider scale, and outside the area being managed (Baillie *et al.* 2000).

14.4.2 Taking account of the requirements of non-bird species

Although various high-profile examples of habitat management have been enacted specifically to benefit individual bird species, these are the exception and have usually been undertaken as "crisis management" for severely threatened birds, such as Kirtland's Warbler *Dendroica kirtlandii*, Corncrake *Crex crex*, Bittern *Botaurus stellaris* or Red-cockaded Woodpecker *Picoides borealis* (Byelich *et al.* 1985; Kulhavy *et al.* 1995; Green and Gibbons 2000; Smith *et al.* 2000). In practice, most habitat management for birds is intended to maintain a characteristic habitat and its associated fauna and flora, taking particular account of the requirements of rare and charismatic species. Conflicts between requirements of birds and other species are rare, not least because most habitat management involves maintaining groups of species that have already coexisted under a previous management regime (see Section 14.2).

Habitat management for plants usually focuses on maintaining or increasing plant species-richness and maintaining populations of rare or otherwise highly valued plants. Both can be irrelevant to the suitability of the habitat for birds and other animals. For plants a major constraint in restoring species-rich terrestrial and aquatic systems is provided by the high levels of phosphorus derived from previous fertilizer application or sewage treatment works. Some types of habitat for birds can often be created at sites with high nutrient levels, although such conditions can hinder successful restoration of aquatic habitat by preventing growth of macrophytes. High nutrient levels can also prevent successful restoration of short, nutrient-poor grassland or heathland until time has allowed these nutrients to leach out.

Birds are larger than invertebrates and occur at lower densities, so usually require larger areas of suitable habitat. Most insects have annual life-cycles, during which they can require a range of different conditions during their larval, pupal, and adult stages. They can therefore only persist at a site if it provides all these conditions every year.

To begin with, new management regimes for birds should be introduced only to a proportion of the existing habitat, to minimize the risk of extinction of invertebrate species of conservation value, should the new management prove unsuitable for them. This is less of a problem for most plants, which can survive at least short periods of unfavorable management as seed or spores. Most birds can move elsewhere if conditions become unfavorable and re-colonize when suitable conditions return. When introducing grazing to a previously ungrazed habitat, it is prudent to start at low stocking levels to determine the effects of light grazing and work up from there to find the optimal grazing level.

14.4.3 Controlling unwanted plants

A frequent issue in habitat management is the control of unwanted plant species, such as Bracken *Pteridium aquilinum* that is invading dwarf-shrub heath or exotic species that are outcompeting native vegetation. The first stage is to evaluate whether the benefits of control will outweigh its costs. Issues to consider are whether the unwanted plant species is spreading, what vegetation is likely to replace it following removal, and the likelihood of the unwanted species re-colonizing from elsewhere. It is also important to decide whether the aim of control is to simply contain the species, or to eradicate it from the site, if indeed the latter is realistic.

There are two methods of reducing the competitive ability of the undesired species relative to that of the surrounding vegetation. The first option to consider is whether the abundance of the unwanted species can be reduced by modifying the existing management. For example, changing the timing of cutting or intensity of grazing might reduce the availability of suitable germination sites for the unwanted species. The second option is to cut just the unwanted species to reduce its vigor relative to that of the surrounding vegetation. Cutting is most effective at reducing vigor if carried out at times of year when the plant has least of its reserves stored underground, that is, when plants have produced leaves but before they have replenished their underground reserves. Repeated cutting of re-growth will further deplete the plant's resources. Some emergent plants, for example, Reed *Phragmites australis*, are most effectively controlled by cutting and immediately flooding their stems, or cutting them underwater. This reduces the plant's strength by cutting off the supply of oxygen from the plant's leaves above the water to its roots in the anoxic mud.

If neither of these strategies proves successful, then the next options are to dig up the unwanted species or use a herbicide. The former will only be feasible on a small-scale, and may be followed by regeneration of the unwanted species from seed in the freshly disturbed soil. Herbicides can be dangerous to people and other nontarget groups, and in general should be used only as a last resort, remembering they can leach into waterbodies. Selective methods include "spot-spraying," weed-wiping (wiping a systemic herbicide against tall, unwanted vegetation without touching shorter desirable vegetation below it), and drilling holes into cut tree stumps and pouring herbicide into them. Manufacturer's instructions and best practice should always be followed.

14.4.4 Taking account of predicted climate change

Impending climate change will necessitate a number of changes in current conservation practises. Shifts in the distribution of species resulting from climate change (e.g. Thomas and Lennon 1999; Harrison *et al.* 2001) will require habitat management to increasingly cater for the requirements of newly colonizing species, rather than just focusing on those already present. The ability of individual species to exploit changes in climate will be largely dependent on their ability to colonize new areas. Less mobile species will be unable to disperse from habitats that become increasingly unsuitable for them (e.g. Warren *et al.* 2001). Shifts in climatic ranges of species can perhaps to some extent be catered for by extending semi-natural habitat to higher altitudes, so that less mobile species have shorter distances to disperse to remain within their climatic ranges. A particularly insidious threat is the increased potential for mobile nonnative species to outcompete native species in semi-natural habitats. Another factor to consider is potential changes in hydrology of wetlands resulting from changes in precipitation and evapotranspiration.

Sea-level rise resulting from climate change and, in some areas, possible increased storm activity are already thought to be responsible for losses of coastal and associated habitat low-lying coastal habitats, such as saltmarshes and brackish and freshwater marshes inland of them. In most cases, similar habitats are preventing from re-forming further inland because of "hard" sea defences, resulting in a net loss of coastal habitat (e.g. Harrison *et al.* 2001). Large-scale habitat creation behind existing seawalls will be necessary to offset these losses.

14.5 Managing grasslands

14.5.1 Introduction

Except in very arid areas, most grasslands require periodic vegetation removal to prevent colonization by scrub and trees, and in the case of some very wet grasslands, succession to fen. This can be done by grazing, cutting, or burning. The latter is likely to be applicable only on drier grasslands, particularly those formerly maintained by natural fires. The primary consideration when managing grasslands for birds is how to create the desired sward conditions at particular times of year, while minimizing the potential damaging effects of management activities on breeding birds. On wet grasslands hydrology is also important in influencing use by wildfowl and waders.

The most important factors influencing bird use on grasslands are usually the height and structure of the sward, and the quantity of litter and bare ground. All these aspects can be manipulated by management, as can vegetation composition. Vegetation composition can itself influence sward structure, and may also directly influence food supply, for example, by providing suitable seeds or palatable grass species.

Vegetation height and structure affect the suitability of nest-sites, abundance and accessibility of prey, and the ability of birds to detect predators (e.g. see review by Vickery *et al.* 2001). Structure can be difficult to define and measure, but generally refers to variation in density and height (see Chapter 11). It is useful to distinguish between fine-scale variation in structure over tens of centimeters (often referred to as "tussockiness") and coarse scale variation (over tens of metres or more). The availability of litter and bare ground can also influence conditions for some birds. Some species, such as Henslow's Sparrow *Ammodramus henslowii*, require dense litter for nesting, while others, such as Stone Curlew *Burhinus oedicnemus*, require bare or sparsely vegetated ground (Green *et al.* 2000). Bare ground may also increase access for birds to soil invertebrates (Perkins *et al.* 2000) and surface-living arthropod prey such as beetles, while a dense litter layer will reduce it. In general, variation in sward conditions will increase the likelihood of suitable conditions for nesting or feeding being present somewhere in the area.

Scattered scrub and trees can increase the numbers of bird species using a grassland, mainly by providing nest-sites and song posts for more generalist species, rather than grassland specialists. They may also provide nest-sites and look-out posts for predatory birds and thereby reduce the breeding success or survival of grassland species (e.g. Green *et al.* 1990*a*).

14.5.2 Effects of cutting and burning on sward condition

Cutting and burning can both remove all or most of the above-ground vegetation at once. Uniformity of vegetation removal encourages uniformity in subsequent vegetation composition, height, and structure. The sudden removal of vegetation is particularly damaging to invertebrates and small mammals, and this might affect the suitability of grassland to birds that feed on them. The lack of litter and tussocks in regularly cut or burnt grasslands will further reduce their suitability for small mammals. Cutting tends to leave more litter than burning. Burning can therefore be more suitable for birds requiring bare ground, such as Upland

Sandpipers *Bartramia longicauda*. By creating more bare ground, burning also favors more annual plants, particularly if carried out frequently. In hay meadows, annual cutting is usually followed by "aftermath" grazing, which is important in maintaining high plant species-richness of agriculturally unimproved grasslands (Smith and Rushton 1994). Removal of cuttings will prevent them smothering re-generating seedlings and other small plants, but leaving them will provide temporary cover for small mammals and some invertebrates until the vegetation has grown again.

The primary considerations when deciding on cutting or burning regimes are their timing and frequency. Timing is influenced by the timing and duration of the breeding season (see below), and by the times of year particular sward heights are required by species of interest. The frequency of cutting and burning influences sward condition in a given area, particularly the quantity of accumulated litter. Thus in dry prairies in Missouri, USA, cutting on a rotation of 1–2 years is considered best for Grasshopper Sparrows *Ammodramus savannarum* that require a light litter layer, while a rotation of 2 years or more is considered better for Henslow's Sparrows (Swengel and Swengel 2001). Cutting on a rotation of more than 1 year, different patches in different years, can also be used to produce coarse-scale variation in sward structure, by creating a mosaic of patches at different stages of re-growth. Rotational management will therefore help maintain a continuity of suitable habitat for invertebrates and small mammals over a given area.

Altering the height at which the sward is cut can influence sward conditions. In agriculturally managed grasslands, cutting is carried out close to ground level to maximize the offtake, and such management also helps to maintain high species-richness of plants by creating gaps for plant regeneration (see above). Regular cutting at a height of 15 cm ("topping") is used to maintain a dense sward of 15–20 cm high to discourage flocks of Starlings *Sturnus vulgaris*, gulls, corvids, and plovers from grassland at airports, in order to reduce bird strikes (Civil Aviation Authority 1998). Altering the height of cutting can also be used to provide preferred sward heights for birds at particular times of year, for example, for wintering geese (Vickery *et al.* 1994).

14.5.3 Effects of grazing on sward condition

Grazing differs fundamentally from cutting and burning in that it removes the vegetation piecemeal, and more selectively, at least at low to medium grazing intensities. Grazing also produces dung, and its associated invertebrates can be important in the diet of some birds, notably Red-billed Choughs *Pyrrhocorax pyrrhocorax* (Roberts 1982). Trampling by stock can create a continuity of bare

Fig. 14.3 Grazing is important in certain habitats to prevent vegetation succession and maintain suitable vegetation structure for birds. A simple but effective way to monitor its effects is to erect grazing exclosures and compare the vegetation inside and outside of them. Yellowstone National Park. (William J. Sutherland)

and disturbed ground, particularly under wet conditions and thereby increase birds' access to soil invertebrates.

The effects of grazing vary depending on the vegetation already present, the densities, and type of stock and the timing and frequency of its access. Grazing at medium stocking levels will selectively remove only a moderate proportion of the sward, and tends to create variation in the vegetation structure. Very high or very low levels of grazing will in time remove, respectively, virtually all, or almost none, of the vegetation, and thereby produce little variation in sward structure. High densities of small mammals can survive under light cattle grazing regimes, which preserve tussocks and maintain a dense litter layer. Grazing creates coarse scale variation in sward structure and composition by accentuating existing variation in plant composition resulting from differences in topography and previous management. If such variation is not present, then grazing is unlikely to create it. The effects of prescribed grazing levels on the sward vary from year to year, primarily due to weather-related differences in vegetation growth. Thus stocking levels often have to be adjusted by eye to achieve the desired conditions. Nevertheless, it is still useful to measure the height and variation in structure of the sward at key times of year, for example, the beginning of the breeding season, to help interpret changes in bird numbers. This can be done by taking measurements of sward height against the graduated side of a Wellington boot as you walk across the field or by using a sward stick (see Chapter 11).

Three types of domestic animal are commonly used to graze grasslands: cattle, equines (horses, ponies, and donkeys), and sheep. Cattle feed by ripping off tufts of vegetation and are the first choice for producing fine-scale variation in sward structure and patches of bare ground. Equines and sheep nibble the vegetation and are more selective in the plants they remove. If they like the vegetation, they nibble it uniformly short, but if they do not, they will ignore it. They therefore produce little fine-scale variation in the sward, but at moderate grazing densities can create coarse-scale variation comprising short, uniform lawns and dense, rank areas. Judicious grazing, particularly by cattle, is therefore better than cutting or burning in providing suitable conditions for birds requiring a close juxtaposition of suitable nest-sites and ranges of feeding conditions. Grazing influences vegetation composition by encouraging unpalatable and low growing plants that can tolerate repeated defoliation, particularly grasses and rosette-forming species.

14.5.4 Minimizing nest and chick loss during management

If done during the breeding season, cutting and burning destroy nests and chicks (Kruk *et al.* 1997). Grazing animals can trample nests (Beintema and Müskens 1987; Green 1988). Vegetation removal should therefore ideally be undertaken only outside the breeding season. This may be impractical on grasslands where management is driven primarily by agricultural requirements and it may allow the sward to grow too tall for chick rearing, for example, by Lapwings *Vanellus vanellus*. One option is to graze or cut fields adjacent to those with nesting birds, so that adults can take their chicks to feed in these fields. Another is to graze fields with nesting birds at low stocking densities, with the aim of offsetting any decrease in nest survival caused by trampling, with an increase in chick survival due to better chick-rearing conditions. The type and age of stock also influence the proportion of nests trampled (Beintema and Müskens 1987). Alternatively, entry of stock or cutting can be delayed at least long enough to increase nest survival to levels sufficient to maintain the population (e.g. Kruk *et al.* 1996). For species whose nests can be easily located, such as Lapwings and Black-tailed Godwits *Limosa limosa*, nest survival can be monitored to determine the proportion lost to trampling. This information can be used to fine tune grazing regimes. Nests of these species can also be protected from trampling using nest protectors (raised metal grilles placed over the nest), or from mowing by marking nests and mowing around them (Guldemond *et al.* 1993). A novel technique is to deter

Fig. 14.4 Mowing regimes can be altered to minimize the loss of nests and chicks of ground-nesting birds. Delaying mowing until after breeding is one method. Corncrakes (inset), though, require uncut grassland throughout the summer, so strips of unmown grassland are left on field edges to retain suitable habitat for them. (RSPB Images)

birds with chicks from entering fields that are about to be mown by erecting "flags" made out of bamboo canes with blue or white plastic bags attached to their tops (Kruk *et al.* 1997).

Chick mortality can also be reduced by altering the pattern of mowing. Fields are normally mown from the outside of the field inwards. This concentrates chicks in an ever-decreasing island of unmown grassland until this is itself cut, killing the chicks. Mowing from the center of the field outward allows chicks to escape to surrounding fields, and together with leaving strips of suitable habitat along field margins has been used to increase productivity of Corncrakes in the United Kingdom (Tyler *et al.* 1998) (Figure 14.4).

14.5.5 Using fertilizer

The attractiveness of grassland to some types of geese can be increased by re-seeding unproductive swards with more nutritious grasses, such as Perennial Rye Grass *Lolium perenne* (e.g. Percival 1993), and by fertilizing the sward with nitrogenous fertilizer (Owen 1975; Percival 1993; Vickery *et al.* 1994). Care should always be taken to minimize run-off and leaching. Reseeding and/or fertilizer application will damage any existing botanical, invertebrate, or breeding bird interest of a grassland.

14.5.6 Hydrology of wet grasslands

Shallow (<50 cm deep) flooding on grassland can be used to attract wildfowl by providing both safe roost sites and suitable feeding conditions. Regular winter flooding followed by grazing or cutting encourages perennial grasses and some other plants important as seed sources for wintering wildfowl, but tends to produce less ruderal vegetation than moist soil management (Section 14.9.2). Regular inundation also encourages some grass species favored by herbivorous wildfowl, notably Creeping Bent *Agrostis stolonifera*. Retention of shallow floods during the breeding season will provide feeding areas for breeding wildfowl, although more densely vegetated ditches are probably more valuable for brood rearing.

Waders on grassland feed on invertebrates in the soil (primarily earthworms Lumbricidae and leatherjackets Tipulidae), amongst vegetation, and in shallow pools. Flooding previously unflooded grasslands creates a short-term "flush" of displaced soil invertebrates, which can attract waders and other species. However, soil invertebrates are slow to re-colonize areas vacated during flooding. This means that flooding large areas of wet grassland, either at the same time or on rotation, is likely to greatly decrease the total abundance of prey for waders (Ausden *et al.* 2001). One effective way of maintaining suitable conditions for breeding waders in the long-term is to maintain a mosaic of unflooded grassland with a high water table (if soils are suitable—see below) on which waders can nest and feed on soil invertebrates, and shallow pools that sequentially dry out and concentrate aquatic prey during the breeding season (Ausden 2001). This is easiest to achieve on sites with varied topography, such as unleveled coastal grazing marsh. Excavation can be used to create shallow creeks and pools at otherwise uniform sites, but it can be difficult to create a natural-looking variation in height. Disposal of unwanted spoil can also be a problem. Retention of surface water is easiest on soils with low rates of water transmission, such as clays.

Maintaining a high water table within fields can benefit Snipe *Gallinago gallinago* and Black-tailed Godwits by keeping the upper soil moist and therefore soft enough for them to probe for soil invertebrates (Green *et al.* 1990*b*). A high water table also concentrates soil invertebrates close to the soil surface, particularly on the margins of shallow floods. It may also retard vegetation growth and thereby help maintain suitable conditions for breeding waders that prefer more open conditions, such as Lapwings (Figure 14.5).

It is usually only possible to maintain a high field water table during the breeding season on soils that have a high rate of water transmission, such as undamaged peat. This can be done by maintaining high water levels in surrounding ditches, particularly if these ditches are closely spaced, so that water can flow laterally

Fig. 14.5 Flooding grassland can increase the accessibility of soil invertebrates to breeding waders such as lapwings by suppressing vegetation growth. If pools remain into late spring and summer, they are rapidly colonized by aquatic invertebrates and provide important feeding areas for wader chicks. (Malcolm Ausden and RSPB Images)

from these ditches through the soil into the field. On soils with lower rates of water transmission, movement of water into the field from surrounding ditches becomes insufficient to replace water lost from the field by evapotranspiration in late spring and summer. On undamaged peat, a water table of 20–30 cm below the soil surface is recommended for breeding Snipe (RSPB, EN and ITE 1997). Again, variation in surface topography is helpful in maintaining high water levels, especially if shallow ditches help to irrigate areas around them.

14.6 Managing dwarf shrub habitats

Dwarf shrub habitats consist of heathland, moorland, maquis, garrigue, and other vegetation comprised of dense, predominantly evergreen low shrubs. Some dwarf-shrub habitats are prevented from succeeding to tall scrub and woodland by edaphic and/or climatic factors, but most are prevented by periodic burning, cutting, and/or grazing. Most experience of managing dwarf-shrub communities to benefit birds stems from management of heather dominated (ericaceous) upland moorlands and lowland heathlands in the United Kingdom.

A large proportion of upland moorlands in the United Kingdom have been traditionally managed to encourage high densities of Red Grouse *Lagopus*

lagopus scoticus for shooting. This involves burning narrow strips or small patches of moorland in winter to provide a mosaic of different aged stands of heather. Management of lowland heathland for conservation also involves small-scale winter burning, typically on a rotation of 15–30 years, in this case to perpetuate the dominance of dwarf-shrubs and produce a patchwork of different growth phases for their associated fauna. Cutting is also used on heathland to produce similar effects, particularly at sites where fire could easily spread out of control. Management of heathland for conservation is usually also accompanied by removal of invading trees and control of bracken, and sometimes also by grazing to prevent regeneration of trees.

Although uplands dominated by moorland support a range of bird species of high conservation value, few of these are associated with pure stands of dwarf-shrubs, the majority being associated with intermixed areas of bog, bracken, and grass (e.g. Brown and Stillman 1993; Stillman and Brown 1994; Tharme *et al.* 2001). Hence, the most important consideration in managing moorland for these species is the retention, and in some cases management, of these associated habitats. All the birds considered characteristic of lowland heathland, such as Dartford Warbler *Sylvia undata*, Stonechat *Saxicola torquata*, Nightjar *Caprimulgus europaeus*, and Woodlark *Lullula arborea*, also prefer mixtures of dwarf-shrubs and scattered scrub and/or trees, bare and disturbed ground and areas of short grassland (e.g. Bibby 1979; Sitters *et al.* 1996; van den Berg *et al.* 2001). On heathlands the main issue to resolve is what proportion of trees and shrubs to remove. The compromise is between removing enough to prevent loss of the heathland, while retaining enough to maintain populations of the key bird and other species.

The length of the burning or cutting rotation influences the suitability of the habitat for the few bird species that use the dwarf-shrub heath, by influencing the proportion of different stages of re-growth present at any one time. For example, Woodlarks and Red-billed Choughs (on maritime heath) prefer recently cut or burnt areas, while Dartford Warblers and nesting Hen Harriers *Circus cyaneus* prefer taller, older, heather. The height and persistence of dwarf-shrub heath can also be influenced by grazing intensity.

14.7 Managing forests and scrub

14.7.1 Introduction

The main factors influencing the avifauna of forest and scrub are their structure and dominant tree species, particularly whether broad-leaved trees, conifers, or a mixture of both. Four methods can be used to influence structure and composition: control of grazing and browsing by domestic animals and deer, burning, felling, and planting. More than any other habitat, management of forests and scrub requires both a short, and long-term perspective. Whereas management can relatively quickly alter the composition and structure of the field layer and structure of the understorey, changes in dominant tree species (except by selective felling) usually take much longer.

Species-richness of breeding and wintering birds and sometimes also overall bird densities, usually increase with age of tree stand (e.g. Manuwal and Huff 1987; Buffington et al. 1997; Donald et al. 1997, 1998). In addition, in Europe the proportion of tropical migrants (mainly warblers) is highest in early successional forest and scrub, particularly in vegetation 1–4 m high, while in eastern North America the proportion of tropical migrants increases with vegetation height, and is greatest in vegetation over 10 m high (Mönkkönen and Helle 1989). This is because most migrants in Europe are scrub birds and most in eastern North America are forest birds. The proportion of cavity-nesting birds also tends to increase with age of stand (e.g. Donald et al. 1998), due to an increase in the number of suitable large and dead trees (Newton 1998).

However, these broad changes in avifauna in relation to age of stand can still be greatly influenced by management. In most countries outside the tropics virtually all forests have been, and still are, managed for wood production, even if not specifically planted for this (Section 14.7.4). Management for wood products aims to maximize production of unblemished wood from a limited range of selected tree species. To this end, trees are harvested prior to maturity. Fallen dead wood is usually taken for firewood, and dead trees and limbs are often removed for safety reasons and to prevent the spread of disease. The overall effect is to reduce tree species diversity, simplify structure, diminish the abundance of mature trees and dead wood, and consequently lower the value for birds. Structural diversity of such forests can be increased by thinning and creating gaps (Section 14.7.5), cavity nest-sites can be provided by adding nest boxes while the quantity of dead wood can be increased as described in Section 14.7.6. By contrast, old-growth forests that have had minimal or no management can support a larger number of bird species including some that are rare or absent in managed forests. These include species that require tree cavities or large-crowned trees for nesting, old trees, standing dead trees (snags) and/or dead limbs and branches for feeding (see Newton 1998; Imbeau et al. 2000; Poulsen 2002). Old-growth forests do not require management, other than perhaps in some cases reintroduction and management of previously exterminated large herbivores and their predators to facilitate more natural forest dynamics.

Forests dominated by broad-leaved trees support a distinctly different avifauna from those dominated by conifers, while mixtures of the two support an intermediate one containing species from both. The effects of individual tree species composition are less well known, although tree species are known to differ in their suitability for foraging (e.g. Peck 1989) and nesting in (e.g. Hågvar *et al.* 1990). Some tree species produce seeds that are much favored by particular bird species: for example, Oaks *Quercus spp.* by Jays *Garrulus glandarius*, Hornbeams *Carpinus betulus* by Hawfinches *Coccothraustes coccothraustes*, and Birch *Betula pendula* by Common Redpolls *Carduelis flammea*, and other finches.

14.7.2 Grazing and browsing

Grazing and browsing can be used to influence the structure and composition of the field layer and the structure of the understorey of forests in the short or medium term. In the longer term, grazing and browsing affect regeneration, and thereby also influence the composition of tree and shrub species. Browsing also removes scrub and leaves from lower branches.

The specific effects of grazing and browsing depend largely on the herbivore involved and its density. High densities of large herbivores reduce the density of vegetation within reach of whichever animals are being used, and encourage low-growing grasses and bryophytes at the expense of taller grasses and forbs. This tends to decrease the densities of birds that forage or nest in low scrub, that nest on the ground or which feed on seeds of herbaceous vegetation. Conversely, high levels of grazing/browsing tend to increase the densities of birds that require an open understorey for feeding, for example, in grazed oakwoods in western Britain, Redstarts *Phoenicurus phoenicurus*, Pied Flycatchers *Ficedula hypoleuca*, Wood Warblers *Phylloscopus sibilatrix*, and Tree Pipits *Anthus trivialis* (see review by Fuller 2001). By reducing the height of herbaceous vegetation, high grazing levels may also reduce the densities of small mammal prey for raptors and owls. Levels of grazing/browsing by wild deer can be reduced by culling and erecting deer-proof fences to prevent re-colonization.

Grazing and trampling produce gaps in which tree seedlings can establish, but prevent them from growing into saplings. These seedlings can be "released" by reducing or excluding grazing, but this tends to create a dense, even-aged understorey of saplings. Regeneration of tree seedlings is often poor once grazing levels have been reduced, because of lack of gaps for germination and because of competition with seedlings from tall grasses and forbs. Medium levels of grazing and variations in grazing levels are most likely to produce patchy and periodic regeneration of trees (Mitchell and Kirby 1990).

14.7.3 Burning

Controlled (prescribed) burning can be used to restore and maintain bird assemblages associated with fire-dependent forest types, particularly open forests dominated by fire-tolerant species of pine or oak (Figure 14.6). Fires which do not reach the forest canopy (ground fires) remove only the field layer, shrubs, and saplings. This simplifies the structure and opens the canopy. Fires which do reach the canopy (crown fires) also kill many mature trees, and can thereby alter tree species composition and forest structure. Crown fires usually occur only when fuel loads are exceptionally high after a period when fires have been artificially suppressed. Fire can also stimulate germination of certain forest tree species, and thereby influences tree species composition.

Fig. 14.6 Burning may seem catastrophic to birds and other wildlife. However, many vegetation types are dependent on periodic fires and suppression of fires by people has resulted in the degradation or loss of these habitats and their associated birds. One example is the fire-dependent scrub of Florida that is the sole habitat of the Florida Scrub Jay *Aphelocoma coerulescens*. (William J. Sutherland)

The short-term reduction in litter, shrubs, and saplings caused by burning usually results in short-term increases in numbers of ground and aerial-foraging birds, but decreases in numbers of ground-nesting species (e.g. Wilson *et al.* 1995; Artman *et al.* 2001). The effects of burning on bird composition have been particularly well studied in forests managed to benefit Red-cockaded Woodpeckers in the southeastern USA. Here, burning on a less than 5-year rotation, often accompanied by thinning, has been used to remove broad-leave trees and to restore and maintain open pine-dominated forests. This procedure increases the densities of birds typical of open pine-grassland habitats, such as Red-cockaded Woodpecker, Northern Bobwhite *Colinus virginianus*, and Blue Grosbeak *Guiraca caerulea*, but decreases the densities of birds associated with broad-leaved trees, such as Tufted Titmouse *Baeolophus bicolour* (Wilson *et al.* 1995, Provencher *et al.* 2002). Prescribed burning can also be used to kill stands of trees in order to encourage desired stages of regrowth, and to stimulate regeneration. In Michigan, USA, burning is used to kill stands of old Jack Pine *Pinus banksiana* to provide young (7–21 years old) stands suitable for Kirtland's Warblers (Byelich *et al.* 1985).

14.7.4 Planting and harvesting regimes

A range of management systems is used for producing and harvesting wood products. Clear-felling consists of periodic harvesting and re-planting of areas of trees. Shelterwood systems involve felling only a proportion of trees at any one time, in order to retain a partial canopy to shelter the following crop. Coppicing involves cutting broad-leaved trees close to the ground to produce regrowth of straight poles for fencing, charcoal production, and other uses. The resulting "coppice stools" are intermixed with "standard" trees, which are left to grow tall and periodically harvested for timber. Conifer trees do not regrow in this way.

The avifauna of harvested blocks of forest changes in relation to age of regrowth. The avifauna of the forest as a whole can therefore be influenced by altering the length of the harvesting rotation, so as to change the proportions of different ages of regrowth present at any one time. The avifauna of regrowth changes in relation to vegetation height as described in Section 14.7.1. However, since trees are harvested prior to maturity, bird species associated with later stages of forest growth are invariably scarce. In coppice, breeding bird densities typically increase during the first 5 years or so of regrowth, remain high during the period of canopy closure, but decline thereafter. Bird densities then remain low during the rest of the coppice cycle and, should coppicing cease, continue to remain low for a considerable time with only a very slow increase in species associated with older-forest growth (e.g. Fuller and Henderson 1992).

Clear-felling systems can be improved for forest birds by increasing tree species diversity, and by leaving a proportion of live and dead trees unharvested. These may be individual trees or groups of trees. However, suitable breeding sites for some important early successional species, such as Woodlark, can only be created by clear felling moderate sized areas (Bowden 1990). Species-richness of birds in conifer plantations can be increased by incorporating a proportion of broad-leaved trees. This can be done by planting, allowing natural regeneration, or retaining existing broad-leaved trees in areas too wet, steep, rocky, and otherwise inaccessible for harvesting conifers. There is evidence that species-richness of birds on conifer plantations is greater if, for a fixed area of broad-leaved trees, these are dispersed throughout the conifers, rather than concentrated in a few large blocks (Bibby *et al.* 1989). Retention of trees during harvesting is common practice in much of the world, but its long-term effects on birds are yet to be fully evaluated. However, in the first few years of regrowth, areas of forest where patches of trees have been retained support higher densities of breeding birds typical of more mature forest, particularly ground and tree-nesting and forest canopy-gap species, than areas of forest which have been clear-felled (e.g. Annand and Thompson 1997; Merrill *et al.* 1998). Retained trees also provide hunting perches for raptors, which might otherwise not be there.

14.7.5 Thinning and creating gaps

The avifauna of structurally simple forest can be diversified by increasing the structural complexity of the forest. Structural complexity increases without intervention, as trees out-compete one-another (self-thinning), create gaps and allow patchy regeneration, but this natural process can be accelerated by felling individual, or groups of, trees. Felling a proportion of trees throughout dense, closed canopy forest (thinning) will encourage regeneration of the shrub layer, allow trees to attain greater size and encourage suppressed broadleaved trees in conifer plantations to mature. Thinning can be carried out selectively to modify tree species composition, while the thinning of dense, multi-stemmed, abandoned coppice (Section 14.7.4) to produce single-stemmed trees (singling). This is thought to increase shrub cover and overall densities of birds, particularly warblers, and probably hole-nesting species as singled trees mature (Fuller and Green 1998). Glades and gaps in the canopy which occur where trees are felled should not always be restored with new trees, as they can be important, albeit often short-lived, habitats for birds and other forest-associated flora and fauna.

Thinning and group felling produce substantial benefits to the woodland avifauna only if done on a large-scale. In practice, this is usually possible only as part of a commercial felling regime. For example, commercially thinned 40–45 year-old Douglas-fir *Pseudotsuga menziesii* stands in Oregon, USA, held higher densities of six bird species compared to unthinned stands, including three characteristic of old-growth forest, namely Hairy Woodpecker *Picoides villosus*, Red-breasted Nuthatch *Sitta canadensis*, and Hammond's Flycatcher *Empidonax hammondii*. In contrast, only Pacific-slope Flycatchers *Empidonax difficilis* were more abundant in unthinned stands, as this species is associated with dense, closed canopy forest (Hagar *et al.* 1996).

14.7.6 Increasing the quantity of dead wood

Dead wood is a particularly important component of forests managed for conservation and, by encouraging woodpeckers, creates more nest-sites for species that use their old nest holes. Standing dead trees (snags) provide nest-sites for cavity-nesting birds and foraging habitat for woodpeckers and other species. Dead wood associated with large populations of ancient trees can also support important assemblages of invertebrates. Trees can rot in two ways—from the outside (sapwood decay) or the inside (heart rot decay). Heart rot decay is generally better for cavity-nesting birds and supports a particularly specialized invertebrate fauna. Standing and fallen dead wood and old trees should not be removed unless they pose unacceptable safety risks.

Dead wood can be increased by killing parts of or whole trees by girdling (making a continuous cut with removal of a ring of bark from around the trunk), or by injecting with herbicide, or in the case of conifers, removing the base of the live crown. These methods can be accompanied by inoculation of fungi to speed up decay. Such techniques should only be used on younger trees, preferably healthy ones, and not on older or partly dead trees that are important in their own right. It is preferable to damage only part of the tree, since whole dead trees fall over more quickly and thereby lose their value to cavity-nesting birds. A recent study found no difference in decay characteristics and woodpecker activity in Douglas Firs killed by girdling, herbicide injection, and cutting off of the base of the live crown with or without inoculation of fungi (Brandeis *et al.* 2002).

14.8 Managing deep water

Few management techniques to benefit birds apply in water over about 1 meter deep. The main ways in which deep waterbodies can be improved for birds are by providing suitable nest-sites by creating islands or providing rafts (Section 14.9.5); by creating shallow areas and emergent vegetation around the margins (see below).

14.9 Managing wetlands

14.9.1 Manipulating the proportions of open water, ruderal vegetation, and swamp

The suitability of shallow ($<ca$ 1m), nutrient-rich freshwater wetlands for birds is primarily influenced by the proportions of open water, swamp, fen, and scrub (Kaminski and Prince 1981; Linz *et al.* 1996 and Section 14.9.6), inputs of nutrients, pesticides, and other pollutants and levels of human disturbance (e.g. see Newton 1998). Each of the different habitats mentioned represents a different stage in the process of vegetation succession and each supports different assemblages of birds. Long-term changes in the extent of open water, swamp/fen, and scrub can be determined from aerial photographs. The effect of nutrients, pesticides, and other pollutants is a complex subject and outside of the scope of this chapter (see Newton 1998 for a review of their effects).

Nutrient-poor wetlands comprise bogs, base-poor fens, and some other types of fen fed by oligotrophic groundwater. Where they persist in the lowlands, their fragile plant and invertebrate assemblages are invariably threatened by low water levels and eutrophication. There is little or no management that could be carried out specifically to benefit birds in these types of wetlands without damaging their existing plant and invertebrate interest. For this reason, their management is not discussed further.

There are two approaches to manipulating the proportions of open water and swamp in a wetland. One involves periodically lowering water levels (drawdowns). The other involves maintaining a more constant water regime, while preventing or reversing succession in specific areas by removing vegetation or lowering the substrate. When installing control structures to manipulate water levels, it is important to consider their potential impact on movement of fish within the site. In particular, water control structures can impede upstream migration of fish such as Common Eels *Anguilla anguilla*, which can be important prey for birds such as Bitterns (Gilbert *et al.* 2003). This problem can be mitigated by installing passes (Knights and White 1998).

Periodic drawdowns during the growing season wetlands expose moist, bare mud on which seedlings of ruderal and emergent plants can germinate. Techniques for optimizing germination of ruderal plants to maximize seed production are discussed in Section 14.9.2. Swamp can be created by re-flooding these emergent plants in autumn, taking care not to completely submerge them, and then allowing them to expand through vegetative growth. The ratio of swamp to open water will be important in influencing bird use and can be manipulated during subsequent drawdowns. For example, in cattail *Typha* spp. and Whitetop Rivergrass *Scolochloa festucacea* swamps in North America, densities of breeding wildfowl are highest where there are equal proportions of swamp and open water (Kaminski and Prince 1981; Linz *et al.* 1996). If the swamp has died back due to herbivory by wildfowl and/or Muskrats *Ondatra zibethicus*, disease, erosion, or other environmental stresses, it can be re-established by lowering water levels in spring and summer to allow germination of emergent plants again. In the Oostvaardersplassen in the Netherlands, a drawdown period of 4 years was needed to re-establish reedbed which had largely disappeared due to erosion and grazing by Greylag Geese *Anser anser* (Ter Heerdt and Drost 1994). If the swamp has expanded too much, then the area of open water can be increased by lowering water levels, cutting, or burning patches of swamp, and then re-flooding. Burning creates more open ground than cutting, and so tends to result in more growth of ruderal vegetation during subsequent drawdowns (de Szalay and Resh 1997). Densities of breeding waterfowl tend to be highest where there is a high level of interspersion of swamp and open water (e.g. de Szalay and Resh 1997; Kaminski and Prince 1981).

Using drawdowns to manage the relative proportions of open water and emergent vegetation is impractical or unacceptable in certain wetlands. In particular, it risks temporary or permanent extinction of less mobile invertebrates in isolated wetlands where it is only possible to dry out all, or most, of the habitat. In these situations it is necessary to control succession by removing vegetation and lowering the ground level. However, this has several disadvantages compared to using periodic drawdowns. It does not create a temporary increase in invertebrate productivity in the shallow water following periodic drying out (see Section 14.9.3), nor does it provide suitable conditions for ruderal vegetation. In addition, permanent flooding maintains anoxic conditions in the sediment, which are thought to exacerbate the effects of eutrophication in causing die-back of reeds in Europe (Van der Putten 1997), unlike periodic drawdowns which allow oxidation of reed litter. Management of reedbeds without periodically drying them out is discussed in Section 14.10.6.

14.9.2 Increasing food abundance for birds in shallow freshwater

The abundance of seeds for wintering wildfowl can be increased by providing suitable bare, saturated mud for prolific seed-producing ruderal vegetation to germinate and grow in spring and summer. This can be done by lowering water levels or irrigating dried out wetlands to keep the soil moist. These seeds can be made available to wildfowl by re-flooding in autumn. This technique is known as "moist-soil management" (Smith and Kadlec 1983; Haukos and Smith 1993).

The most important factors affecting germination and growth of ruderal vegetation are the timing of drawdown and the ability to keep the mud moist enough for germination. For example, in the Playa Lakes Region of North America, drawdowns in April are recommended to maximize seed production of persicarias *Polygonum* spp. (Haukos and Smith 1993). Lowering of water levels in spring can be timed so as to increase food availability to migrants (Section 14.9.4), but may conflict with optimal management for breeding birds. Late summer is considered the best time to re-flood to provide suitable conditions for wintering waterfowl. Disking or ploughing the substrate in autumn prior to re-flooding can be used to increase seed production the following growing season by increasing the abundance of prolific seed-producing annual grasses at the expense of perennial species (e.g. Gray *et al.* 1999). Optimal frequency of drawdowns varies between sites, but is typically once every 5–7 years. If there is a throughput of water, then carrying out drawdowns too frequently might flush out a high proportion of nutrients made soluble during the drawdown (see below), thereby reducing productivity.

Invertebrate biomass tends to be highest in early successional (i.e. recently flooded) wetlands (e.g. Danell and Sjöberg 1982), and can therefore be increased by periodically drying out and re-flooding. The initial high invertebrate biomass is attributed to high overall productivity fueled by release of soluble nutrients from freshly inundated soil and decomposition of flooded terrestrial vegetation, and low levels of predation by predatory invertebrates and fish. As the wetland matures, nutrients released from decaying terrestrial plant material decline, numbers of predatory invertebrates and fish increase, and total invertebrate biomass tends to decline. Drying out will also kill any fish present. Re-colonization by fish following re-flooding is often accompanied by high levels of recruitment of small fish of suitable size for fish-eating birds.

Little is known about the effects of management during a drawdown on invertebrate biomass or bird use following re-flooding. Gray *et al.* (1999) found that soil disturbance during drawdowns reduces the biomass of large invertebrate prey for wildfowl the following winter, probably because it reduces the quantity of above-ground detritus for them to feed on. In rice fields, different treatments of rice stubble (ploughing, burning, chopping, rolling, disking, or cutting and removing) have little or no effect on bird use following re-flooding, although rice harvesting techniques that leave tall rice stems discourage some waterbird species, particularly small waders (Day and Colwell 1998; Elphick and Oring 1998).

Timing of re-flooding following drawdowns can influence the biomass of invertebrate prey for birds. Re-flooding playa wetlands in the southern USA in

September results in a higher biomass of aquatic invertebrates (predominantly ramshorn snails Planorbidae) the following winter than does re-flooding in November (Anderson and Smith 2000). In wetlands where the benthic fauna is dominated by non-biting midge larvae, invertebrate biomass is likely to be higher in winter if re-flooding is done in autumn while adult midges are still active and ovipositing.

14.9.3 Increasing food abundance for birds in shallow brackish and saline water

Invertebrate prey biomass for birds in brackish and saline lagoons can be maximized by maintaining optimum salinities for growth and reproduction of these different suites of prey species. This requires regular monitoring of salinity or conductivity. Measurements should be taken more frequently in summer when high evaporation rates can lead to rapid rises in salinity. Salinity in parts per thousand (ppt or ‰) is approximately equal to $0.64 \times$ conductivity in milliesiemens (mS) (Jones and Reynolds 1996). In temperate lagoons at low salinities ($<ca.$ 8‰) the most abundant invertebrate prey are usually non-biting midge larvae (Chironomidae) in the mud, and water boatmen (Corixidae), and opossum shrimps *Neomysis* spp. in the water column. At higher salinities (above ca. 8‰ and below 40–70‰) the main prey in the mud are polychaete worms, non-biting midge larvae, molluscs and amphipods, with opossum and other shrimps the main prey in the water column (Britton and Johnson 1987; Robertson 1993). Brine shrimps *Artemia* spp. are virtually the only prey present at salinities above ca. 70‰. These animals are restricted to warm climates and can withstand salinities up to ca. 320‰ (e.g. Britton and Johnson 1987). Studies of lagoonal invertebrates in England have revealed that maximum biomass of a non-biting midge larvae/water boatman and ragworm *Hediste diversicolor/Corophium volutator* fauna occur at, respectively, ca. 6‰ and ca. 24‰ (Robertson 1993).

There is little information on the effects of periodically drying out saline waterbodies on the invertebrate food supply for birds. As in freshwater wetlands, drying out kills fish, and high densities of fish can reduce densities of benthic invertebrates (Robertson 1993). Set against this, the small fish found in saline lagoons, such as sticklebacks Gasterosteidae and gobies Gobiidae, are themselves important prey for birds such as herons and egrets. Experimental studies have found that adding dead plant matter to brackish lagoons can increase the biomass of ragworms, but has little or no effect on that of non-biting midge larvae (Robertson 1993). Therefore, invertebrate biomass in ragworm-dominated lagoons could probably be increased by first lowering water levels to allow vegetation to colonize, and then re-flooding to kill this vegetation and increase the quantity of dead plant material available to the invertebrates. Biomass of poly-chaetes and bivalves may be slow to increase following such a drawdown, because they both take several years to reach maximum size. It might be best to retain some pools in the lagoon during a drawdown, from which polychaetes and bivalves can re-colonize the rest of the lagoon following re-flooding. The chironomid life-cycle is very short, and these animals quickly re-colonize from winged adults. Burning patches of emergent vegetation prior to re-flooding to increase interspersion of brackish swamp and open water (Section 14.9.1) has been found to increase the biomass of chironomids, while increasing interspersion by burning had no effect on it (de Szalay and Resh 1997).

The salinity regime will also influence the abundance of plant food for waterfowl. Saline lagoons support only a limited range of macrophytes species, although some of these, notably Beaked Tasselweed *Ruppia maritima* and charophytes, are important waterfowl food. Macrophytes are absent from high salinity lagoons, for example in southern France, those with salinities more than 64‰ (e.g. Britton and Johnson 1987).

14.9.4 Increasing accessibility of food for birds in shallow water

Water levels can be manipulated to provide suitable shallow water as feeding areas for waders, dabbling ducks, herons, and other species. Highest numbers of bird species are typically found in water 15–20 cm deep, with few wading species using water deeper than 40 cm (e.g. Elphick and Oring 1998). Plovers and some other species feed mainly on bare mud exposed by falling water levels. The range of feeding opportunities available at any one time can be increasing by enhancing topographic variation within the area flooded.

Achieving suitable water depths by lowering water levels has the advantage of concentrating aquatic prey of birds, particularly fish, and providing bare mud containing stranded benthic invertebrates on which waders and other birds can feed. For example, periodic lowering of water levels in fish ponds to allow harvesting of commercial fish species attracts large numbers of herons and egrets to feed on stranded non-commercial fish species and shrimps (Young 1998). Creating suitable water depths by raising water levels to flood new habitat may temporarily raise productivity by increasing the availability of detritus (see previous two sections), and provide a short-lived (and probably one-off) abundance of displaced terrestrial invertebrates, particularly on grassland (Section 14.9.2).

Where a number of such waterbodies are under independent hydrological control, feeding conditions for waterfowl can be optimized by sequentially lowering water levels in different waterbodies, to provide a continuity of suitable feeding conditions. It is worth considering lowering water levels at times of year when there is a lack of shallow water available in the surrounding area (Taft *et al.* 2002).

14.9.5 Providing islands and rafts

Nesting areas for terns, gulls, waders, and wildfowl can be created by building islands. These can be covered in shingle to provide suitable conditions for nesting terns and plovers, but such open conditions are difficult to maintain. One option is to design islands so that they are covered in water in winter and then exposed by falling water levels immediately prior to nesting. Winter flooding helps rot down and disperse any vegetation that might have grown on them. Otherwise, vegetation might have to be cleared by hand. Islands can be difficult to construct in deep water and can be subject to rapid erosion by wave action and to flooding during the breeding season should water levels rise. An alternative is to provide suitable nesting habitat, for example, for terns, on anchored rafts (see Burgess and Hirons 1992).

14.9.6 Managing reedbeds

Most experience of managing swamps and fens for birds is from managing reed-dominated vegetation (reedbeds) in Europe, which supports a distinctive assemblage of breeding birds (e.g. Hawke and José 1996; Poulin *et al.* 2002). The avifauna of reedbeds is strongly influenced by the extent of the swamp/open water interface, its physical structure and dominant plant species, the duration and the timing of flooding, and the extent of scrub (Van der Hut 1986; Tyler 1994; Graveland 1998; Poulin *et al.* 2002).

Management of reedbeds usually concentrates on providing wet, open reedbed on the margin of open water (water reed). This is because water reed is the primary habitat for two rare birds in Western Europe, namely Bittern and Great Reed Warbler *Acrocephalus arundinaceus* (Tyler 1994; Graveland 1998), but its extent has declined relative to that of other types of reedbed as a result of succession, die-back of reed margins, and a lack of suitable shallow open water into which early successional reed can spread (Tyler 1994; Van der Putten 1997).

Succession in reedbeds can be slowed by cutting or burning. These reduce the accumulation of litter and raising of the ground surface relative to the water table, and kill or suppress colonizing scrub. Cutting or burning in winter encourages reed at the expense of other tall plants, while cutting in summer reduces the dominance of reed relative to other species. Burning is more effective at reducing litter (Cowie *et al.* 1992), but many site managers consider it more damaging to less mobile invertebrates than cutting, despite research showing little or no difference in invertebrate assemblages between cut and carefully burnt wet reedbed (Ditlhogo *et al.* 1992). However, it is important not to burn in dry conditions in summer or autumn, when the fire will be hotter and burn deeper into the litter, and is likely to be more damaging to invertebrates. Cutting is only practical during dry conditions or if the water is frozen solid enough to allow access. Lowering water levels to enable cutting in winter can compromise attempts to attain suitable water levels for breeding birds in early spring.

Although cutting or burning in winter may be necessary to prevent succession in reedbeds, it can leave areas bare and unsuitable for nesting birds the following spring. In southern Europe, where regrowth of reed is rapid, winter reed cutting only reduces densities of early nesting, resident passerines (Poulin and Lefebvre 2002). In northern Europe, where regrowth of reed is slower, it also reduces densities of later arriving migrant warblers (Graveland 1999). Winter cutting also eliminates moth larvae which overwinter in reed stems, and provide food for some reedbed passerines. However, a recent study in southern France suggested that total invertebrate prey availability for reedbed passerines was actually higher in annually cut reedbeds than uncut ones (Poulin and Lefebvre 2002).

Any deleterious short-term effects of cutting or burning can be reduced by managing areas in rotation, thus increasing the range of conditions in the reedbed as a whole by providing different stages of regrowth. The frequency of winter reed cutting used to arrest succession varies between sites, but is typically once every 5–10 years. Commercial cutting of reed takes place on a 1–2 year rotation to provide high densities of strong, straight reed stems suitable for thatching. Annual cutting of large areas of reedbed is detrimental to some nesting birds and damaging to its invertebrate fauna. Therefore, a compromise between the needs of commercial cutting and conservation is to cut only a proportion of the reedbed for thatching, and to do this on a 2-year rotation (Hawke and José 1996).

Grazing can be used to reduce the dominance of reed, and at high levels can convert fen to mire and wet grassland. Judicious grazing by cattle or ponies can be used to create patchy, open reed interspersed with shallow water and grassland. This is considered the most productive part of some wetlands in northern Europe for breeding waterfowl.

Succession in reedbeds can also be reduced by raising the level of the water relative to that of the surface of the substrate. The timing and duration of flooding influences conditions for feeding and nesting. Maintaining water levels above ground level allows fish to penetrate the margins of reedbeds and thereby provide feeding conditions for birds such as Bitterns (Tyler 1994). Prolonged flooding of reedbeds in summer probably also increases the invertebrate food supply for some nesting passerines (Poulin *et al.* 2002), and might also affect the availability of nest-sites for some species. Consistent high water levels over many years have, though, been implicated in the regress of reeds (Van der Putten 1997). The ideal is to have a mixture of hydrological regimes. Regular monitoring of water levels is important in informing hydrological management and interpreting its effects. Water levels are best monitored using gaugeboards leveled to a benchmark. The relationship between gaugeboard readings and water levels within the reedbed can be determined by mapping water depths on a grid when water levels are at a known, high gaugeboard level. These data can then be used to determine water depths within the reedbed when water levels are lower.

Sparse, early-succesional reed can be encouraged by lowering the ground surface to provide open water into which reed can spread or be planted. Excavation can be difficult in reedbeds, though, and disposal of excavated material can be problematic. However, if excavation is carried out sensitively, it can be used to create a gradient from open water and early successional emergent vegetation, to later successional swamp and fen. It is most frequently used along ditch edges to maximize the length of sparse reed/open water interface, but has been used on a larger scale to lower areas of tens of hectares to provide sparse, open reed for breeding Bitterns in the United Kingdom (Smith *et al.* 2000) (Figure 14.7). The

Fig. 14.7 Bitterns prefer feeding in sparse, early successional reedbed close to open water. One method to set back reedbed succession is to lower the surface of the reedbed relative to the water. Minsmere RSPB Reserve (RSPB Images)

colonizing reed can be kept open by cutting on a short rotation to remove dead stems. Open water can be maintained by cutting reed underwater to reduce its vigor.

14.9.7 Scrub

The presence of scrub in swamps and fens increases the total number of breeding bird species, mainly through addition of generalist scrub species (e.g. Hanowski *et al.* 1999). Extensive colonization by scrub reduces the value of the swamp or fen for scarcer wetland species. Established scrub can be removed by cutting (usually requiring treatment of stumps with herbicide to prevent regrowth), burning, or clearing in winter when the ground is sufficiently frozen using a modified blade on a bulldozer ("shearing"). Burning is most effective in the dry conditions of late summer or autumn, when the fire can burn deep enough to kill the roots of scrub.

14.9.8 Wet woodlands

Wet woodlands can support distinctive assemblages of birds, but are rarely managed specifically for them. An exception is in oak-dominated forested wetlands managed for timber in the Mississippi and associated valleys in the USA (so-called "greentree reservoirs"). Here, water levels are manipulated to mimic natural winter flooding and make acorns and other food available to Mallard *Anas platyrhynchos* and Wood Ducks *Aix sponsa* (Reinecke *et al.* 1989).

14.10 Managing intertidal habitats

Intertidal habitats comprise mudflats, sandbanks, saltmarsh, and mangroves. Saltmarshes can be grazed or burnt to create suitable vegetation structure for birds and encourage preferred food plants for wildfowl. Autumn or early winter burning is commonly used in the southern USA to stimulate succulent new growth of food plants for wildfowl, such as Olney Three-square Bulrush *Scirpus olneyi* for wintering Lesser Snow Geese *Chen caerulescens caerulescens*. Burning also increases access to their nutritious rhizomes, and increases use by loafing and feeding seed-eating ducks (Chabreck *et al.* 1989) and icterids, but decreases use by Marsh Wrens *Cistothorus palustris* and Sedge Wrens *C. platensis* during the season that burning takes place. Use of these areas by passerines returns to pre-burn levels by the following spring or winter (Van't Hul *et al.* 1997; Gabrey *et al.* 1999, 2001). Grazing can be used to create suitable vegetation for nesting Redshank *Tringa totanus* on saltmarshes in the United Kingdom (Norris *et al.* 1997), and to encourage open swards of Common Salt-marsh Grass *Puccinellia maritima* and Red Fescue *Festuca rubra* favored by wintering geese and Eurasian Wigeon *Anas Penelope* at the expense of unpalatable Sea-couch *Elymus pycnanthus* (e.g. Bos *et al.* 2002).

Management of intertidal areas has also involved control of Reed and Common Cordgrass *Spartina anglica*. Along the north Atlantic coast of the USA, disturbance and restriction of tidal influence have encouraged expansion of reed at the expense of cordgrass (*Spartina* spp) meadow vegetation, which supports a more important avifauna (e.g. Benoit and Askins 1999). Common Cordgrass has been controlled in Europe using herbicide or mechanical disturbance where it has colonized mudflats important for feeding waders (e.g. Frid *et al.* 1999).

14.11 Managing arable land and hedgerows

Arable land in Europe supports a high proportion of the population of many threatened and declining bird species (Tucker and Heath 1994), and many of these declines have been associated with agricultural intensification (Tucker and Heath 1994; Chamberlain *et al.* 2000; Donald *et al.* 2001). This has prompted the development of strategies to reverse bird declines, which are capable of being incorporated into arable farming systems, applied on a large scale, and encouraged by agri-environment schemes (e.g. Evans *et al.* in press). Techniques have also been developed to increase the harvest of gamebirds on arable farmland (Potts 1986).

So far, most of the habitat management techniques used to benefit farmland birds have been developed in the United Kingdom. Declines in farmland birds in the United Kingdom have been primarily associated with the following factors:

1. Increased monoculture at a farm and landscape scale, which is thought to have been detrimental to species requiring a mixture of grassland and arable in close proximity during the breeding season (e.g. Galbraith 1988).
2. A change from spring to autumn sowing. In a spring sowing system, stubble is left following harvest in autumn, and only ploughed in the following spring prior to sowing. Weedy stubbles provide an important source of seeds and spilt grain for finches, buntings, sparrows, larks, pigeons, and waterfowl in winter (e.g. Evans 1996). In autumn sowing systems, the stubble is ploughed in immediately following harvest prior to autumn sowing. By spring, autumn sown crops are too dense or tall for some bird species to nest (Hudson *et al.* 1994; Wilson *et al.* 1997).
3. Increased fertilizer (primarily nitrogen) use. This has increased crop growth and density and thereby further decreased the availability of sparse,

open crops for some species to nest in (Hudson *et al.* 1994; Wilson *et al.* 1997). It has also reduced vegetation diversity in grassland.

4. Increased use of herbicides and insecticides, which has decreased the availability of weed seeds and arthropod prey for birds, particularly for chicks of some species (Potts 1986; Campbell *et al.* 1997; Newton 1998).

Management techniques have focused on reinstating spring sowing and providing winter stubbles, reducing herbicide use, planting strips of unharvested crops to provide a winter seed source for birds (wild bird cover) and in particular, providing strips of unsprayed arable (conservation headlands) and/or grassland (grass margins) around the edges of fields to provide invertebrate food for chicks. Restricting such measures to field margins minimizes loss of agricultural production, while maximizing benefits to birds that prefer to feed close to adjacent hedgerows.

The quality of hedgerows adjacent to arable fields also influences their use by birds. The total number of bird species tends to be highest along tall hedges with many trees, although this is mainly due to these supporting more species associated with woodland and woodland edge. Some species though, for example Whitethroat *Sylvia communis* and Yellowhammer *Emberiza citrinella* prefer shorter hedges with few trees (Green *et al.* 1994; Parish *et al.* 1994).

14.12 Conclusions

Although many of the principles of managing specific habitats for birds and other wildlife are fairly well understood, there is clearly much to be learnt in terms of fine-tuning habitat management to different sites. It is only by recording management, monitoring its effects and disseminating this information to others, that it will be possible to optimize the limited resources available for conservation.

As we have seen, habitat management for birds encompasses a wide variety of techniques, ranging from low intensity ones such as light grazing by herbivores to more intensive ones such as lowering the surface of reedbeds. These more intensive techniques are expensive and only ever likely to be practical on a small scale. Within protected areas, the challenge for habitat management in the future is to *minimize* the intensity of management needed to conserve threatened species by increasing the size of existing fragments of semi-natural habitat and restoring, as far as practical, better functioning of natural processes within them. Outside protected areas, the main challenge for habitat management in developed countries will be to identify further practical measures that can be adopted by mainstream agriculture and forestry to benefit birds, other wildlife, and the environment in general. The greatest challenge for habitat management, though, will be to identify practices that prevent long-term degradation of habitats for birds and other wildlife in developing countries, while also incorporating the short- and long-term needs of local people.

References

Anderson, J.T. and Smith, L.M. (2000). Invertebrate response to moist-soil management of playa wetlands. *Ecol. Appl.*, 10(2), 550–558.

Annand, E.M. and Thompson III, F.R. (1997). Forest bird response to regeneration practices in central hardwood forests. *J. Wildlife Manage.*, 61, 159–171.

Artman, V.L., Sutherland, E.K., and Downhower, J.F. (2001). Prescribed burning to restore mixed-oak communities in southern Ohio: effects on breeding bird populations. *Conserv. Biol.*, 15, 1423–1434.

Ausden, M. (2001). The effects of flooding grassland on food supply for breeding waders. *Br. Wildlife*, 12, 179–187.

Ausden, M., Sutherland, W.J., and James, R. (2001). The effects of flooding lowland wet grassland on soil macroinvertebrate prey of breeding wading birds. *J. Appl. Ecol.*, 38, 320–338.

Baillie, S.R., Sutherland, W.J., Freeman, S.N., Gregory, R.D., and Paradis, E. (2000). Consequences of large-scale processes for the conservation of bird populations. *J. Appl. Ecol.*, 37 (Suppl. 1), 88–102.

Beintema, A.J. and Müskens, G.J.D.M. (1987). Nesting success of birds breeding in Dutch grasslands. *J. Appl. Ecol.*, 24: 743–758.

Benoit, L.K. and Askins, R.A. (1999). Impact of the spread of *Phragmites* on the distribution of birds in Connecticut tidal marshes. *Wetlands*, 19(1), 194–208.

Bibby, C.J. (1979). Foods of the Dartford warbler *Sylvia undata* on southern English heathland (Aves: Sylviidae). *J. Zool. Soc. Lond.*, 188, 557–576.

Bibby, C.J., Aston, N., and Bellamy, P.E. (1989). Effects of broadleaved trees on birds of upland conifer plantations in North Wales. *Biol. Conserv.*, 49, 17–29.

Bos, D., Bakker, J.P., de Vries, Y., and van Lieshout, S. (2002). Long-term vegetation changes in experimentally grazed and ungrazed back-barrier marshes in the Wadden Sea. *Appl. Veg. Sci.*, 5, 45–54.

Bowden, C.G.R. (1990). Selection of foraging habitats by Woodlarks (*Lullula arborea*) nesting in pine plantations. *J. Appl. Ecol.*, 27, 410–419.

Brandeis, T.J., Newton, M., Filip, G.M., and Cole, E.C. (2002). Cavity-nester habitat development in artificially made Douglas-fir snags. *J. Wildlife Manage.*, 66(3), 625–633.

Britton, R.H. and Johnson, A.R. (1987). An ecological account of a Mediterranean Salina: the Salin de Giraud, Camargue (S. France). *Biol. Conserv.*, 42, 185–230.

Brown, A.F. and Stillman, R.A. (1993). Bird-habitat associations in the eastern Highlands of Scotland. *J. Appl. Ecol.*, 30, 31–42.

Buffington, J.M., Kilgo, J.C., Sargent, R.A., Miller, K.V., and Chapman, B.R. (1997). Comparison of breeding bird communities in bottomland hardwood forests of different successional stages. *Wilson Bull*, 109, 314–319.

Burgess, N.D. and Hirons, G.J.M. (1992). Creation and management of artificial nesting sites for wetland birds. *J. Environ. Manage.*, 34, 285–295.

Byelich, J., DeCapita, M.E., Irvine, G.W., Radtke, R.E., Johnson, N.I., Jones, W.R., Mayfield, H., and Mahalak, W.J. (1985). *Kirtland's Warbler Recovery Plan*. US Fish and Wildlife Service, Rockville, Maryland.

Campbell, L.H., Avery, M.I., Donald, P.F., Evans, A.D., Green, R.E., and Wilson, J.D. (1997). A Review of the Indirect Effects of Pesticides on Birds. JNCC Rep. No. 227. Peterborough, JNCC.

Chabreck, R.H., Joanen, T., and Paulus, S.L. (1989). Southern Coastal Marshe. In *Habitat Management for Migrating and Wintering Waterfowl in North America*, eds. L.M. Smith, R.L. Pederson, and R.M. Kaminski, pp. 249–277. Texas Tech University Press, Lubbock.

Chamberlain, D.E., Fuller, R.J., Bunce, R.G.H., Duckworth, J.C., and Shrubb, M. (2000). Changes in the abundance of farmland birds in relation to the timing of agricultural intensification in England and Wales. *J. Appl. Ecol.*, 37, 771–788.

Civil Aviation Authority (1998). *Aerodrome Bird Control. CAP 680*. Documedia Solutions Ltd, Cheltenham.

Cowie, N.R., Sutherland, W.J., Ditlhogo, M.K.M., and James, R. (1992). The effects of conservation management of reed beds. II. The flora and litter disappearance. *J. Appl. Ecol.*, 277–284.

Danell, K. and Sjöberg, K. (1982). Successional patterns of plants, invertebrates and ducks in a man-made lake. *J. Appl. Ecol.*, 19, 395–409.

Day, J.H. and Colwell, M.A. (1998) Waterbird communities in rice fields subjected to different post-harvest treatments. *Colon. Waterbirds*, 21(2), 185–197.

de Szalay, F.A. and Resh, V.H. (1997). Response of wetland invertebrates and plants important in waterfowl diets to burning and mowing of emergent vegetation. *Wetlands*, 17, 149–156.

Ditlhogo, M.K.M., James, R., Laurence, B.R., and Sutherland, W.J. (1992). The effects of conservation management of reed beds. I. The invertebrates. *J. Appl. Ecol.*, 29, 265–276.

Donald, P.F., Haycock, D., and Fuller, R.J. (1997). Winter bird communities in forest plantations in western England and their response to vegetation, growth stage and grazing. *Bird Stud.*, 44, 206–219.

Donald, P.F., Fuller, R.J., Evans, A.D., and Gough, S.J. (1998). Effects of forest management and grazing on breeding bird communities in plantations of broadleaved and coniferous trees in western England. *Biol. Conserv.*, 85, 183–197.

Donald, P.F., Green, R.E., and Heath, M.F. (2001). Agricultural intensification and the collapse of Europe's farmland bird populations. *Proc. R. Soc. Lond.*, B, 268, 25–29.

Elphick, C.S. and Oring, L.W. (1998). Winter management of Californian rice fields for waterbirds. *J. Appl. Ecol.*, 35, 95–108.

Evans, A.D. (1996). The importance of mixed farming for seed-eating birds in the UK. In *Farming and Birds in Europe: The Common Agricultural Policy and its Implications for Bird Conservation*, eds. D.J. Pain and M.W. Pienkowski, pp. 331–357. London, Academic Press.

Evans, A.D., Armstrong-Brown, S., and Grice, P.V. (in press). The role of research and development in the evolution of a "smart" agri-environment scheme. *Asp. Appl. Ecol.*

Frid, C.L.J., Chandrasekara, W.U., and Davey, P. (1999). The restoration of mud flats invaded by common cord-grass (*Spartina anglica*, CE Hubbard) using mechanical disturbance and its effects on the macrobenthic fauna. *Aquat Conserv: Mar. Freshwater Ecosyst.* 9, 47–61.

Fuller, R.J. (2001). Responses of woodland birds to increasing numbers of deer: a review of evidence and mechanisms, *Forestry,* 74, 289–298.

Fuller, R.J. and Green, G.H. (1998). Effects of woodland structure on breeding bird populations in stands of coppiced lime (*Tilia cordata*) in western England over a 10-year period. *Forestry,* 71 (3), 199–215.

Fuller, R.J and Henderson, A.C.B. (1992). Distribution of breeding songbirds in Bradfield Woods, Suffolk, in relation to vegetation and coppice management. *Bird Stud.*, 39, 73–88.

Gabrey, S.W., Afton, A.D., and Wilson, B.C. (1999). Effects of winter burning and structural marsh management on vegetation and winter bird abundance in the Gulf Coast Chenier Plain, USA. *Wetlands,* 19, 594–606.

Gabrey, S.W., Afton, A.D., and Wilson, B.C. (2001). Effects of structural marsh management and winter burning on plant and bird communities during summer in the Gulf Coast Chenier Plain. *Wildlife Soc. Bull.*, 29, 218–231.

Galbraith, H. (1988). Effects of agriculture on the breeding ecology of Lapwings *Vanellus vanellus. J. Appl. Ecol.*, 25, 487–503.

Gilbert, G., Tyler, G., and Smith, K.W. (2003). Nestling diet and fish preference of Bitterns *Botaurus stellaris* in Britain. *Ardea,* 91, 35–44.

Graveland, J. (1998). Reed die-back, water level management and the decline of the great reed warbler *Acrocephalus arundinaceus* in the Netherlands *Ardea,* 86, 187–201.

Graveland, J. (1999). Effects of reed cutting on density and breeding success of Reed Warbler *Acrocephalus scirpaceus* and Sedge Warbler *A. schoenobaenus. J. Avian Biol.*, 30, 469–482.

Gray, M.J., Kaminski, R.M., Weerakkody, G., Leopold, B.D., and Jensen, K.C. (1999). Aquatic invertebrate and plant responses following mechanical manipulations of moist soil habitat. *Wildlife Soc. Bull.*, 27, 770–779.

Green, R.E. (1988). Effects of environmental factors on the timing and success of breeding common snipe *Gallinago gallinago* (Aves:Scolopacidae). *J. Appl. Ecol.*, 25, 79–93.

Green, R.E. and Gibbons, D.W. (2000). The status of the Corncrake *Crex crex* in Britain in 1998. *Bird Stud.*, 47, 129–137.

Green, R.E., Hirons, G.J.M., and Kirby, J.S. (1990a). The effectiveness of nest defence by black-tailed godwits *Limosa limosa. Ardea,* 78, 405–413.

Green, R.E., Hirons, G.J.M., and Cresswell, B.H. (1990b). Foraging habits of female common snipe *Gallinago gallinago* during the incubation period. *J. Appl. Ecol.*, 27, 325–335.

Green, R.E., Osborne, P.E., and Sears, E.J. (1994). The distribution of passerine birds in hedgerows during the breeding season in relation to characteristics of the hedgerow and adjacent farmland. *J. Appl. Ecol.*, 31, 677–692.

Green, R.E., Tyler, G.A., and Bowden, C.G.R. (2000). Habitat selection, ranging behaviour and diet of the stone curlew (*Burhinus oedicnemus*) in southern England. *J. Zool.*, 250, 161–183.

Guldemond, J.A., Parmentier, F., and Visbeen, F. (1993). Meadow birds, field management and nest protection in a Dutch peat soil area. *Wader Stud. Group Bull.*, 70, 42–48.

Hagar, J.C., McComb, W.C., and Emmingham, H. (1996). Bird communities in commercially thinned and unthinned Douglas-fir stands of western Oregon. *Wildlife Soc. Bull.*, 24, 353–366.

Hågvar, S., Hågvar, G., and Mønnes, E. (1990). Nest site selection in Norwegian woodpeckers. *Holarctic Ecol.*, 13, 156–165.

Hanowski, J.M., Christian, D.P., and Nelson, M.C. (1999). Response of breeding birds to shearing and burning in wetland brush ecosystems. *Wetlands*, 19, 584–593.

Harrison, P.A., Berry, P.M., and Dawson, T.E. (eds.). (2001). Climate change and nature conservation in Britain and Ireland: modelling natural resource responses to climate change (the MONARCH project). UKCIP Technical report, Oxford.

Haukos, D.A. and Smith, L.M. (1993). Moist-soil management of playa lakes for migrating and wintering ducks. *Wildlife Soc. Bull.*, 21, 288–298.

Hawke, C.J. and José, P.V. (1996). *Reedbed Management for Commercial and Wildlife Interests.* RSPB, Sandy.

Hirons, G., Goldsmith, B., and Thomas, G. (1995). Site management planning. In *Managing Habitats for Conservation*, eds. W.J. Sutherland, and D.A. Hill, pp. 22–41. Cambridge University Press, Cambridge.

Hudson, R.W., Tucker, G.M., and Fuller, R.J. (1994). Lapwing *Vanellus vanellus* populations in relation to agricultural changes: a review. In *The Ecology and Conservation of Lapwings* Vanellus vanellus: *1–33*. eds. G.M., Tucker, S.M. Davies, and R.J. Fuller, UK Nature Conservation No. 9. Peterborough: JNCC.

Imbeau, L., Mönkkönen, M., and Desrochers, A. (2000). Long-term effects of forestry on birds of the eastern Canadian boreal forests: a comparison with Fennoscandia. *Conserv. Biol.*, 15, 1151–1162.

Jones, J.C. and Reynolds, J.D. (1996). Environmental variables. In *Ecological Census Techniques: A Handbook*, ed. W.J. Sutherland, pp. 281–316. Cambridge University Press, Cambridge.

Kaminski, R.M. and Prince, H.H. (1981). Dabbling duck and aquatic macroinvertebrate responses to manipulated wetland habitat. *J. Wildlife Manage.*, 45(1), 1–15.

Knights, B. and White, E.M. (1998). Enhancing migration and recruitment of eels: the use of passes and associated trapping systems. *Fish. Manage. Ecol.*, 4, 311–324.

Kruk, M., Noordervliet, A.A.W., and ter Keurs, W.J. (1996). Hatching dates of waders and mowing dates in intensively exploited grassland areas in different years. *Biol. Conserv*, 77, 213–218.

Kruk, M., Noordervliet, M.A.W., and ter Keurs, W.J. (1997). Survival of black-tailed godwit chicks *Limosa limosa* in intensively exploited grassland areas in the Netherlands. *Biol. Conserv.*, 80, 127–133.

Kulhavy, D.L., Hooper, R.G., and Costa, R. (eds.) (1995). *Red-cockaded woodpecker: recovery, ecology and management.* Center for Applied Studies in Forestry, College of Forestry, Stephen F. Austin State University, Nacogdoches, Texas, USA.

Linz, G.M., Blixt, D.C., Bergman, D.L., and Bleier, W.J. (1996). Response of ducks to glyphosate-induced habitat alterations of wetlands. *Wetlands*, 16, 38–44.

MacArthur, R.H. and MacArthur, J.W. (1961). On bird species diversity. *Ecology*, 42, 594–598.

Manuwal, D.A. and Huff, M.H. (1987). Spring and winter bird populations in a Douglas fir sere. *J. Wildlife Manage.*, 51, 586–595.

Merrill, S.B., Cuthbert, F.J., and Oehlert, G. (1998). Residual patches and their contribution to forest-bird diversity on northern Minnesota aspen clearcuts. *Conserv. Biol.*, 12, 190–199.

Mitchell, F.J.G. and Kirby, K.J. (1990). The impact of large herbivores on the conservation of semi-natural woods in the British uplands. *Forestry*, 63, 333–353.

Mönkkönen, M. and Helle, P. (1989). Migratory habits of birds breeding in different stages of forest succession: a comparison between the Palearctic and Nearctic. *Ann. Zool. Fennici*, 26, 323–330.

Newton, I. (1998). *Population Limitation in Birds*. Academic Press, London.

Norris, K., Cook, T., O'Dowd, B., and Durdin, C. (1997). The density of redshank *Tringa totanus* breeding on the salt-marshes of the Wash in relation to habitat and its grazing management. *J. Appl. Ecol.*, 34, 999–1013.

Owen, M. (1975). Cutting and fertilising grassland for winter goose management. *J. Wildlife Manage.*, 39, 163–167.

Parish, T., Lakhani, K.H., and Sparks, T.H. (1994). Modelling the relationship between bird population variables and hedgerow and other field margin attributes. I. Species richness of winter, summer and breeding birds. *J. Appl. Ecol.*, 31, 764–775.

Peck, K.M. (1989). Tree species preferences shown by foraging birds in a forest plantation in northern England. *Biol. Conserv.*, 48, 41–57.

Percival, S.M. (1993). The effects of reseeding, fertilizer application and disturbance on the use of grasslands by barnacle geese, and the implications for refuge management. *J. Appl. Ecol.*, 30, 437–443.

Perkins, A.J., Whittingham, M.J., Bradbury, R.B., Wilson, J.D., Morris, A.J., and Barnett, P.R. (2000). Habitat characteristics affecting use of lowland agricultural grassland by birds in winter. *Biol. Conserv.*, 95, 279–294.

Potts, G.T. (1986). *The Partridge: Pesticides, Predation and Conservation*. Collins, London.

Poulin, B. and Lefebvre, G. (2002). Effect of winter cutting on the passerine breeding assemblage in French Mediterranean reedbeds. *Biodiversity and Conserv.*, 11, 1567–1581.

Poulin, B., Lefebvre, G., and Mauchamp, A. (2002). Habitat requirements of passerines and reedbed management in Southern France. *Biol. Conserv.*, 107, 315–325.

Poulsen, B.O. (2002). Avian richness and abundance in temperate Danish forest: tree variables important to birds and their conservation. *Biodiversity Conserv.*, 11, 15551–1566.

Provencher, L., Gobris, N.M., Brennan, L.A. Gordon, D.R., and Hardesty, J.L. (2002). Breeding bird response to midstorey hardwood reduction in Florida sandhill longleaf pine forests. *J. Wildlife Manage.*, 66, 641–661.

Reinecke, K.J., Kaminski, R.M., Moorhead, D.J., Hodges, J.D., and Nassar, J.R. (1989). Mississippi Alluvial Valley. In *Habitat Management for Migrating and Wintering Waterfowl in North America*, eds. L.M., Smith, R.L. Pederson, and R.M. Kaminski, pp. 203–247. Texas Tech University Press, Lubbock.

Roberts, P. (1982). Foods of the Chough on Bardsey Island, Wales. *Bird Stud.*, 29, 155–161.

Robertson, P.A. (1993). The management of artificial coastal lagoons in relation to invertebrates and avocets *Recurvirostra avosetta* (L.). PhD thesis, University of East Anglia.

RSPB, EN and ITE. (1997). *The Wet Grassland Guide: Managing Floodplain and Coastal Wet Grasslands for Wildlife*. RSPB, Sandy.

Sitters, H.P., Fuller, R.J., Hoblyn, R.A., Wright, M.T., Cowie, N., and Bowden, C.G.R. (1996). The Woodlark *Lullula arborea* in Britain: population trends, distribution and habitat occupancy. *Bird Stud.*, 43, 172–187.

Smith, K., Welch, G., Tyler, G., Gilbert, G., Hawkins, I., and Hirons, G. (2000). Management of RSPB Minsmere Reserve reedbeds and its impact on breeding Bitterns. *Br. Wildlife*, 12(1), 16–21.

Smith, L.M. and Kadlec, J.A. (1983). Seed banks and their role during drawdown of a North American marsh. *J. Appl. Ecol.*, 20, 673–684.

Smith, R.S. and Rushton, S.P. (1994). The effects of grazing management on the vegetation of mesotrophic (meadow) grassland in Northern England. *J. Appl. Ecol.*, 31, 13–24.

Stillman, R.A. and Brown, A.F. (1994). Population sizes and habitat associations of upland breeding birds in the South Pennines, England. *Biol. Conserv.*, 69, 307–314.

Swengel, S.R. and Swengel, A.B. (2001). Relative effects of litter and management on grassland bird abundance in Missouri, USA. *Bird Conserv. Int.*, 11, 113–128.

Taft, O.W., Colwell, M.A., Isola, C.R., and Safran, R.J. (2002). Waterbird responses to experimental drawdown: implications for the multispecies management of wetland mosaics. *J. Appl. Ecol.*, 39, 987–1001.

Ter Heerdt, G.N.J. and Drost, H. (1994). Potential for the development of marsh vegetation from the seed bank after a drawdown. *Biol. Conserv.*, 67, 1–11.

Tharme, A.P., Green, R.E., Baines, D., Bainbridge, I.P., and O'Brien, M. (2001). The effect of management for red grouse shooting on the population density of breeding birds on heather-dominated moorland. *J. Appl. Ecol.*, 38, 439–457.

Thomas, C.D. and Lennon, J.J. (1999). Birds extend their range northwards. *Nature*, 399, 213.

Tucker, G.M. and Heath, M.F. (1994). *Birds in Europe: Their Conservation Status*. BirdLife International (BirdLife Conservation Series No. 3). Cambridge, UK.

Tyler, G. (1994). Management of Reedbeds for Bitterns and Opportunities for Reedbed Creation. RSPB *Conserv. Rev.*, 8, 57–62.

Tyler, G.A., Green, R.E., and Casey, C. (1998). Survival and behaviour of Corncrake Crex crex chicks during the mowing of agricultural grassland. *Bird Stud.*, 45, 35–50.

Van den Berg, L.J.L., Bullock, J.M., Clarke, R.T., Langston, R.H.W., and Rose, R.J. (2001). Territory selection by the Dartford warbler (*Sylvia undata*) in Dorset, England: the role of vegetation type, habitat fragmentation and population size. *Biol. Conserv.*, 101, 217–228.

Van der Hut, R.M.G. (1986). Habitat choice and temporal differentiation in reed passerines of a Dutch marsh. *Ardea* 74, 159–176.

Van der Putten, W.H. (1997). Die-back of *Phragmites australis* in European wetlands: an overview of the European Research Programme on Reed Die-back and Progression (1993–94). *Aquat. Bot.*, 59, 263–275.

Van't Hul, J.T., Lutz, R.S., and Mathews, N.E. (1997). Impact of prescribed burning on vegetation and bird abundance at Matagorda Island, Texas. *J. Range Manage.*, 50, 346–350.

Vickery, J.A., Sutherland, W.J., and Lane, S.J. (1994). The management of grass pastures for brent geese. *J. Appl. Ecol.*, 31, 282–290.

Vickery, J.A., Tallowin, J.R., Feber, R.E., Asterak, E.J., Atkinson, P.W., Fuller, R.J., and Brown, V.K. (2001). The management of lowland neutral grasslands in Britain: effects of agricultural practices on birds and their food resources. *J. Appl. Ecol.*, 38, 647–664.

Vulink, J.T. and Van Eerden, M.R. (1998). Hydrological conditions and herbivory as key operators for ecosystem development in Dutch artificial wetlands. In *Grazing and Conservation Management*. eds. M.F., WallisDeVries, J.P. Bakker, and S.E. Van Wieren, pp. 217–252. Kluwer Academic Publishers, Dordrecht.

Warren, M.S., Hill, J.K., Thomas, J.A., Asher, J., Fox, R., Huntley, B., Roy, D.B., Telfer, M.G., Jeffcoate, S., Harding, P., Jeffcoate, G., Willis, S.G., Greatorex-Davies, J.N., Moss, D., and Thomas, C.D. (2001). Rapid response of British butterflies to opposing forces of climate change and habitat change. *Nature*, 414, 65–69.

Wilson, C.W., Masters, R.E., and Bukenhofer, G.A. (1995). Breeding bird response to pine-grassland community restoration for red-cockaded woodpeckers. *J. Wildlife Manage.*, 59, 56–67.

Wilson, J.D., Evans, J., Browne, S.J., and King, J.R. (1997). Territory distribution and breeding success of Skylarks *Alauda arvensis* on organic and intensive farmland in southern England. *J. Appl. Ecol.*, 34, 1462–1478.

Young, L. (1998). The importance to ardeids of the Deep Bay fish ponds, Hong Kong. *Biol. Conserv.*, 84, 293–300.

Index